Food Choice, Acceptance and Consumption

Food Choice, Acceptance and Consumption

Edited by

H.L. Meiselman
US Army Natick Research, Development and Engineering Center
Natick, Massachusetts
USA

and

H.J.H. MacFie
Institute of Food Research
Reading
UK

BLACKIE ACADEMIC & PROFESSIONAL
An Imprint of Chapman & Hall
London · Weinheim · New York · Tokyo · Melbourne · Madras

Published by
Blackie Academic & Professional, an imprint of Chapman & Hall,
2–6 Boundary Row, London SEI 8HN

Chapman & Hall, 2–6 Boundary Row, London SE1 8HN, UK

Chapman & Hall GmbH, Pappelallee 3, 69469 Weinheim, Germany

Chapman & Hall USA, 115 Fifth Avenue, Fourth Floor, New York, NY 10003, USA

Chapman & Hall Japan, ITP-Japan, Kyowa Building, 3F, 2-2-1 Hirakawacho, Chiyoda-ku, Tokyo 102, Japan

DA Book (Aust.) Pty Ltd, 648 Whitehorse Road, Mitcham 3132, Victoria, Australia

Chapman & Hall India, R. Seshadri, 32 Second Main Road, CIT East, Madras 600 035, India

First edition 1996

Typeset in 10/12 pt Times by Best-set Typesetter Ltd., Hong Kong
Printed in Great Britain by St Edmundsbury Press Ltd, Bury St Edmunds, Suffolk

ISBN 0 7514 0192 7

A catalogue record for this book is available from the British Library
Library of Congress Catalog Card Number: 96–83685

∞ Printed on acid-free text paper, manufactured in accordance with ANSI/NISO Z39.48-1992 (Permanence of Paper)

Preface

'Does the world need another book about eating?' we asked ourselves when the idea for this book originally surfaced. Our initial reaction to this question was mixed because we were aware of the enormous amount of scientific, technical, clinical, and popular material published about eating. We cautiously studied exactly what was available and what might not be available; we slowly came to the conclusion that another book on eating was not only worthwhile but essential. That would be true if we could organize the type of book which had not yet been written, a truly interdisciplinary book which covered the broad range of human eating.

As the book developed we became more and more excited about its potential. And now that the book is actually finished and published it is clear to us that what we envisaged has been realized. The process of producing this volume involved a series of decisions. First, we selected the topics which we believed needed to be covered in any broad view of human eating. This was neither easy nor obvious, because most books about eating cover the topic from one perspective; for example, abnormal eating or physiological controls of eating. We hoped to cover all aspects from anthropology through marketing to sensory perception. The title of the book developed from the process of selecting chapter topics. These topics are traditionally covered within separate treatments of food choice, food acceptance, and food intake. The goal of this book was an integrated treatment of the whole range of eating in humans, including choice, acceptance, and intake.

Second, we sought to ask the most knowledgeable people to cover specialized material in their areas of expertise. We were amazed at the level of enthusiasm from the authors because we asked the best people to write chapters, people who are always busy and overcommitted. We received a resounding positive response. This was our first realization that this book would be different and that this book would indeed fill a gap in the literature.

Third, and most important, we asked each author not to prepare a traditional review of scientific work in each area (sensory, social, etc.). Such reviews are valuable for forming a collection of what is known about a particular topic, although traditional reviews usually focus on the author's work and usually do not interpret the data reviewed. Rather, we asked each author to present a chapter which would convince the layman that his/her particular discipline was among the most important, if not **the** most important, discipline in the study of eating. On the surface this might not seem like a great challenge, but those familiar with the eating literature (usually called food intake or something else very nonhuman) will realize that authors often leave it to the reader to draw the connection to normal human eating. Directly relating each field to normal human eating was the challenge for all of the authors in this volume. They accepted the challenge enthusiastically and the result is a unique volume which covers a very broad range of material but relates all of that material to one theme of normal human eating.

For the first time, the reader can see presentations of what is known about different specialty areas and what the material means to those interested in eating. What is the relevance of basic sensory research, of physiological research in animals, of research and clinical practice with individuals who are classified as disturbed eaters? Further, how do social research and marketing research relate to understanding eating? Each of the authors in this book is an expert in his or her field. Each has written reviews of his or her particular specialty. For the first time, these experts have pulled together a refreshingly new look at an old topic: eating. But, by reading each chapter and, most important, by integrating the various chapters, the reader is provided with one of the most complete views of normal human eating ever presented. We sincerely hope that this volume stimulates others to work towards a greater understanding of normal human eating and towards a more integrated view of eating.

We wish to thank the publishers for their enthusiasm for the project; the authors for their superb writing jobs; and our office assistants, Jane Johnson and Inge Ellison, for their able support.

Herbert L. Meiselman, Natick, USA
Halliday MacFie, Reading, UK

Contributors

L.L. Birch	Department of Human Development and Family Studies, College of Health and Human Development, The Pennsylvania State University, 110 Henderson Building South, University Park, PA 16802, USA
J. Bonke	Social Forsknings Instituttet, Bogergade 28, DK-1300, Copenhagen K, Denmark
A.V. Cardello	US Army Natick Research, Development and Engineering Center, Natick, MA 01760-5020, USA
J.O. Fisher	Department of Nutrition, The Pennsylvania State University, 110 Henderson Building South, University Park, PA 16802, USA
K. Grimm-Thomas	Department of Human Development and Family Studies, College of Health and Human Development, The Pennsylvania State University, 110 Henderson Building South, University Park, PA 16802, USA
J.L. Guss	New York Obesity Research Center, St. Luke's Roosevelt Hospital Center, 1111 Amsterdam Avenue, New York, NY 10025, USA
C.P. Herman	Department of Psychology, University of Toronto, Toronto, Ontario, M5S 1A1, Canada
H.R. Kissileff	Departments of Psychiatry and Medicine, Columbia University College of Physicians and Surgeons, St. Luke's Roosevelt Hospital Center, 1111 Amsterdam Avenue, New York, NY 10025, USA
H.J.H. MacFie	Consumer Sciences Department, Institute of Food Research, Reading Laboratory, Earley Gate, Whiteknights Road, Reading RG6 2EF, UK
H.L. Meiselman	US Army Natick Research, Development and Engineering Center, Natick, MA 01760-5020, USA

M.T.G. Meulenberg Department of Marketing and Marketing Research, Wageningen University, Hollandseweg 1, 6706 KN Wageningen, The Netherlands

L.J. Nolan Department of Psychology, Columbia University College of Physicians and Surgeons, St. Luke's Roosevelt Hospital Center, 1111 Amsterdam Avenue, New York, NY 10025, USA

J. Polivy Department of Psychology, University of Toronto, Toronto, Ontario, M5S 1A1, Canada

M.M. Raats Consumer Sciences Department, Institute of Food Research, Reading Laboratory, Earley Gate, Whileknights Road, Reading RG6 2EF, UK

P.J. Rogers Consumer Sciences Department, Institute of Food Research, Reading Laboratory, Earley Gate, Whiteknights Road, Reading, RG6 2EF, UK

P. Rozin Department of Psychology, University of Pennsylvania, 3815 Walnut Street, Philadelphia, PA 19104-6196, USA

R. Shepherd Consumer Sciences Department, Institute of Food Research, Reading Laboratory, Earley Gate, Whiteknights Road, Reading, RG6 2EF, UK

D.A.T. Southgate 8 Penryn Close, Norwich NR4 7LY, UK (formerly of Institute of Food Research, Norwich)

H.C.M. van Trijp Department of Marketing and Marketing Research, Wageningen University, Hollandseweg 1, 6706 KN Wageningen, The Netherlands

Contents

1 The role of the human senses in food acceptance

ARMAND V. CARDELLO

1.1 Introduction

1.1.1 The role of the human senses in philosophy and science

Naive realism is the philosophical view that things that exist in the environment truly possess the qualities which they appear to have—a coin is round, an apple is red, peppers are pungent. However, when viewed on edge, a coin appears elliptical, an apple seen under green light appears gray, and peppers may be perceived as pungent to some but not to others. If one accepts the postulates of naive realism, these facts lead to the conclusion that coins are, simultaneously, circular and elliptical, that an apple is both red and gray, and that any single pepper is both pungent and mild. Such irrational conclusions have led most modern thinkers to accept an alternative epistemological view. This latter view is that of *empiricism*.

Both the classical empiricists, e.g. Locke, Hume and Berkeley, and their present day counterparts ascribe to the notion that all knowledge comes by way of the senses. However, this knowledge is not of reality, but of 'phenomena,' literally, *appearances*. This sensory-based philosophy has come to form the foundation for all modern approaches to science, a fact that is best reflected in the use of *observation* as the basic datum of science. The empiricist view also led directly to the philosophy of logical positivism, which maintains that all scientific entities derive their meaning from the operations by which they are *observed*. Thus, the fundamental philosophical assumptions that underlie all of science are rooted in the notion that information obtained through the human senses is the only basis for matters of knowledge, the mind, and human experience. It is with this humbling notion of the importance of the human senses in all human endeavors that the present chapter sets out to examine the role of the senses within one small, yet critically important, aspect of human experience—food acceptance.

1.1.2 Food acceptance: A model

In scientific terms, food acceptance is best classified as a hypothetical construct. Over the years and among different researchers, it has been referred to by such terms as *palatability*, *hedonic tone*, *liking/disliking*, *food prefer-*

ence, and *pleasantness/unpleasantness*. Some of these terms, such as palatability, are merely synonymous constructs; others, such as liking/disliking and pleasantness/unpleasantness, reflect the operational measures by which the construct is commonly measured.

In this chapter, food acceptance is treated as a perceptual/evaluative construct. It is a phenomenological experience, best categorized as a feeling, emotion or mood with a defining pleasant or unpleasant character. Since it is a subjective construct, the measurement of food acceptance relies on the use of psychometric, psychophysical, and/or behavioral methods. In common practice, those investigators who are more concerned with the phenomenology of food acceptance will use verbal report as its primary index; whereas, those investigators who focus more on the consequences of food acceptance will use behavioral measures, e.g. choice and consumption, as its primary index.

Figure 1.1 is a schematic model of the sensory basis of food acceptance, showing the stages, interactions, and measurement levels involved in the processing of sensory and perceptual information about food. At the far left of Figure. 1.1, the physicochemical energies intrinsic to the food are transduced into neurochemical and neuroelectric events in the peripheral nervous system via receptor organs for each sensory system (vision, kinesthesis, olfaction, etc.). At the phenomenological level, we speak of a psychophysical transformation occurring at this stage, which gives rise to the basic sensory dimensions of *quality* (salty, cold, red), *magnitude* (intensity [weak, strong]), and *duration*. As these basic sense data are conveyed through the nervous system, numerous crossmodal sensory interactions (reflected by the gray arrows) modulate the information being mediated along these separate channels.

At the next stage of processing, the sensory information is organized into object-relevant characteristics. This organization draws input from both learning and memory. It is here that taste and odor combine to form a recognizable flavor or that sounds emitted during biting combine with kinesthetic and somesthetic information to result in the perception of the crispness of a food. As these integrated percepts evolve and as object recognition emerges, information from learning and memory creates a framework of contextual information and expectations that serves to modulate these percepts.

At this point in perceptual processing, a stage of hedonic experience is also evoked, in which the perceptual information embodied in the appearance, flavor and texture of the food assumes a hedonic tone (pleasant, unpleasant). Like the percept it accompanies, this hedonic component is subject to a variety of factors unrelated to the stimulus itself. These include previous experience, context, culture, expectations, and physiological status (hunger, thirst). At the phenomenological level, we speak of a *preference* or *liking* response being formed.

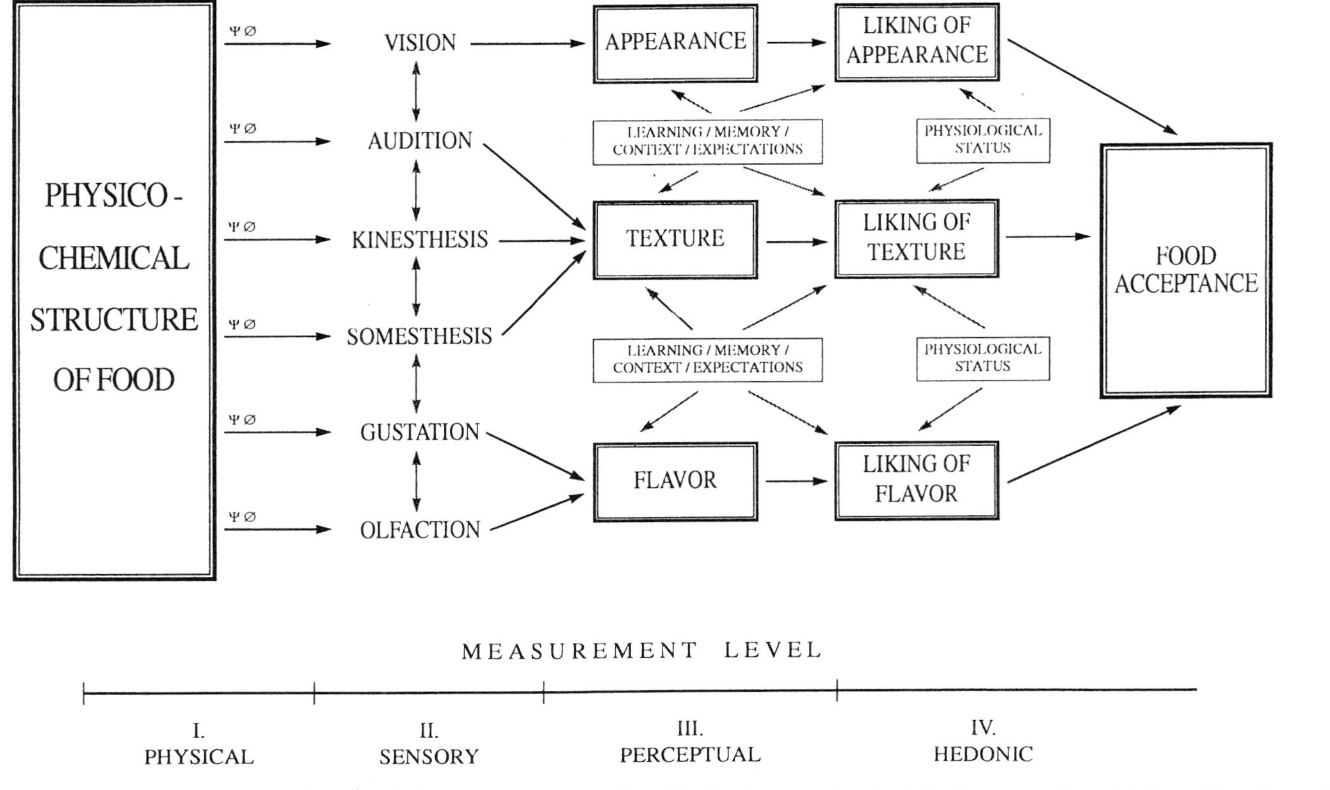

Figure 1.1 Schematic diagram showing the basic sensory, perceptual and hedonic stages involved in the processing of information about the physicochemical structure of food and resulting in food acceptance behavior.

At the final stage of processing, the hedonic information that accompanies the perceptual attributes of the food is integrated to produce a perceptual/evaluative experience that we call food acceptance. It is this experience that is assumed to underlie verbal reports of the overall like/dislike of foods and that contributes, along with social, situational, and marketplace factors, to the choice and consumption of food.

1.1.3 Sensation vs. *perception*

As shown in Figure 1.1, when food is eaten the receptor systems for each of the human senses transduce the physical, chemical and thermal energy of the food into biochemical and physiological events within the nervous system. Through a process still poorly understood, these events give rise to both the experience of specific sensations (salty, firm, red, etc.) and the resultant synthesis of these sensations into a percept of the food being eaten. This duality in human sensory/perceptual experience, i.e. the experience of individual sensory qualities—sweet, sour, cold, etc.—as contrasted with the perceptual gestalt of having eaten an onion, peach, cracker, etc., has been of interest to psychologists for some time.

In 1785, Thomas Reid distinguished perception from sensation by stating that the

> 'conception and belief which nature produces by means of the senses, we call perception. The feeling which goes along with the perception, we call sensation.'

Titchener (1909) commented on this duality by noting the 'curiously unitary character' of the combined sensations experienced while biting into a peach or sipping on coffee, while at the same time noting that each of its sensory constituents must be 'referred to the sense department to which they properly belong.' Titchener's comments underscore a distinction between *synthetic* and *analytic* experience that has become a critical element in characterizing the difference between perception and sensation, and which has also been used to characterize qualitative differences among the senses. (See Boring, 1942, for a review of early views on the distinction between sensation and perception.)

Although early views on the relationship between sensation and perception held that perception could only occur in the presence of sensation, in the mid-1960s James Gibson argued that perception could occur without sensation (defined as conscious awareness of sense data), but not in the absence of *information*. Gibson's view was that the human senses are active perceptual systems—interacting with the environment, seeking to extract information—not simply passive receptors responding to impinging energy from the environment (Gibson, 1966).

The differences between sensation and perception and between passive and active perception can be seen in the different approaches to the assessment of the role of the human senses in feeding behavior. Although a wealth of information on the role of the human senses in feeding behavior has come from psychophysical studies that use model systems to examine the effect of physicochemical structure on basic, passive sensory experience, an equally impressive array of data has come from studies that utilize real foods to examine more global and active aspects of perception and its role in food acceptance behaviors. Only by integrating these two types of data is it possible to develop a fundamental understanding of the complex interactions that occur among human sensory processes, perception, cognition and behavior in everyday eating situations.

1.1.4 Number and nature of sensory systems

Since the time of Aristotle, it has been held that there are five human senses: vision (sight), audition (hearing), gustation (taste), olfaction (smell) and somesthesis (touch). However, Sherrington (1906) showed that, in addition to these five exteroceptor systems (located on the surface of the body), there are three proprioceptor systems (located below the body surface). These are the vestibular sense (balance), the kinesthetic sense (body and limb position) and the sense of deep pressure. In addition, there is a system of interoceptors (located in the gastrointestinal tract and deep body organs) that respond to internal bodily states, such as stomach distention. With the possible exceptions of the vestibular and deep pressure sense, all of the human sensory systems are actively involved in the process of locating, consuming, experiencing and digesting food.

1.1.5 Basic dimensions of sensory experience

Following the terminology of Helmholtz, distinctions between sensations that derive from different sensory systems are referred to as distinctions in *modality*. Within each modality, one may also experience a variety of qualitatively different sensations. For example, within the sense of taste, one can experience 'salty', 'sweet', 'sour' and 'bitter' sensations; while within the visual sense one can experience sensations of 'blue', 'green' or 'red'. These distinct sensations within a single modality are referred to as *qualities*. Similarly, one can experience differences in the *magnitude* and *duration* of the sensation. Thus, the sweetness of an apple may be of low, intermediate, or high intensity; and the pungency of a jalapeno pepper may persist for only a few seconds or for several minutes.

The basic sensory dimensions of quality, magnitude and duration are commonly measured using standardized sensory and psychophysical methods. In the case of sensory quality, the terminology used by untrained individuals to describe their sensations is often poor (Richardson and Zucco, 1989; Desor and Beauchamp, 1974) and varies considerably. Some of this variation may well be due to genetic or physiological variation in the population, as will be addressed in later sections. However, a more common problem is that of psychological concept alignment (O'Mahony, 1991; O'Mahony *et al.*, 1990; Ishii and O'Mahony, 1987, 1990). For example, the odor of a sliced grapefruit might be described by one individual as 'fruity', by another as 'citrus', and by still another as 'orange-like'. All are experiencing the same sensation, but their category labels differ. In order to overcome this problem in defining the sensory properties of food, a variety of descriptive techniques have been developed to standardize terminology in flavor and texture perception. These techniques, which rely on the use of sensory standards and trained panelists, are now an essential element in the practical assessment of the sensory properties of food (Cairncross and Sjostrom, 1950; Caul, 1957; Brandt *et al.*, 1963; Stone *et al.*, 1974; Civille and Szczesniak, 1973; Civille and Liska, 1975; Meilgaard *et al.*, 1991). More recently, free choice profiling techniques, which utilize statistical approaches to enable *untrained* consumers to utilize their own sensory terminology, have also gained popularity as an alternative to traditional profile methods (Williams and Langron, 1984; Steenkamp and van Trijp, 1988).

In addition to being able to describe the quality of sensations, one must also be able to *quantify* these sensations. At the most fundamental level one must be able to state whether or not a sensory quality is evident in a food. The traditional approach to this problem has been to use *threshold* methods, by which either the simple occurrence of a sensation (absolute threshold) or the occurrence of a sensation of a specified sensory quality (recognition threshold) is determined through statistical estimation of the minimal stimulus energy required to elicit the sensation (see Fechner, 1860; Engen, 1971a; ASTM, 1979; Meilgaard *et al.*, 1991). More recently, methods based on signal detection theory (Tanner and Swets, 1954; Green and Swets, 1966) have also been successfully applied to the determination of human sensitivity to food-related attributes (O'Mahony, 1983, 1992; Vie *et al.*, 1991; Irwin *et al.*, 1992, 1993). The latter enable separation of purely sensory effects from those that may be due to contextual biases of the subject.

Although the simple occurrence of a sensation may be critical to off-flavors in foods, the suprathreshold magnitude or intensity of the sensation has a much broader impact on food acceptance. To measure suprathreshold sensory magnitude, a number of psychophysical scaling techniques have been developed, ranging from the simple summation of stimulus differences to ranking and rating procedures, e.g. category scaling and magnitude estimation. Much has been researched and written about these methods and their relative advantages and disadvantages. The reader is referred to the following sources for details on these approaches: Fechner, 1860; Stevens, 1957, 1961, 1975; Engen, 1971b; Anderson, 1981;

Riskey, 1986; Lawless and Malone, 1986; Cardello and Maller, 1987; McBride, 1993; Green *et al.*, 1993.

The last fundamental aspect of sensation, duration, is especially important to food acceptance. Although the duration of any sensation is directly related to the duration of the stimulus, physiological factors also play an important role. For example, *adaptation* is a physiological phenomenon that manifests itself in a reduction in neural activity upon repeated stimulation of the receptor organ. It is this adaptation that is responsible for the rapid diminution in the perceived magnitude of ambient odors encountered in a fish market and the rapid loss of sweetness in lemonade if it is held in the mouth for but a few seconds. In fact, the accumulated effects of adaptation during the act of consuming a food may well be a contributing factor in the termination of eating (see section on *sensory specific satiety*).

The opposite of adaptation, or *persistence*, may also occur in response to foods with strong bitter, sour or pungent characteristics. Such persistence may be due to continued neural excitation in the absence of the stimulus or may be the result of residual stimulus components near the receptor surfaces. In either case, such persistance can produce after-effects of long duration. In order to control for these effects during testing, standard sensory procedures call for the use of water rinses and long interstimulus intervals. Of course, in certain situations, the duration of a sensation produced by a food may be extremely important to its overall acceptance, e.g., the sweetness of fine chocolate. In such circumstances, the measurement of time–intensity relationships becomes essential, and the methods for accomplishing this have also become an important tool in the food industry (Lee and Pangborn, 1986; Overbosch *et al.*, 1986; Dijksterhuis, 1993). Lastly, the *order* in which different sensations are perceived can also be a critical factor in more complex foods and beverages, e.g. wines, and is routinely taken into account by most sensory descriptive profile methods.

1.1.6 The hedonic dimension

Most sensory stimuli, but especially food, elicit a *hedonic* dimension in addition to the basic dimensions of quality, magnitude and duration. The term hedonic means 'having to do with pleasure' and derives its meaning from the philosophy of Hedonism, which holds that pleasure is the ultimate good and the aim of all human behavior. In 1879, Wundt first proposed that a hedonic dimension is inherent in all sensory stimuli, but especially those that give rise to olfactory and gustatory sensations (Beebe-Center, 1932). In point of fact, the hedonic dimension inherent in food lies at the heart of what is commonly meant by 'food acceptance', because, as noted previously, food acceptance is a hypothetical construct inferred from verbal and nonverbal behaviors that reflect the pleasure (or displeasure) aroused by food.

But what is the nature of the hedonic dimension? Is it perceptual or of some other nature? Certainly there is a phenomenological component to pleasure and displeasure. Just as pain is a sensory phenomenon, its opposite, pleasure, might also be construed to be a sensory phenomenon. In fact, one early theorist proposed the existence of 'beneceptive' and 'nociceptive' sensory systems that mediate pleasant and unpleasant sensations (Troland, 1928). However, a fundamental problem with this notion is that the same physical stimulus that arouses pleasure in one individual may arouse displeasure in another individual. Pleasure and displeasure, liking or disliking, are not sensory phenomena, although they accompany most sensory stimuli. Rather, pleasure/displeasure are affective experiences, i.e., emotional responses whose somatic effects are accompanied by a cognitive experience of the emotion.

If one considers the cognitive experience that accompanies the hedonic response to a stimulus, it is clear that it behaves in much the same manner as classical sensory and perceptual dimensions (Helson, 1964). Affective responses have magnitude and duration. Affective responses adapt over time and recover following deprivation. They are subject to contrast and order effects, and they show bipolarity. These similarities between sensory and affective dimensions have led to the use of similar methods and approaches for their measurement. Berlyne (1973), in his discussion of experimental aesthetics, describes how the development of methods for assessing hedonics has paralleled those of classical and contemporary psychophysics. In fact, many of the methods used for assessing hedonics were originally those developed by psychophysicists for scaling the magnitude of sensory experience.

In addition to its similarities to purely sensory dimensions, the hedonic dimension of food serves an essential function in the organization of behavior. Over the years, a vast body of research in the fields of learning and motivation has focused on such concepts as *reward*, *punishment*, *satisfaction*, *arousal*, *reinforcement*, *incentive value*, and *positive and negative feedback*—concepts that all connote the hedonic or affective aspects of anticipated or actual consequences of behavior. It is no coincidence that food, with its ability to evoke intense hedonic reactions, is the primary source of reward and motivation for learned behaviors.

For the measurement of hedonic quality, the 9-point hedonic scale (Peryam and Girardot, 1952; Peryam and Pilgrim, 1957) has become an international mainstay in the field. However, the full range of scalar techniques used to measure sensory intensity is also used to quantify the liking/disliking experience. This includes not only traditional univariate measures, but a wide variety of multivariate measurement techniques as well (see Schiffman *et al.*, 1981 and Martens and Russwurm, 1983). A more detailed discussion of measurement issues in food acceptance can be found in later sections of this chapter.

1.1.7 Sensory–hedonic relationships

Although the sensory and hedonic dimensions of food can be conceptually and theoretically divorced from one another, phenomenologically, they are often confused. Consumers, when asked to describe the taste or texture of foods will often reply with affective descriptions, e.g. 'it tastes bad' or 'the consistency is poor'. Similarly, when multidimensional scaling is applied to judgments of the similarity or dissimilarity of basic taste and/or smell stimuli, the primary dimension that spatially organizes the stimuli is a hedonic one (Yoshida, 1964; Schiffman, 1974). Given the intimate relationship between the sensory and hedonic dimensions of food, it is not surprising that the general function relating sensory experience to hedonic experience is well described. In fact, it was in 1874 that Wundt first proposed that the relationship between sensory magnitude and hedonic tone can be described in the form of Figure 1.2 (Beebe-Center, 1932). This schematic shows that, as the sensory intensity of a given stimulus increases, e.g. the sweetness of sugar, its hedonic tone becomes increasingly pleasant (region A), reaches a maximum level (the bliss point), decreases in pleasantness to the point of neutral hedonic tone (region B), and finally becomes unpleasant at the highest sensory intensities (region C). In spite of

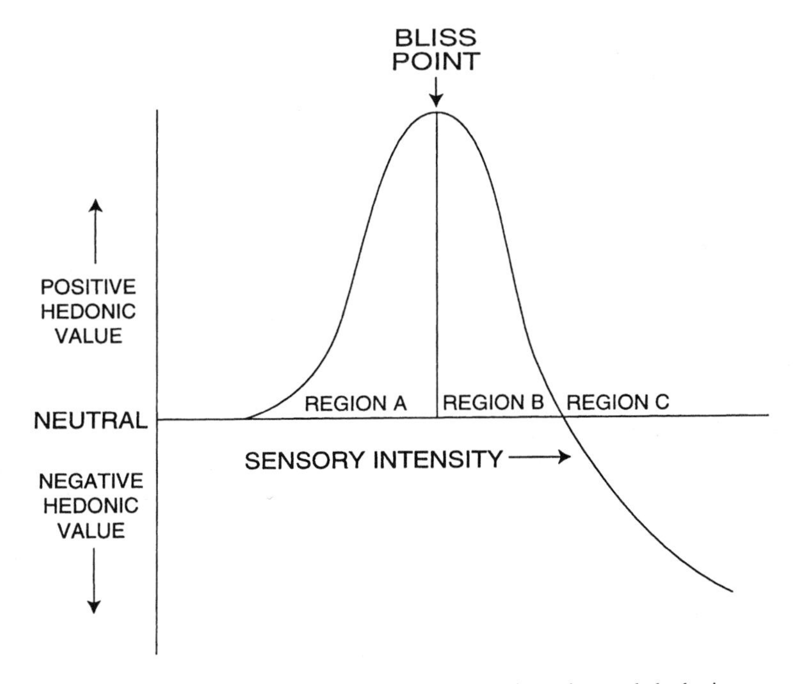

Figure 1.2 The basic relationship between sensory intensity and hedonic response (Wundt curve).

the robust nature of this inverted U function, for certain sets of data, it is better characterized as an inverted L function (Moskowitz *et al.*, 1976). In fact, Booth and co-workers have argued that the inverted U function is really an artifact of averaging data *across* individuals and that the actual function for *individuals* is an inverted V (Booth *et al.*, 1983, 1987; Booth, 1990; Conner *et al.*, 1986, 1988). In either case, it is important to note that this general relationship contrasts sharply with the relationship between perceived sensory magnitude and the physical intensity of the stimulus that elicits it. The latter function has repeatedly been shown to be a monotonically increasing function, e.g. power function (Stevens, 1975; Bolanowski and Gescheider, 1991). As we shall see, the inverted U or inverted V function characterizes the general relationship between the intensities of sensation and their hedonic consequences for a wide range of food attributes. However, the exact form of the function will vary by sensory quality and by conditions of testing.

1.1.8 Modes of sensory influence on food acceptance (liking)

As noted earlier, sensory experience is the ultimate basis for our knowledge of objects in the real world. As such, sensory experience also underlies the emotional and behavioral responses to these objects. In the case of food acceptance, sensory experience functions to influence the emotional and behavioral responses to foods in a number of ways. First is the fact that many forms of sensory experience evoke biologically innate hedonic responses (preferences/aversions), e.g. the innate preference for sweet taste. Secondly, through a vast network of intramodal and crossmodal interactions, many forms of sensory experience can alter the qualitative and quantitative character of other sensory experiences, thereby altering their hedonic and behavioral significance. These numerous and profound interactions are critical to our understanding of how the human senses work in concert to create a percept of multi-ingredient foods and how seemingly singular attributes of food can have complex effects on food acceptance. A third mode of influence is through the ability of food-related gustatory and olfactory stimuli to elicit salivary, gastric and pancreatic secretions (cephalic phase responses) which act to facilitate the perception, consumption and absorption of foods and nutrients. A fourth mode of influence occurs through mechanisms of acquired preference and aversion, whereby the sensory experiences evoked by food serve as a conditioned stimulus for food-related behaviors. Lastly, the totality of sensory experience can serve to establish a context or set of perceptual and hedonic expectations that alters the emotional or behavioral response to any single element of that overall experience.

Although examples of each of the above modes of influence will be seen in the sections that follow, the emphasis in this chapter will be on the fundamental relationships that exist among the physicochemical characteristics of food, their *direct* and *interactive* effects on sensory and perceptual experience, and the hedonic elements of food acceptance behavior. Data that bear more directly on the roles of learning, context, expectations, and/or physiological status will not be discussed in depth, but left to other reviews (see Rozin, 1989; Capaldi and Powley, 1990; Blundell and Rogers, 1991; Rozin and Tuorila, 1993; Cardello, 1994; Bell and Meiselman, 1995), including the chapters by Kissileff and by Meiselman in this volume.

1.2 The role of vision in food acceptance

1.2.1 Basic mechanisms and food-related attributes

In most consumption situations, food is first detected at a distance by the sense of sight. As a result, vision plays a critical role in food acceptance. Like many diurnal species, man has a highly evolved visual system that is designed for efficient color vision. It is estimated that the human eye can distinguish up to 1500 different hues in the visible spectrum of 400–800 nm (Tansley, 1965). Basic visual function is accomplished through the transduction of electromagnetic (light) energy into chemical and electrical energy by visual pigments (rhodopsin) in the cones of the human retina. These photopigments have been shown to possess differential sensitivity to wavelengths of light in the red, green and blue regions of the spectrum (Rushton, 1958; Marks *et al.*, 1964; Wald, 1964, 1968). Although this *trichromatic* theory of color vision has held sway for many years, very recently it has been put into question by the finding of a much larger number of pigment genes than would be expected if there were, in fact, only three photopigments (Neitz and Neitz, 1995). The results of this research have led to the conclusion that there may well be four human photopigments—one blue, one green and two long-wave (red).

Although the exact number of photopigments may be open to question, it *is* well established that for each photopigment, the absorption of light triggers the activation of G-proteins in the photoreceptor cell (Stryer, 1988; Park and Shortridge, 1991). Through a variety of second messenger systems, this activation leads to depolarization of the photoreceptor cell, which, in turn, generates action potentials, which carry the visual information more centrally in the nervous system.

Although color is certainly the most salient aspect of the visual appearance of foods, other visual attributes play an important role in food acceptance. A number of these attributes are related to the degree of transmission, absorption or reflection of light as it strikes the food. For example, the shine or *gloss* seen on many fruits and vegetables (e.g. apples, cucumbers) and viscous foods (e.g. gelatins and glazes) is due to the specular reflection of incident light from their smooth surfaces. The *translucency* of products like soup, jelly and beverages results from the transmission of light through them and is sometimes used as an index of their quality or strength of flavor. The attribute of *turbidity*, seen in many fruit juices, is due to reflection of light from particles suspended in the fluid and contributes to the product's perceived color. *Size* and *shape*, while often taken for granted, are critically important to the consumer's acceptance of such products as cookies, crackers, cereal, candy and fresh produce; while *surface texture* can give important clues to the sensory properties of such foods as vegetables, baked goods and fish.

In spite of the numerous ways by which the appearance attributes of food affect consumer acceptance, the majority of research on the role of appearance in food acceptance has focused on the influence of color. Certainly, no one can deny the importance of color to the perceived quality and acceptability of foods. A banana that has browned, bread that has turned green with mold, and chocolate that has bloomed to a white color are all common examples of how color serves as a signal of the quality (or deterioration in quality) of foods. In fact, in a survey conducted to assess the factors determining food quality among consumer groups in the USA, color/appearance ranked first in importance among sensory variables (Good Housekeeping Institute, 1984).

1.2.2 Innate color preferences

One particularly compelling demonstration of the important role of color in the consumer's response to food was reported in a study by Wheatley (1973). In this study, subjects ate a meal of steak, french fries and peas under color-masking conditions. Halfway through the meal, normal lighting was restored to reveal blue steak, green french fries and red peas. The mere sight of the food was enough to induce nausea in many of the subjects. The stark novelty of the colors used in this study leads one to ask whether certain colors in food are innately preferred or rejected.

Beebe-Center (1932), in numerous studies on the pleasantness of simple colors using standard papers and lights, found that pleasantness varies directly with the degree of saturation of the light. Subsequently, Eysenck (1941), in a study of the color preferences of 40000 individuals, found blue to be the most preferred color and yellow the least. Guilford and Smith (1959) confirmed these results using Munsell chips. They showed that liking was curvilinearly related to wavelength, with liking highest for green-blue hues, intermediate for reds, and lowest for yellow.

In early cultures, many foods were chosen for consumption *solely* on the basis of their color. This was due to the belief that the colors of food imparted physical and emotional properties to the individual who consumed them. As early as the second century, the Greek physician, Galen, urged all young males to eat red food and drink red liquids to become more sanguine, cheerful and confident (Pantone, 1986). More recently, people conjectured that, since there are so few foods whose natural color is blue, blue could be an innately disliked color in foods. While there are no empirical data on this point, the recent popularity of exotic and brightly colored foods and beverages, especially among children (e.g. blue fruit juice and gelatins), suggests that innate color preferences for food are unlikely. Rather, it appears that color preferences for foods are the result of experience, culture, and conditioning. This fact is well-known by food manufacturers and distributors, such as those in the USA, who must cater to the peculiar regional preference for brown eggs among the inhabitants of the northeastern USA, a color preference that exists nowhere else in that country.

1.2.3 Color–taste interactions

As noted in the introduction, understanding the numerous intramodal and crossmodal interactions that occur among sensory qualities is essential to understanding the role of the human senses in food perception and acceptance. One common vehicle by which vision is known to affect food acceptance is through crossmodal interactions with other sensory attributes of food. The effect of color on taste perception is a prime example. Effects of color have been shown to influence measures of taste thresholds, taste recognition, taste discrimination, taste intensity and even the ability of beverages to quench thirst (Clydesdale, 1984, 1993).

Maga (1974) performed the first systematic study of the effect of color on taste *thresholds* in aqueous solutions. Results of his studies showed that:

(i) yellow color increased sweet, sour and bitter thresholds;
(ii) green lowered sweet thresholds but raised thresholds for sour and bitter; and
(iii) red raised sour and bitter thresholds.

Although a later study by Johnson and Clydesdale (1982) found red to also lower sweet thresholds, the interpretation of these results is clouded by the fact that a related study (Johnson, 1982) found a light red strawberry color to increase sucrose thresholds, while a deeper red color had no effect.

In a review of early studies of the effect of color on taste *discrimination*, Pangborn (1960) found no effect of red, green or yellow coloring on the discrimination of sweetness in aqueous solutions. However, she did find a decrease in sweetness discrimination in green-colored samples of pear nectar. Pangborn and Hansen (1963), also working with pear nectars, found sweet and sour discrimination to be diminished in colored *vs.* uncolored nectars. This latter effect is consistent with the threshold findings of Maga (1974), which showed that taste thresholds are generally raised by the addition of color to aqueous solutions.

Although the above studies provide useful information about the effects of color on taste, the relationship is best seen in a series of studies examining the effect of color on suprathreshold taste *intensity* conducted by Clydesdale and coworkers (Kostyla, 1978; Kostyla and Clydesdale, 1978; Johnson and Clydesdale, 1982; Johnson *et al.*, 1982, 1983; Gifford and Clydesdale, 1986; Gifford *et al.*, 1987; Roth *et al.*, 1988; Clydesdale *et al.*, 1992). In this lengthy series of studies, the effect of red color on sweetness intensity was examined using aqueous solutions (Johnson and Clydesdale, 1982), strawberry-flavored beverages (Kostyla, 1978; Johnson, 1982), cherry-flavored beverages (Johnson *et al.*, 1982) and fruit punch (Clydesdale *et al.*, 1992). The general results of these studies showed that increasing the intensity of red color produced increasing sweetness, as shown in Figure 1.3 from Johnson *et al.* (1982). Other studies, such as those by Roth *et al.* (1988), who examined the effects of green and yellow color on the sweetness of lemon and lime beverages, and Fletcher *et al.* (1991), who examined the effect of visual masking and blue color on the sweetness of

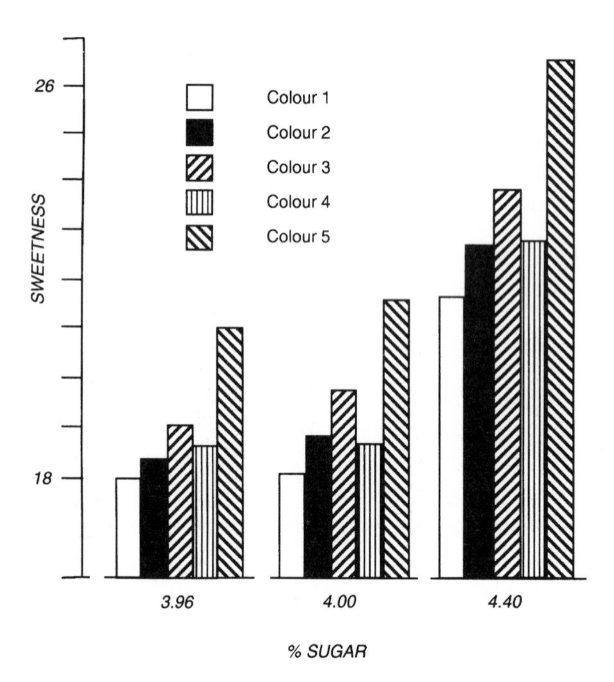

Figure 1.3 Judged sweetness at three sucrose concentrations as a function of increasing red color intensity (colors 1–5) in cherry-flavored beverages (from Johnson *et al.*, 1982).

lemonade, have shown more complex effects of color intensity on sweetness, with few generalizations possible.

Although numerous studies have examined the effects of varying color on the perceived sweetness of foods/beverages, studies of the effect of color on suprathreshold judgments of other taste qualities have been few in number. Notable exceptions are the studies by Gifford and Clydesdale (1986) and Gifford *et al.* (1987) in which the intensity of a yellow-brown color was examined for its effect on the saltiness of aqueous solutions and chicken broths. While some effects of color were observed, the effects were not monotonic with color intensity and could not be generalized.

1.2.4 Color–odor interactions

In one of only a few studies that have looked at the effect of color on odor perception, Christensen (1983) examined the effects on odor intensity of normally colored and uncolored/miscolored margarines, gelatin, bacon strips (soy analog), orange breakfast drink and cheese. Her results are shown in Figure 1.4 for the colored and uncolored samples. As can be seen, with the exception of the bacon analog, there was a reduction in odor intensity in the uncolored/miscolored samples. In addition, the uncolored/

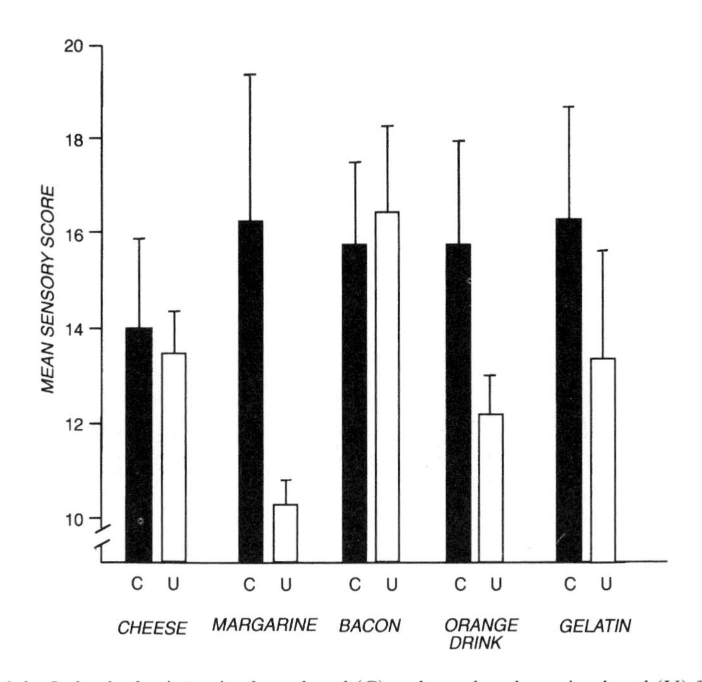

Figure 1.4 Judged odor intensity for colored (C) and uncolored or miscolored (U) food and beverage samples (from Christensen, 1983).

miscolored samples had lower aroma quality ratings than did the colored samples (bacon excepted). In a more recent study, Zellner *et al.* (1991) examined the effect of color on both odor identification and odor liking of appropriately colored, inappropriately colored and uncolored (subjects blindfolded) fruit odorants. Results showed significantly fewer identification errors when the odorant solutions were *appropriately* colored than when they were *inappropriately* colored or uncolored. A similar effect was observed for identification response times. For liking ratings, the appropriately colored and uncolored samples produced significantly higher ratings than did the inappropriately colored samples.

1.2.5 Color–flavor interactions

It is well known that flavor is the integrated perceptual response to the combined taste and odor of foods. Since color has been shown to interact with both taste and odor, it is only logical to expect color to influence flavor perception. In the study cited earlier by Christensen (1983), in which real foods were used as test stimuli, subjects also judged the overall flavor intensity and flavor quality of the foods. Although the results were generally consistent with the results for odor, i.e. there was a reduction in judged flavor intensity and judged flavor quality for the uncolored/miscolored samples, the results were more variable and failed to reach statistical significance. However, in a series of studies by Dubose *et al.* (1980), using noncarbonated fruit beverages as the test stimuli, dramatic effects of color on flavor identification and flavor intensity were observed. In one such study, combinations of four different color levels and four different flavor levels of cherry and orange beverage were presented to consumers, who rated them for overall flavor intensity. Significant effects of color intensity were found for the orange beverages, with flavor intensity increasing with increasing color intensity. Similar trends were found for the cherry beverages, but the effect did not reach statistical significance (a similar finding to Clydesdale *et al.* (1992) who used red colorant in fruit punch).

Of additional interest in the study by Dubose *et al.* (1980) were the results obtained from the study of flavor *identification*. Here, the stimuli consisted of cherry, orange, lime and flavorless beverage bases prepared in red, orange, green and colorless versions. These 16 stimuli were presented to consumers who were asked to identify their flavors. The results of this study showed a dramatic effect of the color of the beverage on perceived flavor identity. In the colored but flavorless samples, the greatest percentage of induced flavor responses was for the flavor category normally associated with the beverage color. In the flavored samples, the highest percentage of incorrect flavor judgments was for flavor categories normally associated with the color of the beverage, e.g. when cherry-flavored beverage was colored green, 26% of flavor identifications were 'lime'; similarly, when lime beverage was colored red, 33% of flavor identifications were for fruit flavors associated with the color red (cherry, strawberry, raspberry). These results are consistent with those found earlier by Hall (1958), who used miscolored sherbets and found that inappropriate colors influenced flavor identification in the direction of the flavor associated with the sherbet's color, as well as those by Stillman (1993), who found inappropriate colors to significantly degrade the identification of the flavor of raspberry and orange-flavored beverages.

1.2.6 Appropriate vs. *inappropriate colors*

To summarize the data on the crossmodal effects of color on taste, odor and flavor, it is safe to say that one of the most important variables is the *appropriateness* of the color to the food and/or its flavor constituents. In general, *appropriate* colors will increase flavor identification and recognition and will increase the perceived intensity (magnitude) of color-associated flavors. *Inappropriate* colors, on the other hand, will increase chemosensory thresholds and decrease taste discrimination, presumably by adding an irrelevant dimension to the basic detection/discrimination problem. Similarly, inappropriate colors will decrease flavor identification and flavor recognition.

1.2.7 Color–acceptance relationships

The issue of color appropriateness has relevance to the effects of color on liking. Christensen (1983) suggested that:

> 'visual cues permit the identification of food and through repeated dietary experience evoke an *anticipated* set of oral sensations.' (p. 787)

More recently, Cardello (1994) has proposed that the color of food, like many other sensory and ideational variables, creates a set of sensory and hedonic expectations for the food. The extent to which these expectations are subsequently confirmed or disconfirmed can, in turn, have a profound effect on the acceptance of the food by the consumer.

For *appropriately* colored foods, the effect of color on acceptability is well described. Figure 1.5 shows data from two studies examining the effect of systematic color changes to food acceptability. The data on the left are from Johnson *et al.* (1983) and show color acceptability ratings as a function of physical color intensity in strawberry beverages. The data on the right are from Dubose *et al.* (1980) and show the relationship between color acceptability and color intensity in lemon cake. In spite of the differences in both product categories and product colors, both sets of data are similar in showing that at low color intensity levels, increasing color intensity increases the judged acceptability of the color of the product. At intermediate

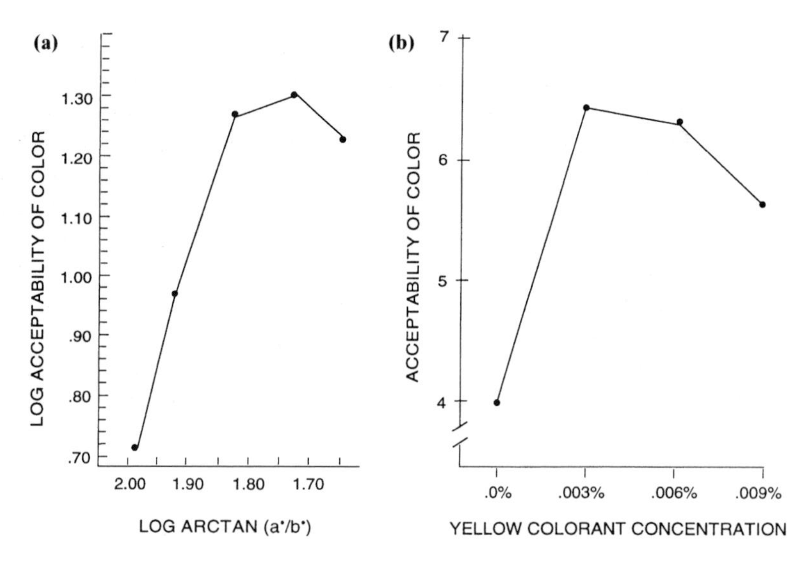

Figure 1.5 (a) Color acceptability (geometric mean magnitude estimate) as a function of red color intensity (arctan [a*/b*]) in strawberry beverages. Data are plotted in log–log coordinates (from Johnson *et al.*, 1983). (b) Color acceptability (9-pt hedonic scale) as a function of yellow color intensity (% concentration) in lemon cake (from Dubose *et al.*, 1980).

color intensity levels, acceptance peaks; and at the highest color intensity levels, acceptance declines. This fundamental relationship has been shown in numerous other studies, including those of Trant and Pangborn (1983) and Johnson *et al.* (1982).

An important point to be made is that the color-acceptability functions shown in Figure 1.5 differ at the high intensity levels from the classical sensory-hedonic function of Figure 1.1. This is because they are truncated at the upper end. While the inverted U function in Figure 1.1 describes the hypothetical relationship between sensory and hedonic responses over a wide range of intensities, it must be remembered that studies that examine color (or other) sensory dimensions of food over a restricted intensity range will produce results that appear to differ from those in Figure 1.1. The specific results will depend upon where the tested product intensities lie on the acceptance curve, as well as the relative salience of that attribute dimension to other attribute dimensions in the product.

For *inappropriately* colored foods, the effect of color on acceptance can be more dramatic, as found in Wheatley's (1973) study, described earlier. In general, inappropriately colored foods have a negative effect on acceptance. This is true both for foods that have colors normally associated with spoilage (brown meat, green bread, black tomatoes), as well as for inappropriate colors not associated with spoilage (blue steak, red peanut butter, etc.). One plausible mechanism by which inappropriate colors have a negative effect on food acceptance is through the violation of cognitive expectations. This thesis draws support from several studies on the role of disconfirmed sensory expectations on food acceptance (Cardello *et al.*, 1985; Cardello and Sawyer, 1992; Cardello, 1994), as well as studies by Zellner *et al.* (1991) in which inappropriately colored odorants were found to be much less acceptable than appropriately colored ones, and by Ringo (1982), in which inappropriately colored milk shakes received lower acceptance ratings than appropriately colored ones. Although expectations play an important role in food acceptance, the mechanism by which the effects are mediated is clearly cognitive in nature. The reader is referred to several recent volumes on the role of expectations and other cognitive and psychological variables on food acceptance (Shepherd, 1989; McBride and MacFie, 1990; MacFie and Thomson, 1994).

1.3 The role of somesthesis and kinesthesis in food acceptance

Physical contact with a food or beverage comes through biting or sipping it. This results in a perception of both its physical structure (firmness, viscosity, etc.) and its thermal properties. The primary sensory systems involved in these events are the kinesthetic and somesthetic senses, and their combined influence results in our perception of the *texture* and *temperature* of food (*cf.* Figure 1.1).

1.3.1 Basic mechanisms and food-related attributes

The kinesthetic sense, which literally means 'feeling of motion', mediates sensation of limb position and limb movement. Passive limb movement is mediated by receptors in joints, while voluntary limb movement and resistance to movement is mediated by receptors in muscles. When foods are placed in the mouth, they create resistance to active jaw movements during mastication (biting, chewing). Such resistance activates receptor organs in the intrinsic and extrinsic muscles of the tongue and the masticatory muscles that move the mandible (Frank, 1964; Gill, 1971). Receptors in the temporomandibular joint, which connects the mandible to the skull, also contribute to kinesthetic perception in the mouth (Frank, 1964; Halpern, 1977).

The somesthetic sense, which is a skin sense, mediates sensation of touch, temperature and pressure from skin structures located in or near the oral cavity. The receptors mediating simple pressure or touch lie in the oral mucosa itself (Grossman and Hattis, 1967; Ringel, 1970). However, when pressure is applied to the teeth, the force is transmitted to the periodontal membrane, where the same type of mechanoreceptors that are found in the mucosa mediate sensations of pressure (Pfaffmann, 1939; Cash and Linden,

1982). In the case of temperature sensitivity, free nerve endings that can be found throughout the mucosa serve as the receptors (Hensel, 1977). In general, for touch, temperature, and pressure sensitivity, the lips, tip of the tongue and hard palate have greater sensitivity than do other oral/facial areas. Table 1.1 from Heath and Lucas (1987) shows the sites of sensory nerve endings involved in mastication and food texture perception, the type of sensations mediated, and the relevance of the sensation to food mastication and texture.

1.3.2 Trigeminal (irritant) sensitivity

The perception of pain from the oral cavity is mediated by high-threshold mechanical, thermal and chemical receptors (Burgess and Perl, 1977). However, chemical irritant sensitivity in the oral cavity is part of a more generalized sensitivity that encompasses the nasal mucosa and other skin areas (Scheuplein, 1976). This generalized chemical sensitivity is mediated by the trigeminal nerve and is referred to as the *common chemical sense.* It is responsible for sensations commonly described as burning, cooling, stinging and pungent (Cliff and Heymann, 1992). While the common chemical sense plays an important role in the perception of many spicy foods, e.g. ginger, hot peppers and onions, the nature of the receptor mechanisms that are

Table 1.1 Sources of sensory information relevant to the perception of food texture and the control of mastication

Sites of sensory nerve endings	Modality of sensation	Phase of chewing cycle when active	Relevance of sensation
Periodontal membrane	Displacement Force	Occlusal loading	Fine discomminution of texture Fine and coarse control of comminution
Muscle, spindle-system in mandibular elevators	Stretch Position Displacement	All movements	General control of mandibular movement and position
Golgi tendon-organs	Tension Vibration	Late closure	? Protection against unexpected load
Oral mucosa	Light touch	All movements	Perception of food texture and movement
Temporo-mandibular joint	Position Displacement	All	General control, particularly wide gaps
Teeth (pulp)	Pain	Extreme occlusal loading only	Protective reflex unloading by muscle inhibition
Bone	Displacement	? As pulp	? As pulp
Ear	Sound	Occlusal loading	Perception of food comminution

involved and the relationship between the common chemical sense and the cutaneous senses is poorly understood (Green, 1990). What is clear, though, is that the common chemical sense has a profound capability to influence behavioral reactions to food and is tied directly to basic human defensive reflexes, e.g. coughing, sneezing, sweating, salivating and mucus flow.

1.3.3 Somesthesis, kinesthesis and food texture

Although the kinesthetic and somesthetic systems are separate modalities, common practice in food research is to measure their combined perceptual effect, i.e. food *texture* and perceived *temperature*. At the perceptual level, food textures can be classified into three types:

(i) *Mechanical* those related to the responses of foods to applied forces, e.g. hardness, chewiness, gumminess, viscosity;
(ii) *Geometric* those related to the geometry (size, shape, orientation) of the food, e.g. grainy, gritty, flaky, puffy, fibrous; and
(iii) *Moisture/fat-related* those related to the water and/or fat content of the food, e.g. moist, oily, greasy (Szczesniak, 1963, 1991).

Although the texture of food has a profound influence on the acceptability of such categories of foods as meat, fish, fruits and vegetables, until recently, texture has been one of the least studied sensory attributes of food. However, important inroads have now been made in assessing the role of texture and the kinesthetic/somesthetic complex in food acceptance. Many of these inroads are a direct result of the pioneering work of Szczesniak (Szczesniak, 1963, 1971, 1972, 1991; Szczesniak *et al*., 1963; Szczesniak and Kahn, 1971, 1984) and co-workers (Civille and Liska, 1975; Civille and Szczesniak, 1973), who developed the first system for quantifying texture in foods, a method whose offshoots are now used throughout the international food industry, e.g., Hough and Contarini (1994).

1.3.4 Texture preferences

Oral texture perception is much more subject to postnatal developmental processes than are the senses of vision, taste or smell. When an infant is born, his/her masticatory and deglutory development prevents consumption of anything other than liquids. However, by 4 months, sufficient ability to mechanically break down food in the mouth has developed to allow consumption of semi-liquid foods, e.g. soft cereals and mashed fruits and vegetables (Gesell *et al*., 1956). With the development of teeth at around 10 months, lateral chewing movements enable the consumption of solid foods. From this time through adolescence, the child goes through a developmental pattern of preferences for food textures that begins with soft, smooth, unidimensional textures and progresses to firm, rough and complex textures

(Szczesniak, 1972). Szczesniak has suggested that this developmental pattern is innate, since evidence exists that proper development of the jaws and teeth is dependent upon specific chewing loads at certain times during development (Moulton, 1955). With aging, loss of dentition may result in still other changes in texture preference, commonly resulting in a reversion back to the soft food preferences of youth. Such changes in texture preference with aging may well require the tailoring of food products with softer textures to meet the future demands of a rapidly aging population (Peleg, 1993).

In a series of studies with both children and adults, Szczesniak and co-workers (Szczesniak and Kleyn, 1963; Szczesniak, 1971, 1972; Szczesniak and Kahn, 1971, 1984) examined the awareness of texture and its role in food acceptance, using word association and interview techniques. In the earliest of these studies (Szczesniak and Kleyn, 1963), the awareness of food texture was found to be dependent upon such factors as gender (women were found to be more texture conscious than men); experience (individuals who worked with food were more texture conscious than those who did not); and type of food (texture awareness was greater for bland, crispy and crunchy foods). These data were subsequently confirmed (Szczesniak, 1971) and extended to show the effects of socioeconomic class on texture awareness (consumers of higher socioeconomic class were more texture conscious than were those of lower socioeconomic class).

Concerning generalized and/or innate texture preferences, Szczesniak and Kahn (1971) concluded that both crispness and crunchiness are universally liked textures. On the other hand, such texture attributes as soggy, watery, lumpy, sticky, slimy, crumbly and tough, all of which give a sense of lack of control in the mouth, are generally disliked.

The early work of Szczesniak and co-workers is important for two other reasons. First, it showed that the role of texture in food acceptance is highly product-dependent, with foods like celery, peanut butter, bacon, mashed potatoes, tapioca, apples, scrambled eggs, rice and meat being cited most often for the contribution of texture to their overall acceptance. Second, this work showed that socially and culturally learned associations have a significant influence on consumer evaluations of texture (Szczesniak and Kahn, 1971). In their words:

> 'people's awareness of texture is accentuated when expectations are violated or when non-food associations are triggered.' (p. 286)

This important influence of consumer expectations on texture perception has been cited in subsequent work by Vickers (1991) (see Cardello, 1994).

1.3.5 Viscosity–taste interactions

As with vision, the kinesthetic and somesthetic senses can influence food aceptance through a variety of crossmodal interactions. Perhaps the most well-studied of these involves the role of viscosity on taste perception. In the first of a long series of studies on this topic, Mackey and Valassi (1956) and Mackey (1958) showed that thresholds for the four basic tastes were higher in solid foods, foams and gels than in water solutions. They also found that sensitivity for bitter and sweet compounds continuously decreased when the tastants were dispersed in water, gel or oil. In related work, Stone and Oliver (1966) found decreasing sweetness discrimination when sucrose was dissolved in water, cornstarch or carboxymethylcellulose (CMC).

This early work, focusing on taste *sensitivity* and *discrimination*, was subsequently followed by a series of studies examining the effect of viscosity on perceived taste *intensity*. In studies by Moskowitz and Arabie (1970) and Arabie and Moskowitz (1971), the method of magnitude estimation was used to scale the perceived taste intensity of citric acid, glucose, quinine sulfate and sodium chloride dissolved in CMC. The perceived tastes of these compounds were shown to decrease continuously with increasing viscosities of CMC from 1 to 1000 cps (Moskowitz and Arabie, 1970) and again from 1 to 10 000 cps (Arabie and Moskowitz, 1971). Pangborn *et al.* (1973), working with extremely low viscosities of gum, found that the sweetness of sucrose, the bitterness of caffeine, and the sourness of citric acid all decreased at viscosities greater than 16 cps. However, she also found that the sweetness of sodium saccharin increased with increasing viscosity in gums that contained sodium ions. These results reinforced the notion that viscosity–taste interactions are dependent upon the nature of the thickening agent, a suggestion made earlier by Marshall and Vaisey (1972), who observed that sweetness intensity was diminished most in those gels that broke down slowly in the mouth. Although the results of Pangborn *et al.* (1973) showed significant effects of viscosity on taste perception at very low viscosity levels, subsequent studies by Christensen (1977, 1980a,b) found these effects to be more subtle. In Christensen's studies, significant depression of sweetness and saltiness occurred only at the highest viscosities tested (1296 cps) and only for the lowest tested concentrations of sucrose (0.06 M) and NaCl (0.05 M). Christensen's (1980a) effects were also dependent upon the type of thickener (CMC) used, with the effects occurring only with the high viscosity form of CMC. In constrast, Izutsu *et al.* (1981) were able to find significant depression of sweetness with increasing viscosity levels for solutions prepared with low, medium *and* high viscosity forms of CMC.

The combined results of the above studies show that, if there is no physicochemical interaction of the thickening agent and the tastant, higher viscosities will decrease perceived taste intensity. However, different thickeners (gums *vs.* oils *vs.* gels) will have differential effects. Whether these results are due to an interference in the diffusion of the tastant, reduced access to the receptors (as in the reduced taste response created by an oily mouthcoating (Lynch *et al.*, 1993)), or some other mechanism is yet to be resolved (Kokini, 1985; Baines and Morris, 1987). Nevertheless, since thickening agents are commonly used in processed foods, the implications of the data for commercial food formulation are obvious. One example of these practical implications can be seen in the data of Pangborn *et al.* (1978), who demonstrated that increasing the viscosity of beverage products via the addition of hydrocolloid gums decreased the sourness and saltiness of *tomato juice*, decreased the bitterness of *coffee*, and decreased the sourness of *orange drink*. Still other examples of the important effects of thickening agents can be seen in data on viscosity–odor/flavor interactions, below.

1.3.6 Viscosity–odor/flavor interactions

Although much more research has been conducted on the effects of viscosity on taste, some data do exist on the effects of viscosity on food odors and flavor. For example, Pangborn and Szczesniak (1974) showed that a variety of hydrocolloids, when used as carriers for aromatic flavor compounds (acetaldehyde, butyric acid, dimethyl sulfide and acetophenone) reduced their *odor* intensities. A similar depression of their *flavor* intensities was also observed (acetaldehyde excepted). Similarly, in the work described earlier on viscosity–taste interactions by Pangborn *et al.* (1973), increasing viscosity reduced the overall flavor intensity of the samples, as well as their odor intensity. Clearly, such viscosity-related decrements in odor and flavor intensity, when combined with similar effects on taste, can have a deleterious effect on the consumer's acceptance of gum-stabilized products, unless these effects are compensated by other formulation changes.

1.3.7 Temperature–taste interactions

The perception of viscosity is a kinesthetic sensibility, because its evaluation requires the application of oral forces to a fluid and the perception of resistance to these forces. On the other hand, the perception of the temperature of a fluid is a somesthetic sensibility. In one of the earliest studies to examine the effect of temperature on taste thresholds, Hahn and Gunther (1932) found that NaCl (salty) and quinine hydrochloride (bitter) thresholds increased with increasing temperature of solution. However, HCl (sour) thresholds did not change with temperature. Dulcin (sweet) thresholds were lowest at 34°C and increased at higher and lower temperatures. This same V-shaped function was reported many years later by Griffin (1966) to describe the threshold changes with temperature for a variety of sapid compounds. However, Griffin found minimum thresholds at the much lower temperature of 22°C. Pangborn *et al.* (1970) provided general support for the findings of Griffin, showing that sensitivity to NaCl concentrations was greatest in the range from 22–37°C and lower at 0° and 55°C. McBurney *et al.* (1973) further confirmed this relationship for NaCl,

HCl, dulcin and quinine sulfate, finding lowest thresholds between 22°C and 32°C.

In that same year, Moskowitz (1973), using the method of magnitude estimation to scale suprathreshold taste intensities, found that the taste intensities of compounds representing the four basic taste qualities were highest at 35°C, a temperature corresponding to the temperature of the human tongue. However, the rate of growth in perceived magnitude as a function of concentration (exponent of the psychophysical functions) for these compounds did not change with temperature. More recently, Bartoshuk *et al.* (1982) and Calvino (1986) used magnitude estimation to scale the sweetness of sucrose at temperatures ranging from 4–44°C and 7–50°C, respectively. The results of these studies were consistent in showing that increasing temperature increased the perceived sweetness of the solutions, although the effect was most pronounced at the lowest sucrose concentration and converged at about 0.5 M sucrose.

In a review of the area of temperature effects on taste, Green and Frankmann (1987) concluded that, with the exception of the observed effects for sucrose, temperature effects for other compounds were quite variable. In their own studies, Green and Frankmann (1987) found that the temperature of the tongue exerts greater control over the perceived magnitude of the sensation than does solution temperature, suggesting that temperature has a greater effect on the sensory transduction process than on the thermomolecular properties of the solutions. One implication of this finding is that a cold, sweet beverage will become progressively less sweet as the tongue is cooled by the solution; however, as the beverage itself begins to warm from exposure to the ambient air temperature, its cooling effect on the tongue will diminish and sweetness will increase. Such crossmodal time–intensity modulation of perceived taste qualities as a function of both changing food and oral temperatures is an important factor in the determination of the preferred temperature of foods eaten in real life dining situations.

1.3.8 Temperature–oral trigeminal interactions

In addition to the effects of temperature on taste, temperature has been shown to affect sensations mediated by the common chemical sense. Common chemical irritants, e.g., capsaicin, piperine and ethanol, are perceived as burning or hot (Stevens and Lawless, 1988). Further, it is known that these irritants stimulate temperature-sensitive receptors (Croze *et al.*, 1976; Kumazawa *et al.*, 1987). Green (1986) showed enhancement of the perceived chemical warmth of capsaicin by thermal warming of capsaicin solutions, as well as the inverse—the enhancement of perceived thermal warmth by the addition of capsaicin. Stevens and Lawless (1988) expanded upon these results by showing that thermal cooling reduces the perceived irritation of a wide range of irritants, including capsaicin, piperine and ethanol. This research was later confirmed by Green (1990), who showed that thermal cooling reduces the perceived irritation of these compounds, as well as high concentrations (3.0 M) of NaCl. Moreover, Green (1990) demonstrated that pressure applied to the oral mucosa enhances the chemical irritant effect of ethanol, while vibration reduces the irritant effect of capsaicin. The above findings suggest that a wide variety of complex and mutual interactions occur between the somesthetic and common chemical senses. Of course, it must be kept in mind that these interactions are best classified as *intramodal* rather than crossmodal, since the common chemical sense is, itself, a somesthetic sensibility.

1.3.9 Temperature–acceptance relationships

The practical aspects of the role of temperature on food acceptability led Blaker *et al.* (1961) to examine the optimal serving temperatures for hot entrées, beverages, vegetables and potatoes. They found the most preferred temperature of hot entrées and beverages to be 130–170°F and of vegetables and potatoes to be 140–165°F. Similarly, Thompson and Johnson (1963) identified the acceptable range of meats to be 150–160°F and of vegetables and potatoes to be 160–170°F.

While each of these studies examined temperatures within the range of normal serving temperatures, a laboratory study by Cardello and Maller (1982) examined the acceptability of water, selected beverages, and foods at temperatures ranging from 38–135°F. Results of this study showed that:

(i) foods/beverages typically served cold, e.g. fruit beverages, milk, decreased in acceptability with increasing temperatures;
(ii) foods/beverages typically served hot, e.g. beef stew, hashed brown potatoes, increased in acceptability with increasing temperature; and
(iii) foods/beverages normally consumed both hot and cold, e.g. coffee, had maximum acceptance at high and low temperatures and minimum acceptance at room temperature.

The results obtained for foods typically served hot were later replicated by Lester and Kramer (1991) in an institutional foodservice setting and were extended to show that consumption is increased when hot foods are served at increasingly higher temperatures than room temperature.

Although the temperature of a food or beverage can have a profound effect on its perceived taste and acceptance, the effects of temperature on product acceptance are primarily the result of learned associations for its proper temperature. A good example is beer, which is commonly drunk cold in the USA, but at room temperature in most other countries. In a study of the preferred temperature of a novel tropical juice, Zellner *et al.* (1988) instructed one group of subjects that the juice was normally consumed at room temperature, while another group of subjects was told nothing about the juice. Ratings of the acceptability of the beverages revealed that those subjects who were told that the appropriate drinking temperature was room temperature liked the room temperature samples significantly more than did subjects in the control group. Thus, as we have seen before in vision and kinesthesis, learned associations and the expectations that they engender about the sensory properties of food can have a pronounced impact on both the perceived attributes of the food and food acceptability itself.

1.3.10 Other influences of texture on food acceptability

Due primarily to the difficulty in identifying the underlying physical dimensions responsible for most food textures, the development of acceptability functions in texture has progressed only slowly. In fact, in a 1976 review paper on acceptability functions in texture, Moskowitz (1976) cited only the theoretical work of Drake (1974) and a paper on the pleasantness function for the roughness of sandpaper (Ekman *et al.*, 1965). More recent reviews of sensory food texture (Szczesniak, 1987; Moskowitz, 1987) have similarly been devoid of citations on univariate or bivariate acceptability functions in texture (work of Okabe, 1979, excepted). Yet, this is not to say that the role of texture in food acceptance has been ignored.

For example, the *meat* industry was one of the first to focus significant attention on the role of texture in food acceptance. Early work by Cambell (1956) showed that toughness and dryness were the two most undesirable characteristics of meat, while *tenderness* and *juiciness* were ranked high among desirable characteristics. Similarly, Rhodes *et al.* (1955) identified *tenderness* as the most important attribute in consumer preference for meats. Forty years later, Hodgson *et al.* (1992) included three texture attributes among the four sensory attributes used to develop a quality grading system for beef carcasses: *tenderness*, *juiciness*, and *perceived amount of connective tissue*. In the intervening years, numerous studies have examined the effect of carcass characteristics, feed, processing variables, storage conditions, cook times, etc. on meat texture, quality and acceptability. With the advent of restructed meats, extended meat products and meat analogs, the study of meat particle size and the role of additives on sensory texture have become especially important to the consumer acceptance of meat products. Reviews of the role of sensory texture in meat products can be found in Harries *et al.* (1972), Howard (1976), Larmond (1976), Lawrie (1991), Munoz and Chambers (1993) and the technical series on *Developments in Meat Science* (Lawrie, 1989).

Other major food categories in which sensory texture plays an important role in product acceptance are fruits and vegetables, fish products and foods containing fats and oils.

Although a great deal of research on sensory–instrumental relationships of the texture of *fruits and vegetables* has been conducted (Bourne, 1979, 1980; Szczesniak and Ilker, 1988), research on acceptability relationships has been resigned to variety-to-variety comparisons, storage effects, and the like. Similarly, research on *seafood* texture has primarily focused on instrumental measures of quality, with few systematic studies examining the role of perceived texture on consumer acceptability. The latter is partly due to the fact that texture, while an important characteristic of fish products, is considered to be secondary to flavor in the consumer's eye (Connell and Howgate, 1971; Laslett and Bremner, 1979; Wesson *et al.*, 1979; Moskowitz, 1992). Of some interest, though, is the fact that in at least two studies on consumer awareness and acceptability of fish (Szczesniak, 1972; Sawyer *et al.*, 1988), texture was found to be a much more important variable for consumers who *dislike* fish than for those who like it. For the fisheries manager trying to increase the public's consumption of underutilized species, the latter information may be critical for identifying new species to introduce to the marketplace.

Concerning the role of texture in foods containing *fat*, several studies have shown that texture plays an important role in the perception, acceptance, and/or rejection of these foods (Cooper, 1987; Shepherd and Stockley, 1985). Mela (1988) and Mela and Christensen (1988) have established that oral texture sensations play a critical role in the perception of the oiliness of high fat foods, while other studies have shown that such textural attributes as creaminess, thickness and smoothness are the defining characteristics for many high fat foods, e.g. mayonnaise, peanut butter, cream and margarines (Pangborn and Dunkley, 1964; Kokini *et al.*, 1977; Cussler *et al.*, 1979). Unfortunately, for the nutritional well-being of many consumers, the complex of texture and flavor sensations inherent in high-fat foods is irresistably appealing. In a recent study of consumer expectations for regular fat and fat-free products, consumers not only *expected* the acceptability of regular fat products to be better, but they also rated these products to be more acceptable when they were merely *labeled* that way (Tuorila *et al.*, 1994).

In an interesting set of studies designed to uncover the factors controlling the preferences for fatty foods, Drewnoski and co-workers have shown a strong relationship between the perceived texture of fatty foods, their perceived sweetness, and acceptance (Drewnowski and Greenwood, 1983; Drewnowski, 1987, 1991, 1992). Figure 1.6 is taken from Drewnowski (1987) and shows a two-dimensional MDS solution for 10 dairy products. The arrows reflect the direction from the origin for each of the bipolar sensory descriptors used to describe the products. As can be seen, the two major dimensions that underlie the perception of these foods are:

(i) *sweet–sour* (which in these data corresponds to a pleasantness/unpleasantness dimension and, as noted previously, is a common dimension that falls out of MDS analyses of food); and

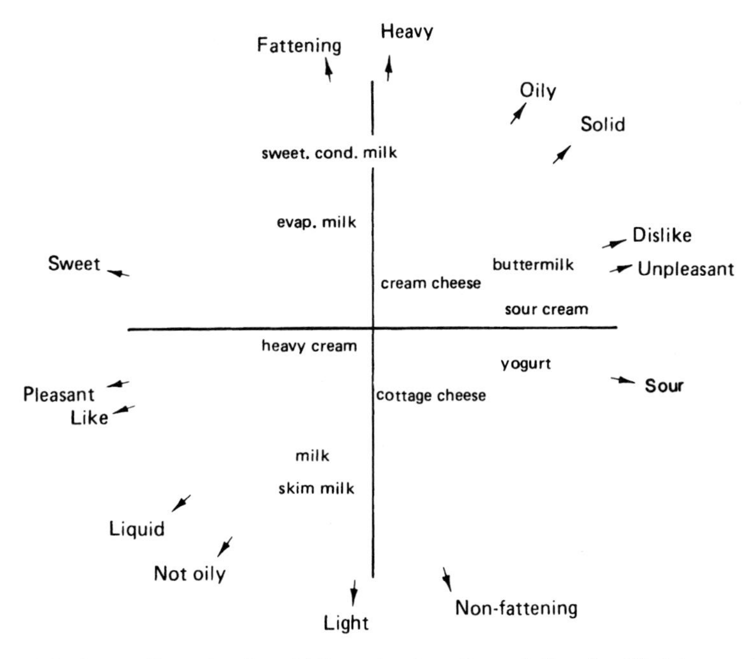

Figure 1.6 A two-dimensional multidimensional scaling solution for 10 dairy products. Arrows with verbal descriptors indicate the direction from the origin along which bipolar attribute vectors pass (from Drewnowski, 1987).

(ii) *fattening–nonfattening* (or light–heavy).

Based on these and other data showing that obese subjects prefer a higher fat-to-sweet ratio than do normal weight subjects, and that anorectic subjects prefer a higher sugar-to-fat ratio, Drewnowski has postulated that the *sweet–fat* complex of foods may have special significance for both the perception of high caloric foods and the development of obesity. While there is, in fact, sufficient evidence that preference for fats is greater in obese individuals (Drewnowski *et al.*, 1985; Mela and Sacchetti, 1991), it appears unlikely that there exists an innate preference for fat (Rogers, 1990). Nevertheless, Drewnowski's data provide convincing evidence of the important role of taste and texture in not only food acceptance, but choice and consumption behavior as well.

1.4 The role of audition in food acceptance

1.4.1 Basic mechanisms and food-related attributes

Audition (hearing) is the sensory experience produced by stimulation of auditory receptors in the human cochlea, a cone-like structure located deep within the inner ear. The most common stimulus for hearing is sound waves, which are funneled into the external auditory meatus (ear canal) by the cartilage of the outer ear (pinna) and then strike the tympanic membrane (ear drum). In response, the tympanic membrane moves in concert with the frequency/intensity of the incoming soundwave. These movements are transmitted to fluids in the cochlea through a set of small, connecting bones. As the fluid in the cochlea moves, it sets up traveling waves of the basilar membrane in the cochlea (Bekesy, 1960). Depending upon the frequency of the incoming sound, these waves cause movement of the auditory receptor hair cells (auditory receptors) on distinct areas of the basilar membrane (Wever, 1962). It is this movement of auditory hair cells that produces depolarization in the receptor cell and initiates the chain of events leading to the transmission of auditory information to the brain.

Three physical attributes of the stimulus waveform—its *amplitude*, *wavelength* and *purity*—define three psychological dimensions of sound: *loudness*, *pitch* and *timbre*. Pitch is considered to be the primary qualitative dimension of sound, and humans can perceive variations in pitch from 20 to 20000 Hz. Unlike vision, where there are only a few distinct receptor types and primary colors, in audition there is a continuous series of qualitatively different pitches. While pitch is encoded in the peripheral nervous system by the location of maximum stimulation on the basilar membrane, this 'tonatopic' coding of pitch is actually preserved as the information travels more centrally in the nervous system, so that even in the brain there is a specific spatial distribution of sound frequencies across the cortical surface (Gulick, 1971). Audition also differs from vision in that it is an *analytic*, rather than a *synthetic*, sensory system. That is, when two sounds of different frequency are presented simultaneously, they are perceived as two distinct sensations. This is in contrast to vision, where, when two colored lights of different wavelengths are presented simultaneously, e.g. red and green, they are synthesized into a single sensation of yellow.

Although hearing is not commonly considered to be a food sense, if one considers the sounds produced by such foods as celery, apples, potato chips, popcorn and crackers, the important role of sound in the perception and acceptance of many foods becomes obvious.

Early research by Drake (1963, 1965) showed that the sounds emitted by different foods during biting and chewing could be differentiated on the basis of their amplitude, frequency and duration. More recently, Vickers and co-workers (Vickers, 1975, 1979, 1988; Vickers and Bourne, 1976) have shown a direct relationship between the sounds emitted by foods during biting and the perception of the *crispness* and *crunchiness* of these foods. Vickers has shown that both crispy and crunchy foods are characterized by irregular sound amplitude as a function of time, but that crispy foods produce a higher frequency (pitch) than crunchy foods. The notion that both total sound level and pitch are critical sensory characteristics for crispy foods was confirmed in a set of studies by Lee *et al.* (1988, 1990). Figure 1.7 presents data from Lee *et al.* (1990) showing the total sound level emitted during the chewing of a fresh (top) and stale (bottom) potato chip as a function of time and number of chews. As is evident from the data, the total sound amplitude emitted upon each chew was greater for the fresh chip than the stale chip. Moreover, correlation of the peak sound amplitude with sensory judgments of the crispiness of the chips were uniformly high ($r >$ 0.87). Figure 1.8, from Lee *et al.* (1988), shows sound spectra obtained from the same subject on the first (a), third (b), and sixth (c) mastication of a potato chip. Again, a large decrement in sound amplitude can be seen from the first to the sixth chew, as the product becomes less crisp. In addition, the higher frequencies of sound (right side of abcissa) can be seen to be greatly diminished by the sixth chew.

The sound spectra/sound pressure data shown in Figures 1.7 and 1.8 are reflections of the fact that crispy and crunchy foods often possess a stiff cellular or porous structure. As dental forces crush this structure, repeated fracturing of the cellular matrix occurs, with each fracture contributing to the total sound spectrum. It is for this reason that the sound spectra in Figures 1.7 and 1.8 are so jagged. This same jaggedness occurs in force–deformation curves of crisp/crunchy products (Gibson and Ashby, 1988). Since jaggedness is amenable to fractal analysis (Normand and Peleg, 1988; Peleg, 1993), recent research (Barrett *et al.*, 1992, 1994) has applied Fourier and fractal analysis to the characterization of the compression curves of crisp and crunchy foods. Such fractal parameters have been shown to correlate well with the perceived crispiness and/or crunchiness of these products,

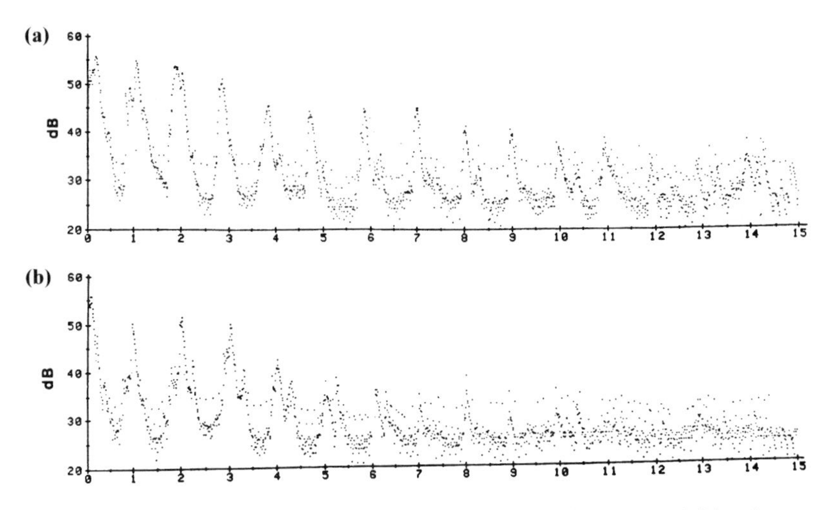

Figure 1.7 Total sound level (dB) as a function of time for (a) a fresh and (b) stale potato chip. Data are for a single subject who chewed each chip at a rate of 1 chew per second (from Lee *et al.*, 1990).

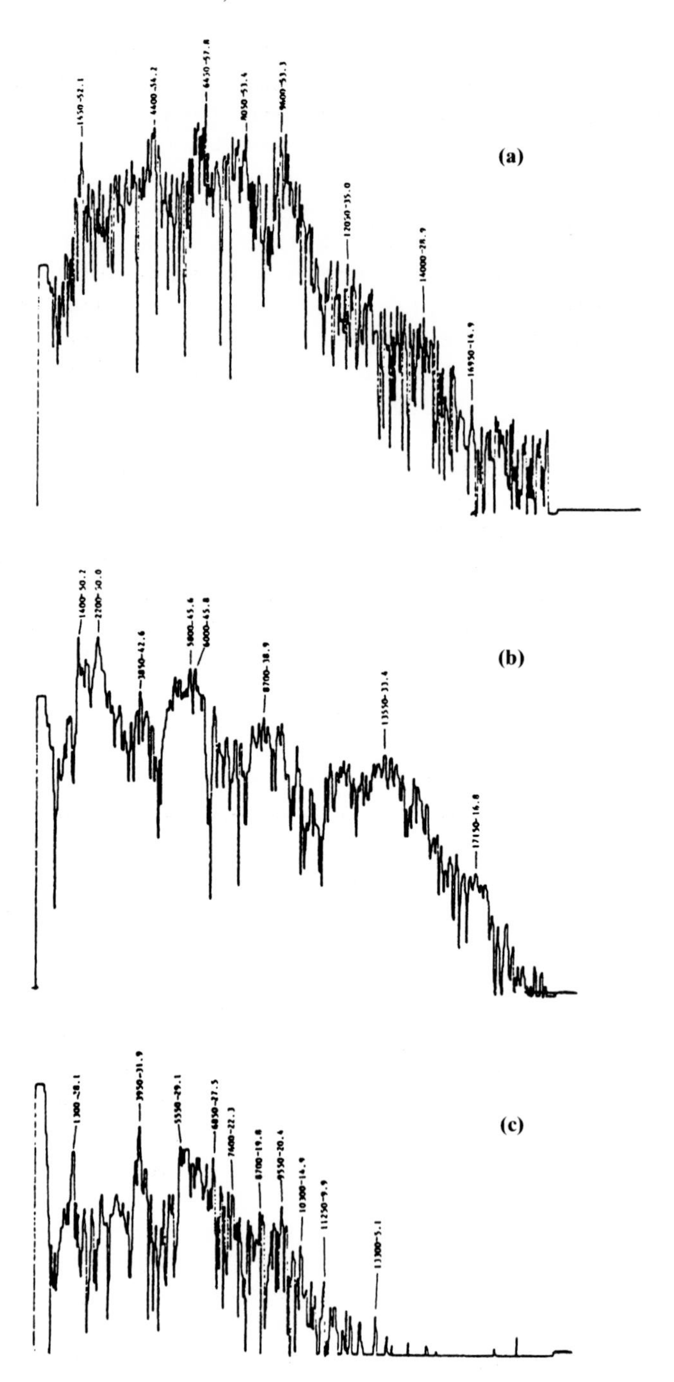

Figure 1.8 Sound spectra for the (a) first, (b) third, and (c) sixth chew of fresh potato chips. Data are from the same subject. The ordinate is sound level (dB) and the abscissa is frequency (Hz), with higher frequencies to the right (from Lee *et al.*, 1988).

suggesting that information encoded in the jaggedness of both the force–deformation curves and sound spectra of these foods may be better predictors of crispiness and crunchiness than are more traditional, rheological (structural) parameters.

1.4.2 Innate auditory preferences

Although the issue of innate preferences for sounds emitted by food is purely speculative at this time, an indirect argument can be made that certain food sounds are universally well-liked. This argument follows from Vickers' (1975, 1979) contention that crispiness and crunchiness are auditory rather than kinesthetic phenomena. This argument, combined with Szczesniak and Kahn's (1971) data showing that crispiness (and to a lesser extent, crunchiness) are universally liked food attributes, can be taken as evidence that the sounds emitted by crispy and crunchy foods constitute a generalized preference in the population. Such an argument has merit from a survival standpoint, since the diet of many herbivores is composed of crispy and crunchy foods, e.g. fruits, nuts and vegetables; while a primary source of nutrition for carnivores during times of survival is the bone marrow of prey, obtained by crushing bones with their teeth (Valkenburgh and Hertel, 1993).

1.4.3 Auditory–acceptance relationships

Concerning the role of audition in hedonics and food acceptance, the work of Iles and Elson (1972) demonstrated that sensory crispness and texture preference are highly correlated. Subsequent research by Vickers (1983) showed that crispness is associated with the pleasantness of biting sounds, giving support to the notion that audition has a direct effect on the acceptance of many foods. In the study by Vickers (1983), standard tape recordings were made of the sounds produced during biting and chewing of 16 foods. Fifty-two subjects judged 9 auditory qualities and the pleasantness of these tape-recorded sounds, as well as the pleasantness of sounds produced while they actually ate samples of the 16 foods. Pleasantness ratings for both types of food sounds (recorded and actual) varied systematically by food type. With the exception of two outlier foods (turnips, crackers), the correlation between the pleasantness ratings of the tape-recorded food sounds and the food sounds accompanied by oral sensations of eating was 0.87. Table 1.2, taken from Vickers (1983), shows the correlations among the ratings of the 9 auditory qualities and pleasantness of the tape-recorded sounds. As can be seen, sounds characterized as 'crisp' and 'crunchy' were the most positively correlated with pleasantness; while sounds characterized as 'tearing' had a negative correlation with pleasantness.

In summary, the effect of the auditory sense on food acceptance is closely

Table 1.2 Correlation matrix for ratings of 9 sensory descriptors and the pleasantness of tape recorded chewing sounds for 16 foods

	Crackly	Pitch	Snap	Light and delicate	Loud	Tearing	Crisp	Brittle	Crunchy
Pitch	0.58								
Snap	0.65	0.53							
Light and delicate	–0.01	–0.01	–0.14						
Loud	0.60	0.54	0.65	–0.30					
Tearing	0.01	0.16	0.10	–0.01	0.15				
Crisp	0.63	0.56	0.69	–0.15	0.69	0.01			
Brittle	0.62	0.52	0.61	–0.11	0.56	0.01	0.62		
Crunchy	0.64	0.52	0.66	–0.17	0.65	0.01	0.77	0.63	
Pleasant	0.30	0.16	0.26	0.01	0.18	–0.12	0.36	0.28	0.31

Source: Vickers (1983).

tied to the perception and acceptance of food texture, most notably crispiness and crunchiness. Research on the role of sound and hearing on other textural qualities, e.g. juiciness (Szczesniak and Ilken, 1988) will likely extend the established effects of hearing on food texture and acceptance. However, the role of hearing on other food attributes, e.g. taste and odor, is far more speculative at the present time.

1.5 The role of gustation in food acceptance

1.5.1 Basic mechanisms and food-related attributes

Gustation (taste) is the sensory experience produced by the stimulation of chemoreceptors located on the tongue, palate and other areas of the oral cavity by chemicals in solution with saliva or with other liquid constituents of food. Taste receptors in man are organized into groups of 50–150 to form specialized structures known as taste buds. The microvilli at the apical end of the taste cells make contact with the oral environment through an opening in the taste bud, known as the taste pore (see Miller, 1995, for a review of taste anatomy). Molecules of food entering the oral cavity are dispersed in the saliva and make contact with the microvilli at the taste pore. Here, the tastant molecules interact with ion channels located on the microvillar membrane (see Roper, 1989 and McLaughlin and Margolskee, 1994 for reviews), causing depolarization of the taste cell and initiating the transmission of taste information along one of three cranial nerves (facial, glossopharyngeal or vagus). A fourth nerve, the trigeminal, also innervates the oral cavity, but, as has been noted previously, subserves the common chemical sense.

On the tongue, taste buds are located on structures called papillae, of which there are four types: fungiform, foliate, circumvallate and filiform. However, only the first three mediate taste sensations. Fungiform papillae are mushroom-shaped structures located on the anterior two-thirds of the tongue, while foliate papillae appear as folds of tissue on the lateral surfaces of the tongue. Although each fungiform papilla houses from 1 to 10 taste buds, foliate papillae contain as many as 120 taste buds each. Circumvallate papillae are large, raised mounds that form a chevron pattern on the posterior dorsal surface of the tongue. Each circumvallate papilla contains upwards of 200 taste buds. Although early studies involving the chemical and electrical stimulation of single papillae provided evidence of taste specificity in individual papillae, i.e. the notion that each papilla is sensitive to only a single taste quality (Bekesy, 1964, 1966), later research demonstrated the multiple sensitivity of individual papillae to a large range of qualitatively different compounds (McCutcheon and Saunders, 1972; Bealer and Smith, 1975; Cardello, 1978, 1981; Kuznicki, 1978; Kuznicki and Cardello, 1986). More recently, a direct tie has been shown between taste papillae density on the tongue and taste sensitivity for fungiform papillae (Zuniga *et al.*, 1993).

Although, phenomenologically, it appears that there are a large number of distinct taste qualities, the bulk of research supports the existence of only four basic taste qualities: salty, sweet, sour and bitter (McBurney, 1974; McBurney and Gent, 1979). Nevertheless, the issue is far from being resolved, because there is lingering concern that these four taste qualities may only be points along a continuous taste spectrum (Erickson and Covey, 1980; Schiffman and Erickson, 1980; Erickson, 1982; Scott and Giza, 1995). In fact, a potential 'fifth' taste quality has received significant attention during the past decade. This sensation has been termed 'umami' taste (Yamaguchi, 1979) and is described as 'savory' or 'delicious' by most subjects. Umami taste is elicited by certain L-amino acids, e.g. monosodium glutamate, and derivatives of 5′-ribonucleotides, e.g. disodium 5′-inosinate. Common foods with significant umami components include mushrooms, fish, tomatoes and seaweed. The volume by Kawamura and Kare (1987) provides an excellent summary of basic research on umami taste.

For each of the four generally accepted taste qualities, a set of adequate stimuli and mechanisms of molecular interaction with the taste receptor cell has been proposed. For example, the adequate stimulus to elicit the *sour* taste has long been known to be proton-donating molecules. Thus, the sourness of such foods as citrus fruits, pickles and yogurt are all directly attributable to the presence of hydrogen ions. However, the number of hydrogen ions required for perception of a sour taste is different for weak acids than for strong acids; e.g. acetic acid tastes more sour than mineral acid at the same pH. Moreover, some amino acids are sweet and others are bitter. These facts suggest that the anion and any undissociated acid modifies the taste of the compound (Ganzevles and Kroeze, 1987). Another modulating factor may be the lipophilicity of the compound, which affects access of the compound to the receptor (Gardner, 1980). Although several models of sour receptor stimulation have been proposed (Beidler, 1967, 1971; Makhlouf and Blum, 1972; Price and DeSimone, 1977), many assumed that taste transduction requires absorption of the chemical stimulus to the receptor membrane. However, several lines of evidence now suggest that sour taste transduction occurs through blockage of potassium conductance in potassium channels on the taste cell membrane (Kinnamon and Roper, 1988) or through direct penetration of the hydrogen ion through sodium channels on the membrane (Gilbertson *et al.*, 1992).

Like sourness, *saltiness* results from stimulation by ions in solution. Electrophysiological evidence (Beidler, 1954, 1978), as well as human psychophysical data (Bartoshuk, 1980; Murphy *et al.*, 1981), have established that the sodium ion (Na^+) and certain cations of other salts are responsible for eliciting the salty taste, while the anions play an inhibitory role. In general, low molecular weight salts have a predominant salty taste, while high molecular weight salts have a bitter taste. Moreover, at low concentrations, NaCl and other salts taste sweet (Renqvist, 1919; Dzendolet and Meiselman, 1967; Cardello and Murphy, 1977; Bartoshuk *et al.*, 1978). As concentration increases, the taste may change to salty, sour and/or bitter. It has been suggested that these taste quality shifts are caused by physicochemical changes that occur in these salts as a function of concentration (Dzendolet, 1968; Cardello, 1979), converting the effective stimulus for one quality to that for another.

The initial event in the perception of salt taste was believed to be the movement of sodium ions across the sodium/hydrogen uniport on the apical membrane of taste cells (DeSimone *et al.*, 1981). However, more recent evidence based on the taste effects of amiloride (which blocks salty taste in humans) suggested that modification was needed to this 'ion transport theory' of salt taste (Schiffman *et al.*, 1983; Desor and Finn, 1989). This led to a series of studies showing that salt taste transduction occurs by passive diffusion of sodium ions through amiloride-sensitive sodium channels (DeSimone and Farrel, 1985; Schiffman *et al.*, 1986).

Unlike sourness and saltiness, *sweet* taste stimulation is more complex. Sweet taste can be elicited by a variety of organic compounds, as well as by inorganic compounds, like lead and beryllium salts. While sugars are still the most common sweeteners, many non-nutritive sweeteners such as saccharine, aspartame, acesulfame-K and the dihydrochalcone sweeteners are potent stimuli for eliciting the sweet taste (O'Brien-Nabors and Gelardi, 1986). Numerous theories have been proposed over the years to explain the diversity of compounds that elicit the sweet taste. The most generally accepted is the hydrogen-bond theory (Shallenberger and Acree, 1967). This theory proposes that sweet-tasting substances possess an AH–B system, where AH^+ is a hydrogen ion bonded to an electronegative atom, and

B is an electronegative atom in sufficient proximity to permit the formation of a hydrogen bond with the receptor molecules on the taste cell surface. Complicating the development of an adequate theory of sweet stimulation is evidence that shows that there may be more than one type of sweet sensation (Boudreau, 1986) and perhaps as many as six different receptor sites for sugars alone (Jakinovich and Sugarman, 1988).

Whereas salt and sour taste transduction occurs through interactions with receptor membrane channels, sweet taste transduction occurs through interactions of the sweet molecule with membrane receptor proteins, resulting in the activation of G-proteins and cyclic nucleotides inside the taste cell (Tonosaki and Funakoshi, 1988; Striem *et al.*, 1989). Through second messenger pathways, this activation of G-proteins ultimately results in the blockage of potassium channels and depolarization of the taste cell.

The last of the four primary taste qualities is the *bitter* taste. Bitterness is important for its role in alerting the organism to potentially dangerous constituents of foods. Common bitter-tasting compounds include the alkaloids, e.g. quinine, caffeine and nicotine, heavy halide salts and certain amino acids. Of course, many quite safe, nutritious and appealing foods, like lettuce, coffee, tea, cucumber, beer and chocolate, are also bitter. Not surprisingly, it has been shown that intake of such bitter tasting foods and beverages is inversely related to one's sensitivity to bitter compounds (Tanimura and Mattes, 1993).

Curiously, some compounds, such as phenylthiocarbamide (PTC) and 6-n-propylthiouracil (PROP), are extremely bitter to some individuals, but tasteless to others (Fox, 1932; Kalmus, 1971; Bartoshuk, 1979). The percentage of tasters *vs.* nontasters of these compounds has been shown to vary by ethnic group (Allison and Blumberg, 1959), reflecting the fact that this taste ability is related to a Mendelian recessive characteristic among nontasters. Moreover, it has been shown that tasters *vs.* nontasters of these compounds also differ in their perception of the bitterness of certain other bitter compounds, e.g. urea and KCl (Bartoshuk *et al.*, 1988), but not all, e.g. caffeine (Mela, 1989; Schifferstein and Frijters, 1991). The apparent difference in the results of these studies may well be due to differences in the criteria used to define PTC/PROP taster status (Bartoshuk *et al.*, 1992; Marks *et al.*, 1992). Bartoshuk *et al.* (1992) have, in fact, proposed that there are three categories of PTC/PROP tasters and that only the most sensitive of the supertasters show the classic taster–nontaster differences.

While it is known that there are at least three different mechanisms of bitter taste transduction in humans (Herness and Pfaffmann, 1986; Yokomukai *et al.*, 1993), the similarity in chemical structure between many bitter and sweet-tasting compounds (e.g. the α anomer of D-mannose is sweet, but the β anomer is bitter), has led to proposals that the physicochemical feature common to bitter-tasting compounds is an AH–B system, like that proposed for sweet taste, but with a different AH–B distance (Kubota and Kubo, 1969; Temussi *et al.*, 1978; Tancredi *et al.*, 1979). Also, as in the case of other taste qualities, the lipophilicity of the compound has been implicated in bitterness perception (Gardner, 1978, 1979). The similarity in chemical structures between bitter and sweet-tasting compounds has led to the search for debittering compounds among those compounds known to suppress sweetness (Roy, 1990).

In terms of transduction processes for bitter, one well-established mechanism involves interaction of the bitter substrate with receptor surface cells, thereby activating G-proteins and leading to the release of calcium ions. The release of the calcium ions causes hyperpolarization of the cell membrane by activation of potassium channels (Hwang *et al.*, 1990; Akabas *et al.*, 1988), leading to the generation of action potentials. A variety of other transduction processes for bitter taste have also been proposed, but they are more controversial.

1.5.2 Innate taste preferences

Of all sensory modalities, taste is the one for which the best argument can be made for the existence of innate preferences. In a classic series of studies examining the facial expressions of infants, Steiner (1973, 1977, 1979) found hedonically toned facial responses to sweet, sour and bitter compounds in infants only a few hours old. However, these responses were elicited by extremely high concentrations of tastants. When weaker stimuli were used (Ganchrow *et al.*, 1983), results were less clear. Steiner's work was partially replicated in a study by Rosenstein and Oster (1988), who also found that infants responded with positive facial responses to sweet tastes and with negative facial responses to bitter tastes. However, unequivocal responses to salty and sour stimuli were not discernible.

In a series of studies examining the intake by infants of solutions representing the four basic taste qualities, Desor *et al.* (1973, 1975) and Maller and Desor (1973) found a strong preference for sweet solutions, but little difference in the preference for sour, salty and bitter solutions. In a related study, Desor *et al.* (1975) found suppression of intake to sweet solutions in infants by the addition of citric acid, leading to the conclusion that the sour taste of citric acid was innately unpleasant. In a recent review of this area, Cowart and Beauchamp (1990) have concluded that there is good evidence of an innate preference for sweet taste and reasonable evidence supporting an innate rejection of sour tastes. Evidence for the innate rejection of bitter tastes is more problematic, because some bitter compounds, e.g. quinine, produce a strong negative response in infants, whereas other bitter compounds, e.g. urea, produce little effect. With regard to salty taste, it seems clear that infants are relatively insensitive to salt at birth. However, a preference for salt does emerge as early as four months after birth

(Beauchamp *et al.*, 1986). Although this preference has been attributed to nervous system maturation (Cowart and Beauchamp, 1986), children as young as six months old may have their preference for salt modified by experience (Birch, 1979; Birch and Marlin, 1982; Harris and Booth, 1987). A more detailed review of the ontogeny of human taste preferences can be found in Ganchrow (1995).

1.5.3 Effects of aging on taste and food acceptance

While there appears to be a strong biological influence on taste preference at birth, biological influences also come into play during aging. Numerous investigators have demonstrated large decrements in taste sensitivity (thresholds), discrimination, and suprathreshold intensity judgments with increasing age (see Murphy, 1986; Murphy and Gilmore, 1990; and Schiffman and Warwick, 1991, for reviews). Although near-threshold taste loss occurs to some extent for all four taste qualities, at suprathreshold levels it is more quality-dependent, with bitterness showing the greatest loss and sweetness the least (Weiffenbach *et al.*, 1986; Murphy and Gilmore, 1989, 1990). Age-related shifts in hedonic preferences for tastants have also been known for some time. Laird and Breen (1939) showed an increase in the preference for sour *vs.* sweet tastes in older individuals. Similarly, Desor *et al.* (1975) showed that adults preferred lower levels of sweetness than did children, while Enns *et al.* (1979) reported a reduced preference for sucrose in elderly *vs.* college-age students. Figure 1.9 from Murphy and Withee (1986) shows the effect of age on the growth of pleasantness with sucrose concentration in water (a) and beverage base (b). As can be seen, pleasantness is in inverted U-function for all test populations. However, the effect of age is to shift the concentration of maximal preference from a lower to a higher concentration. A similar preference shift to higher concentrations of salt was also observed, confirming earlier work by Pangborn *et al.* (1983) in which they showed that older subjects add more salt to broths in order to maximize liking.

Schiffmann (1991) has made the argument that losses in the ability to sense saltiness and sweetness can create severe nutritional problems in elderly diabetic and hypertensive populations, because these populations are likely to compensate for losses in taste intensity by increasing the levels of salt and sugar in their food. However, since bitter taste appears to show the greatest taste decrements, the more common effect is likely to be a loss of the ability to detect the bitter flavors of herbs, seasonings, coffee, tea and similar bitter-tasting foods and ingredients. Age-related losses in taste are an important contributory cause to the decline in the enjoyment of food by the elderly. It is one of the most telling examples of the role of human sensory function on food acceptance, choice and consumption behaviors.

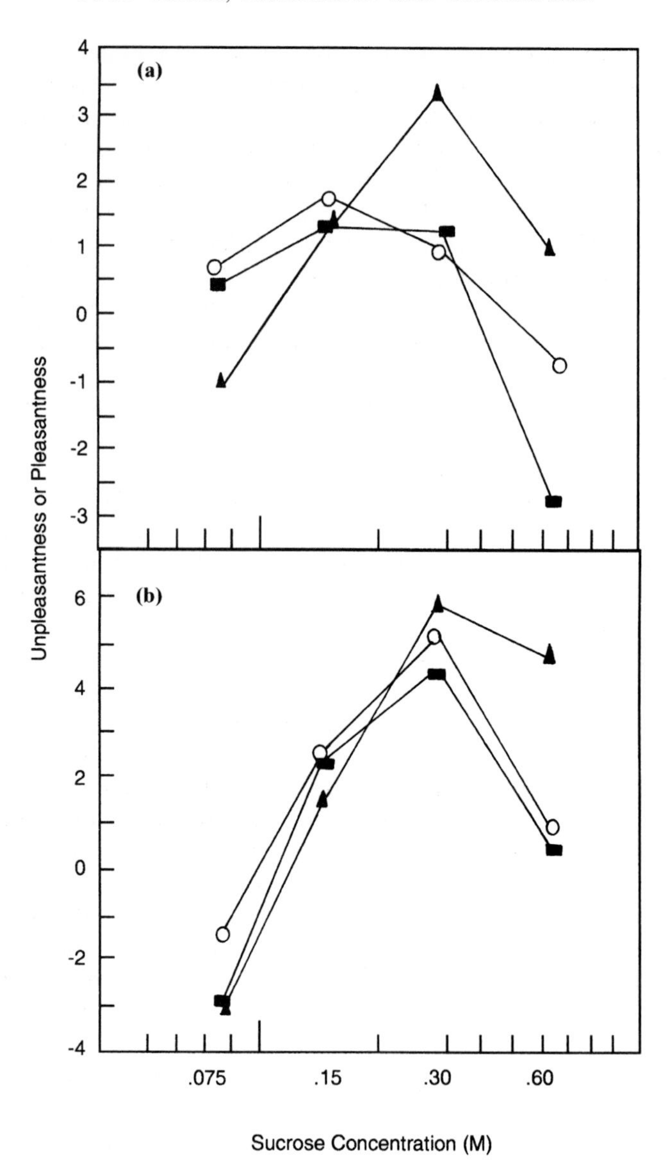

Figure 1.9 Plot of pleasantness/unpleasantness ratings for sucrose (a) in water and (b) in beverage base as a function of age of subjects (from Murphy and Withee, 1986). ○, Young; ■, middle; ▲, elderly.

1.5.4 Intramodal taste interactions

Almost all foods and beverages are comprised of a complex mix of ingredients, each with its own unique taste profile. But what happens when separate tastes are mixed? Are each of the components still perceived in the mixture (providing support to the notion that taste is an analytic sense) or does a new, singular taste emerge (support that taste is synthetic)? The majority of research to date supports the notion that the components of the mixture are still perceived separately and that taste is an analytic sense. However, a number of studies have shown that taste mixtures are not always perceived as true mixtures of independent taste qualities, nor are single tastants always perceived as singular (Erickson, 1982; Schiffman and Erickson, 1980; Erickson and Covey, 1980; O'Mahony *et al.*, 1983).

In addition to possible *qualitative* changes, most taste compounds exhibit a wide variety of *quantitative* interactions when mixed with compounds of either the same or different taste qualities. These interactive effects may differ, depending upon whether the two compounds act simultaneously or sequentially on the taste receptors.

All sensory stimuli *adapt* after a period of sustained stimulation. This is why, for example, the salt in saliva cannot be tasted. When adaptation produced by one stimulus reduces the perceived intensity of a different stimulus having the *same* taste quality, e.g. two different salts, we speak of *cross*-adaptation as having occurred (Bartoshuk *et al.*, 1964; Meiselman, 1968; McBurney, 1972; McBurney *et al.*, 1972). When stimuli of *different* qualities are presented one after another, an increase in the perceived intensity of the second stimulus may occur. This is known as cross-*enhancement*. Perhaps the most ubiquitous example of this phenomenon is 'water taste' (Bartoshuk *et al.*, 1963; McBurney and Schick, 1971). The latter refers to the bitter taste of distilled water, which results from the pre-adaptation of the oral receptors to the salt in saliva. In fact, a simple water stimulus can take on a variety of tastes depending upon the qualitative nature of the pre-adapting stimulus (McBurney and Schick, 1971).

Although cross-enhancement can occur when stimuli of different qualities are presented sequentially, taste *suppression* occurs if one combines the two tastants simultaneously in a single solution. In such a case, the perceived intensity of the mixed solution will be less than the total of the perceived intensities of the unmixed solutions. This suppression of the taste intensity of mixture components is ubiquitous, resulting in the intensity of any component in the mixture being less than the intensity of that same component at the same concentration in an unmixed solution (Lawless, 1987; Kroeze, 1990).

It has been argued that the physiological mechanism for mixture suppression is located centrally in the nervous system (Kroeze, 1979; Lawless, 1979). This argument is based on studies of the release of mixture suppression by pre-adaptation and the demonstration that mixture components presented to different sides of the tongue still produce the suppression (Lawless, 1978; Kroeze and Bartoshuk, 1985). The important point to keep in mind about mixture suppression is that it occurs in almost all cases where two tastants are combined. For multi-ingredient foods, mixture suppression is a critical concern. For example, take the simple case of adding an acidulant to a sweet beverage base formulation. As we shall see, the acceptability of the sweetness of sucrose as a function of concentration is an inverted U function, like that in Figure 1.2. If an acidulant is added to the sucrose base, mixture suppression will reduce the perceived sweetness of the beverage, shifting its acceptance higher or lower, depending upon where on the acceptability function the beverage base lies. Thus, slight formulation changes in multi-ingredient foods can have cascading effects, not only on the food's overall flavor profile, but, most importantly, on its acceptance to the consumer.

Other forms of intramodal taste interactions are those that result from the action of taste modifiers, e.g. compounds like monosodium glutamate (MSG) and inosine 5′ monophosphate (IMP), which can raise sour and/or bitter taste thresholds (Yamaguchi, 1987); *gymnema sylvestre* and *ziziphus jujuba*, which can block or reduce the perception of sweet taste (Bartoshuk *et al.*, 1969; Meiselman and Halpern, 1970a,b; Bartoshuk, 1974; Meiselman *et al.*, 1976); and miracle fruit, which can change the sour taste of foods to a sweet taste (Bartoshuk *et al.*, 1969; Bartoshuk, 1974, 1980). While MSG and 5′ ribonucleotides have found wide application in the food industry, most notoriously in Chinese food, *gymnema sylvestre*, miracle fruit, and other, more exotic taste modifiers have yet to be adopted for commercial food use.

1.5.5 Taste–viscosity interactions

It would seem logical from data presented earlier in this chapter that if color and texture can influence taste, then taste must influence color and texture. While few research studies have addressed the effects of taste on the visual appearance of food, the available data on taste–viscosity relationships support the notion that mutual interactive effects may occur between taste and the oral kinesthetic/somesthetic senses. For example, research by Christensen (1977) showed that increasing both the salty and sweet tastes of simple solutions resulted in an increase in the judged viscosity of those solutions. A subsequent study by Christensen (1980b) confirmed this effect for sweetness. However, NaCl was found to decrease perceived viscosity, as did citric acid. No effects were observed for caffeine. A likely explanation for the effect observed for sweetness is that subjects have come to associate high sugar levels with high viscosity in solutions/beverages. Thus, when they taste a solution/beverage with a high sweetness level, the judged viscosity is raised to match the learned sweetness–viscosity relationship. Effects for other taste qualities appear elusive, because there are few foods that contain sufficiently high concentrations of bitter, sour or salty compounds that viscosity levels would be affected.

1.5.6 Taste–oral trigeminal interactions

A common technique to cool the burning sensation that accompanies foods with a very hot, spicy 'taste' is to take a drink of water. While water can be somewhat effective in this situation (Stevens and Lawless, 1986; Green, 1986), a variety of foods, e.g. rice, butter, crackers, sugar, and beverages, e.g., pineapple juice and beer, are more effective (Stevens and Lawless, 1986; Hutchinson *et al.*, 1990). This leads to the question of how various taste substances interact with trigeminal sensations to reduce or augment those sensations. In three separate studies, data have clearly shown the effectiveness of sucrose in reducing the burning sensation that results from both capsaicin (Sizer and Harris, 1985; Stevens and Lawless, 1986; Nasrawi and Pangborn, 1989) and piperine (Stevens and Lawless, 1986). In only one study did citric acid reduce the burn of trigeminal stimuli (Stevens and Lawless, 1986). Salt and quinine (bitter) had no effect.

In the same way that taste has been shown to affect the sensory response to trigeminal stimulation, so has trigeminal stimulation been shown to affect taste. Although early work by Szolcsanyi (1977) showed no effect of capsaicin on taste thresholds, later research on suprathreshold taste function has, indeed, shown such effects. Lawless and Stevens (1984) showed that a rinse of either capsaicin or piperine reduced the perceived intensity of compounds representing three of the four taste qualities (no effect of capsaicin was observed on NaCl intensity, although piperine did show an effect). However, subsequent work by Lawless *et al.* (1985) found decrements in the taste of sucrose, quinine, citric acid and NaCl following a 2 ppm rinse of capsaicin. Although Cowart (1987) was unable to demonstrate similar effects using 2 ppm capsaicin, either as a rinse or in a mixture with these solutions, Prescott *et al.* (1993) found suppression of sucrose intensity by capsaicin in solution with the sucrose, as well as some minor suppression of saltiness in high capsaicin/low NaCl mixtures. In a still more interesting study, Gilmore and Green (1993) examined the effect of a capsaicin rinse on the saltiness, sourness and irritation produced by high concentrations of NaCl and citric acid. Their results showed a significant desensitization effect of the capsaicin on both the perceived taste intensity *and* perceived irritation of these compounds. Thus, it seems that the common chemical sense has broad interactive effects with both basic taste function, as well as with the irritant effects produced by compounds other than capsaicin. These broad interactive effects are consistent with the wide distribution of free nerve endings subserving the common chemical sense throughout the oral cavity.

1.5.7 Taste–smell/flavor interactions

The most well-studied crossmodal sensory interaction with taste is the one that occurs with smell. Since taste and smell combine to produce our perception of food flavors, it is reasonable to assume that taste and odor show mutual interaction. Interestingly, the available data do not support the type and extent of interactions that one might expect.

Research by Murphy *et al.* (1977) and Murphy and Cain (1980) examined perceived taste intensity, perceived odor intensity and the overall intensity of taste–odor mixtures of saccharine and ethyl butyrate, sucrose and citral, and NaCl and citral. Their results showed a high degree of independence of taste and smell, i.e. overall intensity of the tastant–odorant mixture was perceived to be a simple additive function of the taste and odor components separately. However, when samples of the odorants were presented alone, subjects commonly perceived the sensory effect to be one of taste, not odor. This illusory effect is reminiscent of the early report by Hollingworth and Poffenberger (1917) that the uncertainty about the locus of stimulation of odor/taste stimuli are commonly resolved in favor of taste. Although the effect is totally illusory (pinching the nose to block retronasal passage of the volatile compound to the olfactory receptors will prevent the effect), it highlights the role that odorant stimulation plays in the perceived taste of foods in the mouth. Concerning the basic finding of additivity of taste and smell, it is interesting that other researchers who have presented tastants directly into the oral cavity and odorants directly into the nares have shown less than total additivity (Burdach *et al.*, 1984; Enns and Hornung, 1985; Hornung and Enns, 1986). In at least one case, this suppression amounted to 33% less than the predicted value based on additivity (Enns and Hornung, 1985). Whether the differences between the studies are due to cognitive or task-dependent differences has yet to be resolved, but Hornung and Enns (Hornung and Enns, 1986, 1989; Enns and Hornung, 1988) have offered a model that accounts for these taste–smell mixtures in terms of cognitive factors.

In addition to studies which have looked at taste–smell interactions in model mixtures, a variety of studies have looked at taste–odor/flavor interactions in real foods and beverages. For example, in a study of the effect of sweetness (sucrose) on the odor attributes of fruit juice, Von Sydow *et al.* (1974) showed that ratings of the intensity of pleasant odors increased, while ratings of the intensity of unpleasant odors decreased in response to added sugar. Working with lemon juice, McBride and Johnson (1987) found enhancement of the overall flavor of the juice with increasing levels of both sucrose and citric acid. In a similar study of blackberry juice, Perng and McDaniel (1989) showed that sucrose enhanced the flavor ratings of the juice, but, in contrast to McBride and Johnson, they found increasing levels of acidity to lower ratings of flavor. In research using aspartame as a sweetener, Wiseman and McDaniel (1989), Baldwin and Korschgen (1979), and Bonnans and Noble (1993) showed that aspartame enhanced and prolonged the fruitiness of fruit-flavored solutions more than did sucrose.

ship. As seen earlier for the effect of appropriate *vs.* inappropriate colors on acceptance, taste–acceptance relationships are greatly influenced by the appropriateness of the taste within the overall food/taste context. Since 'appropriateness' is a learned phenomenon, independent of innate taste preferences, these data again point to the interdependence of food acceptance behavior on both sensory factors and learned associations.

1.5.9 Taste preference/aversion learning

As suggested by the data on the liking of sweetness in eggs, the liking of the taste of any food may well be controlled as much by learned associations and contextual influences as by innate preference mechanisms. This point has been made clearly in the case of sweetness by Booth *et al.* (1987). Such learned associations can have profound effects on both learned preferences and aversions for food. Although the wide variety of mechanisms by which learning and conditioning affects human food acceptance goes beyond the scope of this review of *sensory* mechanisms in food acceptance, the critical importance of the sensory stimulus in all conditioning paradigms warrants comment here.

As noted earlier, food acceptance is a construct measured solely through verbal and nonverbal responses of subjects. Like any other response, liking/disliking responses are subject to conditioning phenomena. In the case of food *aversion* learning, numerous studies in both animals (Garcia *et al.*, 1966; Garcia and Koelling, 1966) and humans (Garb and Stunkard, 1974; Bernstein and Webster, 1980; Logue *et al.*, 1981; Pelchat and Rozin, 1982) have shown that a single pairing of food with illness is sufficient to establish an intense and prolonged aversion to the food. In fact, in one large-scale survey (Midkiff and Bernstein, 1985), up to 57% of respondents reported to have at least one learned food aversion. Interestingly, a majority of these aversions appear to be for protein foods. Concerning the relative role of the senses in learned aversions, Bartoshuk and Wolfe (1990) have suggested that the aversion that forms to foods in these cases is by association with the food's odor, not its taste. This argument derives strength from a survival perspective, since to be of future benefit, the aversion to illness-producing food should occur at the earliest sensory contact with the food. The reader is referred to the chapter by Chambers and Bernstein (1995) for a recent review of this area.

In the case of learned food *preferences*, mere exposure to a novel or disliked taste, odor or food can increase its acceptance (Torrance, 1958; Capretta and Rawls, 1974; Domjan, 1976; Cain and Johnson, 1978; Balogh and Porter, 1986; Davis and Porter, 1991). This effect serves as a potent strategy for overcoming food neophobia, the reluctance to consume novel foods (Birch and Marlin, 1982; Birch *et al.*, 1987). In what may be the most interesting example of learned preferences, Rozin and Schiller (1980) have shown that repeated exposure to increasing levels of chili in food will reduce even the innate aversion to chili peppers. This effect has been shown to shift the sensory–hedonic function for the active ingredient in chili peppers, capsaicin, from a monotonically decreasing function to an inverted U function (Stevenson and Yeomans, 1993).

Other research has shown increases in the preference for foods when satiety is used as the unconditioned stimulus (Booth, 1972, 1981, 1982, 1990; Mehiel and Bolles, 1988; Birch *et al.*, 1990). In these studies, the pairing of a neutral or unique flavor with the sensation of satiety has resulted in an increase in the preference for that flavor. Still other studies in both animals and man have demonstrated that a neutral taste paired with a sweet taste will increase the preference for the neutral flavor (Fanselow and Birk, 1982; Zellner *et al.*, 1983; Breslin *et al.*, 1990).

Based on these representative findings, it seems clear that taste/food aversions are highly susceptible to conditioning, especially through the mechanism of postingestional illness. On the other hand, the conditioning of taste/food preferences is more subtle, with evidence pointing toward positive effects from mere exposure, flavor–flavor associations, and postingestive satiety.

1.5.10 Sensory specific satiety

In addition to the influence of learned associations on food acceptance, food acceptance has also been shown to be susceptible to a form of sensory/perceptual adaptation. When a food is eaten, its hedonic tone (pleasantness) declines relative to foods that have not been eaten. This effect has been well described and documented (Siegel and Pilgrim, 1958; Cabanac, 1971; Rolls *et al.*, 1981, 1982, 1988a,b). Although this phenomenon was first viewed to be related to physiological satiation, the fact that the effect has been shown to occur in response to the taste of the food even when it is not consumed (Stang, 1975; Murphy, 1982; Drewnowski *et al.*, 1982) has led to the conclusion that the effect is sensory or perceptual in nature. This type of adaptation has also been shown to vary by food type (Rolls *et al.*, 1981; Drewnowski *et al.*, 1982; Johnson and Vickers, 1991), with some foods, e.g. sweet beverages, showing a strong influence of sensory specific satiety, and other foods, e.g. bread, showing little effect. Although some evidence has shown that certain sensory attributes are more or less susceptible to sensory specific satiety (Drewnowski *et al.*, 1982; Johnson and Vickers, 1991), a complete understanding of the mechanism underlying the phenomenon and its generalizability is lacking. Nevertheless, this phenomenon is a powerful one, influencing not only the acceptance of food, but also its intake, as determined by the cessation of eating (Wisniewski *et al.*, 1992; Swithers and Hall, 1994).

1.6 The role of olfaction in food acceptance

1.6.1 Basic mechanisms and dimensions

Smell is the sensory experience produced by the stimulation of receptors located in the olfactory epithelium by airborne chemical compounds. While eating, volatile components of food reach the receptors in the olfactory epithelium through both the anterior and posterior nares. The latter route enables odorant molecules to pass directly from the mouth to the olfactory epithelium (see Lanza and Clerico, 1995, for a review of nasal anatomy). Like taste receptor cells, olfactory receptor cells have hair-like structures (cilia) which project into the mucous layer covering the epithelium. Olfactory molecules interact with the receptors at the surface of these cilia (Getchell, 1986; Bruch, 1990). Of some importance is the fact that olfactory molecules that reach the epithelium are spatially distributed in a nonrandom fashion, as a result of a variety of physical mechanisms (Mozell and Hornung, 1985). There is also some evidence that receptor cell sensitivities are nonrandomly distributed across the epithelium (Kubie and Moulton, 1980; Mackay-Sim *et al.*, 1982; Kauer, 1991). The combination of both these inherent and imposed spatial distributions provides the basis for a spatial mechanism of olfactory quality coding, similar to that found in audition.

Unlike taste, where there are only four qualitatively distinct sensations, there are literally thousands of different odor qualities. In order to perceptually categorize this large a number of distinct qualities, a variety of odor classification schemes have been developed. These have been described and reviewed in a number of excellent sources (Moncrieff, 1967; Harper *et al.*, 1968; Lawless, 1988). Perhaps the most useful of these classification schemes is the one proposed by Amoore (Amoore *et al.*, 1964; Amoore, 1965, 1975). This classification scheme garners its strength from the fact that it is tied to a viable theory of olfactory functioning. In Amoore's system, eight fundamentally distinct categories of odors (odor primaries) have been identified through the study of specific anosmias, i.e. the clinical absence of the ability to smell a particular odorant (Amoore and Steinle, 1991). These primary odor qualities include *sweaty*, *spermous*, *fishy*, *camphoraceous*, *minty*, *malty* and *urinous*. For each of these primaries, Amoore has proposed the existence of a specific stereochemical arrangement of stimulant molecules and receptor sites. This 'lock and key' hypothesis presumes that the initial event in the olfactory transduction process is the fitting of an odorant molecule into a conformationally similar receptor site (Amoore, 1965, 1975). While the lock and key theory is a viable hypothesis for the mechanism of odorant–receptor interaction, several authors have noted that some odorant molecules which are almost identical in size and shape have vastly different odor quality (Schiffman, 1974; Wright, 1982). Such anomalies pose problems for Amoore's theory and suggest that alternative mechanisms may well be operating.

Regardless of the exact mechanism of odorant interaction with the receptor surface, biochemical research has established the fact that this initial binding event is followed by second messenger activity involving G-proteins and cyclic AMP (Gold *et al.*, 1989; Firestein *et al.*, 1991). It is these second messenger events that open ion channels in the olfactory cell membrane to cause depolarization of the receptor cell (Lancet, 1986; Lancet and Pace, 1987). Other second messenger systems have also been proposed (Bruch and Teeter, 1989), and the volumes by Brand *et al.* (1989) and Doty (1995) include excellent summaries of these different mechanisms.

Recently, Buck and Axel (1991) have isolated members of a large multigene family that encode specific olfactory receptor proteins. Based on their research, it is believed that upwards of 1000 different olfactory genes may exist. Such a large number of receptor proteins would make it possible to account for the extremely large number (>10 000) of perceptibly different odors in human olfaction. It also suggests that almost all olfactory information processing may occur in the peripheral receptor system, not in the brain. Such a possibility is consistent with the fact that many animals who have a small brain still possess a highly evolved sense of smell.

1.6.2 Innate odor preferences

Evidence for the existence of innate odor preferences parallels research demonstrating innate taste preferences. For example, early research by Steiner (1977) examined the facial responses of infants during the first few hours after birth. By presenting stimuli consisting of butter, banana, vanilla, shrimp and rotten egg odors to these infants and recording their facial expressions for scoring by judges, Steiner showed a much higher proportion of negative facial responses to the shrimp (fishy) and rotten egg odors than to the other samples presented. Steiner concluded that the infant facial expressions that were classified by observers as reflecting acceptance are elicited by the same odors that adults find pleasant. Similarly, those infant facial expressions that reflect odor aversions are elicited by the same stimuli which adults find pungent or disgusting. In more recent work (Makin and Porter, 1989; Porter *et al.*, 1992), evidence for an innate preference for the odor of breast milk has been demonstrated in 2-week-old infants. Such a biologically important preference is consistent with other innate preferences that have been observed, all of which facilitate behaviors that foster survival of the species (see Porter and Schall, 1995, for a detailed review of innate olfactory preferences).

In contrast to the above finding with infants, Engen (1982) found young children to be relatively unresponsive to odors and argued that learning accounts for the vast majority of negative hedonic responses to odors in these subjects. More recently, Schmidt and Beauchamp (1988) tested three-year-old children and found distinct and reliable preferences and aversions for odors. In fact, most of the children's preferences/rejections matched those of adult populations, with the exception of androstenone, which children found more aversive than did adults. The results with androstenone are interesting, because this compound, which is found in celery, pork and truffles, is perceived quite differently among individuals. It can be described as either urinous, musky or floral by different people. Yet, upwards of 50% of the population are anosmic to the compound (Labows and Wysocki, 1984). Wysocki and Beauchamp (1984) have shown a strong genetic basis for the perception of androstenone and have since demonstrated the multiple involvement of genetic, developmental *and* experiential factors in its perception (Wysocki and Beauchamp, 1991).

Although stimulation of olfactory receptors by volatile food aromas plays an important role in eating, through its purely sensory effects, it also has an important effect on the utilization and absorption of the food. Both olfactory and taste stimulation mediate what is known as the 'cephalic phase response'. This response refers to the increased volume of salivary, gastric and pancreatic secretions in anticipation of consummatory behavior and has even been shown to be correlated with the quality/acceptability of the foods initiating the response (Jonowitz *et al.*, 1950). The role of the cephalic phase response in food consumption and nutrient utilization has been reviewed in several reports (Mattes, 1987; Mattes and Mela, 1988) and is an important innate, biological influence of olfaction (and taste) on food consumption and acceptance.

1.6.3 Effects of aging on olfaction

As in the case of taste, large decrements in olfactory sensitivity, discrimination and perceived intensity occur with aging (see Murphy, 1986 and Schiffman, 1986 for reviews). In fact, the depression of perceived suprathreshold intensity levels is even greater for olfaction than for taste (Murphy and Gilmore, 1990). These profound olfactory losses are manifested in neural morphology, where the olfactory bulb may actually look 'moth eaten' (Schiffman *et al.*, 1979). The losses that accompany aging are so pervasive that Murphy *et al.* (1994) caution that all efforts to establish normative data on olfactory functioning must incorporate age as a variable. The importance of age-related olfactory losses to food acceptance is magnified by the fact that the vast majority of what we commonly describe as food flavor is attributable to the olfactory components of foods. This has led Schiffman (Schiffman, 1987; Schiffman and Warwick, 1988, 1991) to propose the addition of commercial food odorants to foods for the elderly, in order to amplify their olfactory impact. Furthermore, aging has also been shown to cause decrements in the perceived intensity of oral irritants, e.g. menthol (Murphy, 1983).

Although age-related shifts in odor *preferences* have been reported (Moncrieff, 1966), systematic data are lacking. However, given the large age-related losses in olfactory functioning, including the loss of odor identification (Murphy and Cain, 1986) and recognition (Murphy *et al.*, 1991), significant shifts in food acceptance resulting from these odor losses are to be expected.

1.6.4 *Intramodal olfactory interactions*

Like taste mixtures, simple mixtures of odorants can be analyzed into their components. However, complex mixtures combine to form holistic sensations. The latter phenomenon is commonly assumed to support the fact that olfaction is a synthetic sense (Burgard and Kuznicki, 1990).

As is the case in taste mixtures, the perceived intensity of the mixture of two odorants is typically less than the sum of the perceived intensities of the unmixed components, although greater than their average intensity (Jones and Woskow, 1964; Berglund *et al.*, 1973; Cain, 1975; Laing *et al.*, 1995). A vector model for the interaction between two single odorants was proposed by Berglund *et al.* (1973) and has had fair success in the prediction of simple mixture phenomena. However, while mixture suppression is a common finding and has a physiological basis in olfactory receptor functioning (Kurahashi *et al.*, 1994), synergism has been observed occasionally in cases where the components of the mixture are of low intensity (Engen, 1982; Laska *et al.*, 1990). Another problem associated with developing a complete model of odor mixtures is that the normal suppressive effect of odor occurs only in mixtures of up to 3 or 4 components. After that, overall odor intensity tends to stay the same (Moskowitz and Barbe, 1977).

The problem of multiple component odors is clearly important to the food industry, where multi-ingredient foods are commonplace. Figure 1.11 shows data on the percentage of correct identifications of the constituents of mixtures containing up to five odorant compounds by untrained, trained and expert judges. While training certainly improved the identification of odors in mixtures of up to three or four odorants, identification rates fell to near-zero with five odorants. These results suggest a physiological limit on olfactory processing of odor mixtures (Laing and Livermore, 1992).

Although the suppression of odorant components in a mixture can be used to distinct advantage in the development of aerosol deodorants to mask undesired food and cooking odors, the mutual suppression of volatile food components can create difficult problems for the flavorist attempting to achieve a desired flavor profile in a multi-ingredient food. Combine this with the varied interactions within and between other sensory modalities and one may well begin to appreciate both the scientific and artistic talents

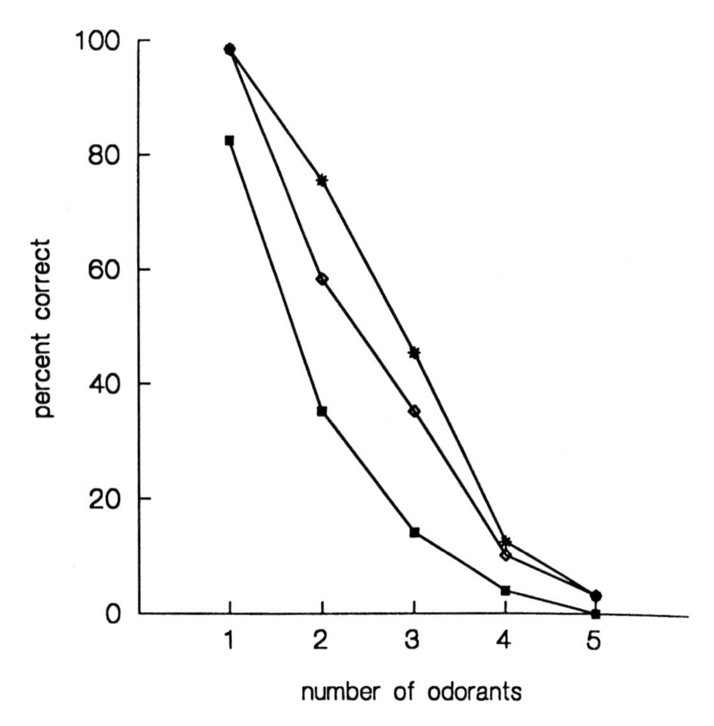

Figure 1.11 Percent correct identifications of the constituents of mixtures of up to 5 odorants by untrained, trained, and expert judges (from data of Laing and Livermore, 1992; after Laing, 1995). ■, Untrained; ◇, trained; ∗, expert.

of those individuals charged with creating the flavor complexes of even the most simple foods.

1.6.5 Oral–nasal irritant interactions

It is of no surprise to anyone who encounters ground horseradish for the first time that the common chemical sensitivities extend to the nasal mucous membranes. While horseradish and other compounds, like ammonia and halogens, can cause pungent, even painful, sensations in the nasal cavity, many relatively benign odorants also have a trigeminal component (Walker *et al.*, 1990). Cain (1974) has pointed out that such compounds are problematic when it comes to assessing their olfactory contribution alone, since it is likely that the observed response to these compounds is the result of a combined effect of the compound's odor and its irritant effect.

When odorant compounds are mixed with irritant compounds, mutual suppression occurs, much like with mixtures of odors. Cain and Murphy (1980) demonstrated this effect by mixing the nasal irritant, carbon dioxide, with various concentrations of amyl butyrate. The result was a mutual suppressive interaction, i.e. increasing concentrations of carbon dioxide resulted in decreasing judgments of the odor of amyl butyrate, and increasing concentrations of amyl butyrate resulted in decreasing judgments of the pungency of carbon dioxide. Moreover, by presenting the odorant compound to one nostril and the irritant compound to the other, it was possible to show that the locus of interaction was located centrally in the nervous system (Cain and Murphy, 1980).

1.6.6 Olfactory–acceptance relationships

Like taste, olfaction induces strong hedonic reactions. Yoshida (1964) showed that odors are often perceived as being liked or disliked before they are even identified. Indeed, one sniff of a long-forgotten odor can evoke powerful emotion-laden memories, suggesting that odor memory is unique among the senses (Lawless and Cain, 1975). Yet, for many common odors there is good agreement among individuals as to their pleasantness/unpleasantness. Figure 1.12 shows data from Cain (1979) on the odor pleasantness/unpleasantness of different odorants. The points to the right

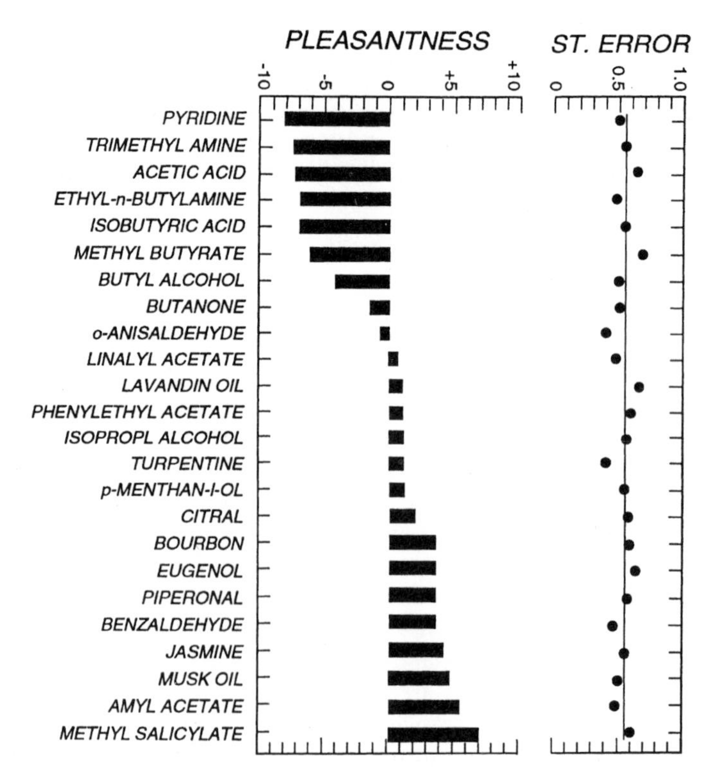

Figure 1.12 Means and standard errors of the mean for the pleasantness of 24 common odorants, expressed as deviations from hedonic neutrality (from Cain, 1979).

of each bar show the standard error of the pleasantness/unpleasantness ratings. As can be seen, there is little evidence of variation in the standard errors as a function of the hedonic nature of the stimulus, suggesting that there is similar agreement among subjects for hedonically extreme odors and hedonically neutral ones (Cain, 1979).

Henion (1971) examined the pleasantness of amyl acetate (banana odor) and found that pleasantness grew as a negative power function of intensity. However, Doty (1975) found this odorant to exhibit an inverted U-function, as did the odorants geroniol (rose), benzaldehyde (almond), eugenol (clove), benzyl acetate (fruity), and anethole (anise). As would be expected, the only odorants exhibiting a monotonic decrease in pleasantness with increasing concentration were unpleasant odorants, e.g. furfural and methyl ethyl ketone. A direct correlation between ratings of unpleasantness and the concentration of odorants that are primarily unpleasant was subsequently demonstrated by Piggot and Harper (1976).

In still other studies involving the pleasantness/unpleasantness of a wide range of odorants, Moskowitz (1977) and Moskowitz *et al.* (1976b, 1977) showed that pleasantness ratings for a wide range of odorants took on one of three distinct relationships with concentration. The most common relationships were a flat function, reflecting little change in hedonic tone with change in concentration, and a monotonically decreasing function, reflecting a steady decline in pleasantness with increasing concentration. The least common relationship was an inverted U-function. The latter characterized the hedonic functions for such compounds as lemon oil, chocolate and clove. Interestingly, many food odors were characterized by a flat hedonic function (Moskowitz, 1981).

Overall, the data on odor pleasantness show that, like the unpleasant taste qualities of bitter and sour, unpleasant odors grow increasingly unpleasant as a function of increasing concentration. For neutral and/or pleasant odors, pleasantness follows the inverted U-function, but with a broad optimum. Like the hedonics of other sensory modalities, olfactory pleasantness/unpleasantness is susceptible to the effects of adapting stimuli (Cain, 1979) and context (Murphy, 1982; Lawless *et al.*, 1991) and to the influences of culture (Davis and Pangborn, 1985). As a whole, olfactory mechanisms involved in determining acceptance follow closely the perceptual and learning mechanisms involved in taste.

1.7 Food acceptance measurement: The relative importance of the senses

The foregoing sections of this chapter have reviewed the various mechanisms by which the human sensory systems, alone and in concert, convert the physical characteristics of food stimuli into percepts of specific sensory attributes. In addition, it has delineated many of the known relationships between sensory experience and hedonic response. However, the human hedonic response to the complex, multi-attribute foods that are eaten in the increasingly rich variety of consumption situations that exists in today's societies is never as simple or direct as one might predict from the established relationships between one or two sensory attributes and liking. Food acceptance measurement is a complex problem, with numerous pitfalls. A consideration of some of these problems is necessary before attempting to evaluate the relative importance of the senses to food acceptance.

1.7.1 Measures of food acceptance

The first problem to consider is how to measure food acceptance. As noted earlier, there are two general classes of measures that are commonly used. One class consists of language-based measures. The other consists of non-verbal, behavioral measures. (See the volumes by Meilgaard *et al.*, 1991; Thomson, 1988; and MacFie and Thomson, 1994 for an overview of these methods and their applications.) Behavioral measures include, most prominently, choice or purchase behaviors and intake (consumption) measures. While these latter measures are commonly used to provide an overt index of the relative preference for different foods, they are often impractical to assess the relative contribution of different sensory attributes within the food to its overall liking. Part of this reason is the fact that it is time-consuming and costly to obtain consumption and/or marketplace choice data for the numerous combinations of ingredient and processing variables that may be investigated for their role in food acceptance. In addition, these measures fail to address the phenomenology of food acceptance and the analytic aspects of the sensory experience that accompanies food acceptance behaviors. For these reasons, most researchers interested in the relative contribution of sensory attributes to food acceptance use other indices of food acceptance. Most notably, these include direct psychophysical scaling of the hedonic element of foods.

As evident from any sampling of the research cited in previous sections, the most commonly used scale to index liking and acceptance is the 9-point hedonic scale developed by Peryam and co-workers at the US Army Quartermaster Institute in the 1950s (Peryam and Girardot, 1952; Peryam and Pilgrim, 1957). This scale consists of a series of 9 labeled scale points ranging from 1 = 'dislike extremely' to 9 = 'like extremely', with a neutral category of 5 = 'neither like nor dislike'. Although this scale is used throughout the world as a simple and direct index of the liking/disliking for food and/or its sensory attributes, it suffers from a number of well-known problems common to category scaling techniques. These include the fact that the category labels do not constitute equal intervals, that the end-categories are underutilized, and that the neutral category reduces the efficiency of the scale (Moskowitz, 1980). For these reasons, other hedonic scaling techniques have evolved, e.g. linear graphic rating and magnitude estimation. While the intent here is not to elucidate the specific advantages and disadvantages of various scaling techniques (see the section on sensory scaling for references related to these points), it is important to realize that different methods will produce different results. Table 1.3, from Moskowitz (1980), shows a comparison of the numerical ratings of liking that correspond to the verbal labels of the 9-point hedonic scale and those obtained using a magnitude estimation scale. The problem involved in interpreting the relative importance or salience of sensory attributes that are rated as 'liked slightly' versus 'liked very much' are clear from an examination of the numeric values in Table 1.3.

Another approach to hedonic scaling is that proposed by Booth (Booth *et al.*, 1983, 1986, 1987; Booth, 1990; Conner *et al.*, 1986, 1987, 1988). This approach utilizes a *reference to ideal* scale. The use of this type of scale is grounded in a linear model of preference psychophysics that argues that preferences are based on perceptual differences in the underlying sensory dimensions. Such a model implies that differences in preference scores should be proportional to discriminable differences in the sensory properties that underlie the preference. One consequence of this theoretical model is the notion that the inverted U function that has commonly been used to describe sensory–hedonic relationships is actually an inverted V function, at least at the level of the individual.

1.7.2 Individual variability in sensory–hedonic relationships

The problem of drawing conclusions about the relative importance of sensory attributes to overall liking from grouped data is made especially difficult by the fact that differences exist in both the shape and the location of peak magnitude in single-peaked preference functions (see Moskowitz,

Table 1.3 A comparison of the numerical values associated with the verbal labels of the 9-pt hedonic scale and magnitude estimation ratings of these same labels

Hedonic scale label	Hedonic scale point value	Magnitude estimation value[a]
Like extremely	9	+163
Like very much	8	+121
Like moderately	7	+82
Like slightly	6	+48
Neither like nor dislike	5	0
Dislike slightly	4	–41
Dislike moderately	3	–80
Dislike very much	2	–117
Dislike extremely	1	–148

[a] By magnitude estimation scaling of the words. *Source*: Moskowitz (1980).

1994). While some of these differences may be attributable to simple random variations, other differences reliably differentiate segments in the population. Figure 1.13 from Moskowitz (1991) shows group functions for two different segments of consumers, i.e. likers and dislikers of a product. For likers (segment 1), the data show a monotonic increase in liking with increasing sensory intensity. For dislikers (segment 2), the data show a monotonic decrease in liking. The third function (total panel) shows the averaged response curve across both segments of consumers. Clearly, the single-peaked preference function for the combined data is an artifact of averaging across different subject populations. This argument against grouped preference functions has been made previously (Pangborn, 1981) and is a recurring theme in the approach to food preference measurement by Booth and co-workers (Booth, 1987, 1990; Conner and Booth, 1992).

1.7.3 The hedonics of tasting vs. *swallowing*

Another source of variability in the assessment of sensory hedonics and food acceptance is the role of swallowing *vs.* expectorating samples. When foods are consumed in real-life situations, the pleasure or displeasure asso-

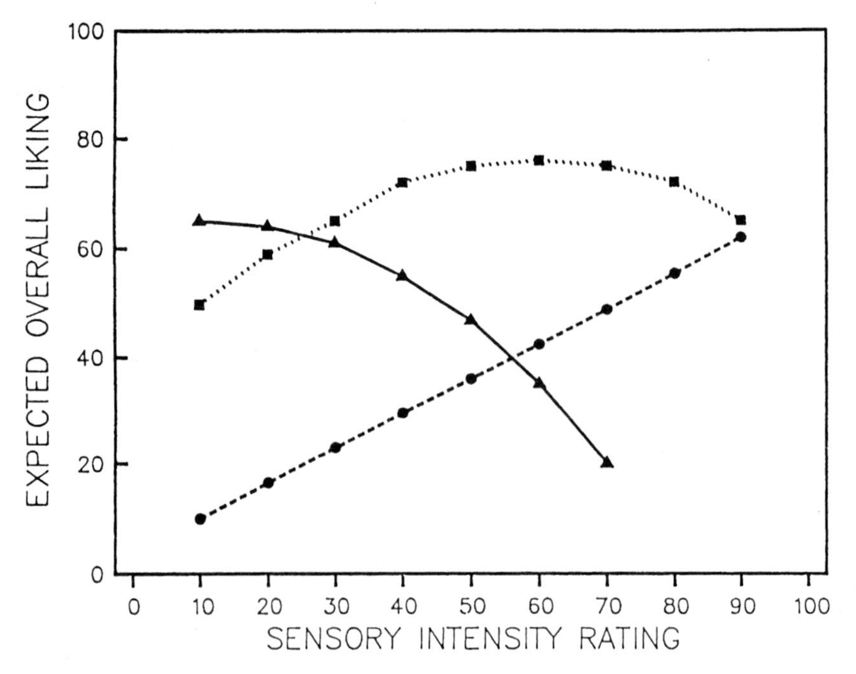

Figure 1.13 Group intensity-liking functions for likers (segment 1; ●) and dislikers (segment 2; ▲) of a product, showing monotonic changes in liking as a function of sensory intensity. The function for the total panel (■) can be seen to be an inverted U, but is merely an artifact of averaging the data from the two different consumer segments (from Moskowitz, 1991).

ciated with the food begins almost immediately and continues throughout the consumption of the food. However, in many sensory tasting situations, the samples are expectorated rather than swallowed. In a study by Lucas and Bellisle (1987), using sweetened yoghurt samples, differential effects were observed for the liking of the samples, depending upon whether the samples were swallowed or expectorated. Lower sucrose concentrations were more pleasant following consumption, while higher sucrose concentrations were less pleasant following consumption. However, Kelly and Heymann (1989), using milk varying in milkfat and beans varying in salt, and Taylor and Pangborn (1990), using chocolate milk varying in milkfat, found no effect of ingestion *vs.* expectoration on either sensory or hedonic judgments.

Figure 1.14 shows data from the study by Taylor and Pangborn (1990). The figure shows time–intensity curves for the degree of liking/disliking of chocolate milk samples as a function of time and percent milkfat. The top curve is for subjects who swallowed the samples, while the bottom is for those who expectorated. As can be seen, maximum liking or disliking (36% fat sample) occurs at approximately 20 seconds, both after swallowing and expectorating, and then gradually returns to a neutral baseline. Although the similarity in the time course of hedonic sensations in Figure 1.14 supports the notion that little difference is to be expected between studies that use swallowing or expectorating procedures, the shift in absolute hedonic rating as a function of time underscores a strong potential for unwanted variability in food acceptance testing. For any subject in either condition, at any given instant, both his/her instantaneous liking and his/her average liking up to that point in time is different from any other instant in time. The overt hedonic response will depend on the time at which the subject chooses to assign a cognitive, internalized rating for the pleasantness/unpleasantness of the sample. Unfortunately, very little is known about the introspective processes involved in making hedonic judgments of food or model stimuli, and common sensory procedures do not define either the exact time or the internal process by which hedonic judgments are to be made.

1.7.4 Relative importance of the senses to food acceptance

Although problems exist in the measurement of food acceptance, one can still ask what the *relative* importance of the senses is to food acceptance, when measured by any one technique. The problem can be addressed in several ways. One common approach is to conduct a multiple regression of individual sensory attribute ratings against overall liking (e.g. Rasekh *et al.*, 1970; Powers *et al.*, 1977; Azanza *et al.*, 1994). If the attribute ratings have been normalized, the obtained coefficients of the equation will index the degree of importance of the attribute to overall liking. Thus, we can view acceptance as in the following equation:

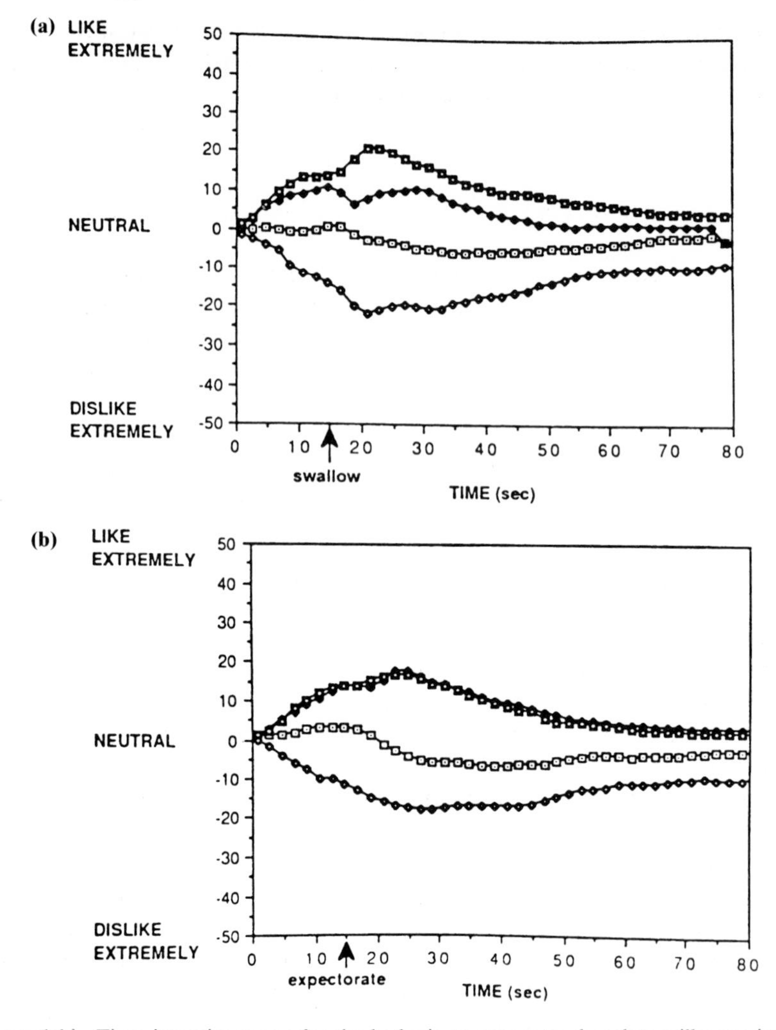

Figure 1.14 Time–intensity curves for the hedonic responses to chocolate milks varying in levels of milkfat. Data are based on two replicates with each of 17 subjects. (a) Curves obtained from subjects who swallowed the samples after 15 seconds. (b) Curves obtained from subjects who expectorated the samples after 15 seconds (from Taylor and Pangborn, 1990). □, 0% fat; ◆, 3.5% fat; □, 10.5% fat; ◇, 36% fat.

$$\text{Acceptance} = k_0 + k_1(A_1) + k_2(A_2) \ldots k_n(A_n)$$

where k_0 is a constant, A_{1-n} are the perceived intensities of each of the judged attributes, and k_{1-n} are the coefficients of the function relating overall liking to sensory attribute intensities. In such an equation, coefficients with a negative value reflect undesirable attributes in the product, whereas positive coefficients reflect desirable attributes. This basic equation can be further modified to include interaction terms:

$$+\mathrm{k}_{12}(A_1 \times A_2)\ldots$$

to account for the combined influences of two (or more) attributes in a product.

One problem with this approach is that some sensory dimensions may have several attributes, while others may have only one. Another problem is that, since liking is nonlinearly related to attribute ratings, the relative importance of the attributes will vary over the entire intensity range, with the importance of an attribute becoming increasingly greater as the attribute moves away from its optimal value (Moskowitz, 1984). Thus, changing the level of ingredients in a product will shift the relative importance of the attributes in the product. Such a situation calls for the use of a quadratic equation, such as:

$$\text{Acceptance} = \mathrm{k}_0 + \mathrm{k}_1 A_1 + \mathrm{k}_2 A_1^2 + \mathrm{k}_3 A_2 + \mathrm{k}_4 A_2^2 + \mathrm{k}_5 A_1 \times A_2$$

for a two-attribute model.

Another approach to the problem of the relative importance of sensory dimensions is to obtain judgments of their importance via questionnaires (Schutz and Wahl, 1981; Szczesniak, 1972) or from direct ratings of their importance (e.g. Tuorila-Ollikainen *et al.*, 1984). However, while direct ratings of importance may correlate with attribute importance determined by statistical regression (Moskowitz, 1984), concordance is not a logical necessity.

Still another approach to this problem is to look at the relationships between overall liking and ratings of the *liking* of the appearance, flavor and/or texture of the food (*cf.* Figure 1.1). In many studies, the correlation coefficients between overall liking and the independent ratings of attribute liking are taken as an index of the relative importance of the sensory dimensions to overall liking. For example, in a study of the relative importance of flavor *vs.* appearance (color) in two fruit-flavored beverages and cake, Dubose *et al.* (1980) concluded that flavor played a more important role in the overall acceptance of these products. This conclusion was reached by virtue of the fact that flavor liking ratings were more highly correlated with overall liking ($r = 0.86$, 0.89 and 0.76 for cherry beverage, orange beverage and lemon cake, respectively) than were color liking ratings ($r = 0.38$, 0.58 and 0.51, respectively). Using a similar approach, Tuorila-Ollikainen *et al.* (1984) concluded that taste (and sweetness) were more important than either odor or appearance in the overall acceptance of raspberry and pear drinks.

Although the approach of correlating attribute liking to overall liking is a commonly used one, Moskowitz (1992) and Moskowitz and Krieger

(1993, 1995) have argued that it is the *slope* of the function relating attribute liking to overall liking that is the best index of the importance of the attributes to overall liking, not the correlation coefficient. Moskowitz and Krieger (1993, 1995) argue that if one takes the simple linear equation:

$$\text{Overall liking} = M(\text{Attribute liking}) + \text{k}$$

the slope, M, of the function will be an index of the importance of that attribute to overall liking. If, for example, overall liking is related to *texture* liking by an equation of this type with a slope of 2.0, it suggests that for every unit increase in texture liking, a two unit increase in overall liking is effected. If the slope for the equation relating overall liking to *flavor* liking is only 0.5, then we can conclude (all things being equal) that texture is a far more important driver of the overall liking of the product than is flavor.

In order to index the relative importance of multiple attributes, Moskowitz and Krieger (1993) propose that the ratio of any single attribute slope to the sum of all attribute slopes be used. This relative attribute importance for each subject can then be plotted as in Figure 1.15 to graphi-

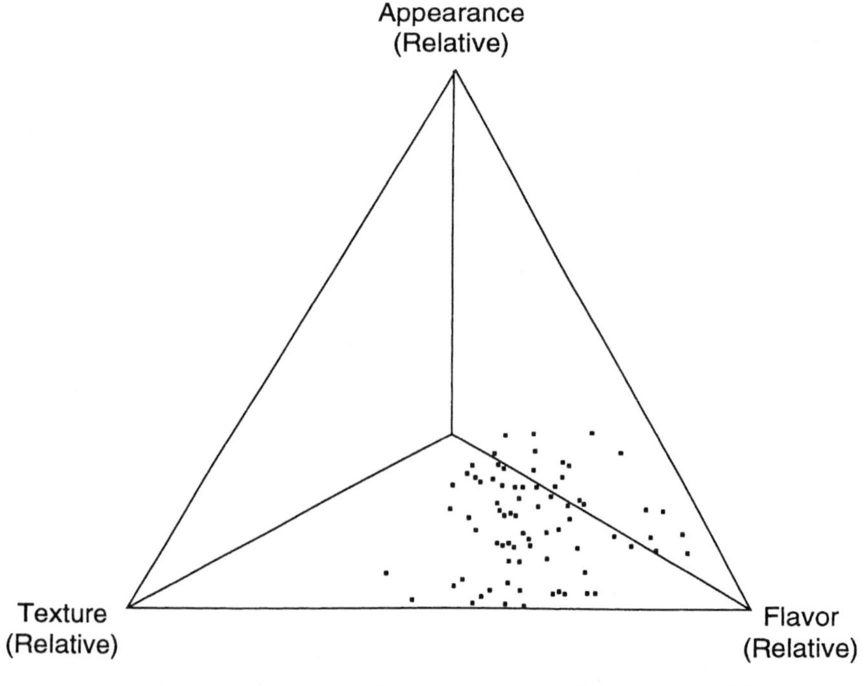

Figure 1.15 Triangular plot of the relative slopes of attribute liking to overall liking for fruit pies for the attributes of appearance, texture, and flavor. Data are for 98 consumers (from Moskowitz and Krieger, 1993).

cally depict the sensory drivers of overall acceptance in the product. The data in Figure 1.15 are for ratings of blueberry pie filling (Moskowitz and Krieger, 1993) and show that ratings of flavor liking are the primary drivers of the overall acceptance of this product. Based on this type of analysis, the authors have offered methods for classifying consumers into different types based on the relative importance of sensory dimensions to their overall liking of food products, e.g. those who pay primary attention to flavor, those who divide their attention between flavor and texture, etc. Moreover, based on research spanning several years and including numerous product categories, the authors have concluded that the predominant pattern of relative importance to overall acceptance is *taste/flavor > texture > appearance* (Moskowitz and Krieger, 1995).

One caveat that should be noted with regard to the above analyses is that it assumes that flavor liking, texture liking, and appearance liking are all determined solely by inputs from the sensory channels corresponding to these liking categories. However, as much of this chapter has shown, there are strong crossmodal influences at work in food perception and acceptance, making it impossible to assume that flavor liking is not partly due to appearance and/or texture variables in the product.

1.8 Future perspectives

As has been shown in this chapter, the human senses play a ubiquitous role in food acceptance. The human senses, alone and in combination, are responsible for the appearance, flavor and texture of food. Each sensory system exerts its influence on the acceptance of food, either through direct mechanisms of innate preference/rejection, through crossmodal interactions with other sensory systems, or through learned associations with the reinforcing properties of food. Although a variety of relationships between individual sensory attributes and acceptance have been demonstrated, the general relationship that re-occurs under numerous circumstances is the inverted U or inverted V relationship. And while a variety of methods and approaches reveal flavor (odor and taste) to be more important to the overall acceptance of foods than either appearance or texture, the relationship varies both by food item, by individual, and by population segment.

In a recent survey of the sources of pleasure from eating (Westenhoefer and Pudel, 1993), it was observed that the 'excellence of taste' in a food was the single most important aspect of the pleasure associated with eating in the more affluent industrialized region of West Germany. In less affluent East Germany, other factors played a more important role, e.g. a pleasant atmosphere. As improvements are seen in the economies of nations, consumers will begin to place even greater influence on the sensory quality of the foods they purchase. For international food manufacturers, meeting the demand for improved sensory quality will require a better understanding of what contributes to the sensory acceptability of foods. Reliance on databases that include only regional or national consumers or that derive from a context of only regional and national brands will not suffice. The innate sensory preferences that appear to exist across populations will be seen to become less important when compared to the culture-based, learned preferences/aversions for foods and the effects of widely varying food/brand contexts across nations.

In such circumstances, a far better understanding will be needed of how innate sensory preferences/aversions, especially neophobic responses, are changed by experience, culture, socialization, and cognition. This will require greater collaboration among researchers whose primary interest is the phenomenology and antecedents of sensory and hedonic experiences and those researchers whose interest is the behavioral consequences of those experiences. Only by such collaborative efforts will we come to know the various mechanisms by which the human senses come to influence food acceptance and the behavioral responses to food.

References

Akabas, M.H., Dodd, J. and Al-Awqati, Q. (1988) A bitter substance induces a rise in intracellular calcium in a subpopulation of rat taste cells. *Science*, **242**, 1047–1050.

Allison, A.C. and Blumberg, B.S. (1959) Ability to taste phenylthiocarbamide among Alaskan eskimos and other populations. *Human Biology*, **31**, 352–359.

Amoore, J.E. (1965) Psychophysics of odor. *Cold Spring Flavour Symposia on Quantitative Biology*, **30**, 623–637.

Amoore, J.E. (1975) Four primary odor modalities of man: Experimental evidence and possible significance, in *Olfaction and Taste* (eds D.A. Denton and J.P. Coghlan), Academic Press, New York, pp. 283–89.

Amoore, J.E. and Steinle, S. (1991) A graphic history of specific anosmia, in *Chemical Senses, 3, Genetics of Perception and Communication* (eds C.J. Wysocki and M.R. Kare), Dekker, New York, pp. 331–351.

Amoore, J.E., Johnston, J.W., Jr. and Rubin, M. (1964) The stereochemical theory of odor. *Scientific American*, **210** (2), 42–49.

Anderson, N.H. (1981) *Foundations of Information Integration Theory*, Academic Press, New York.

Arabie, P. and Moskowitz, H. (1971) The effects of viscosity upon perceived sweetness. *Perception and Psychophysics*, **9**, 410–412.

ASTM (1979) Standard practice for determination of odor and taste thresholds by a forced-choice ascending concentration series method of limits, in *Annual Book of ASTM Standards*. E-679. ASTM, Philadelphia, PA.

Azanza, F., Juvik, J.A. and Klein, B.P. (1994) Relationships between sensory quality attributes and kernel chemical composition of fresh-frozen sweet corn. *Journal of Food Quality*, **17** (2), 159–172.

Baines, Z.V. and Morris, E.R. (1987) Flavour/taste perception in thickened systems: the effect of guar gum above and below c*. *Food Hydrocolloids*, **1**, 197–205.

Baldwin, R.E. and Korschgen, B.M. (1979) A research note: intensification of fruit flavors by aspartame. *Journal of Food Science*, **44**, 938–939.

Balogh, R.D. and Porter, R.H. (1986) Olfactory preferences resulting from mere exposure in human neonates. *Infant Behavior and Development*, **9**, 395–401.

Barrett, A.H., Normand, M.D., Peleg, M. and Ross, E.W. (1992) Characterization of the jagged stress–strain relationships of puffed extrudates using the fast Fourier transform and fractal analysis. *Journal of Food Science*, **57**, 227–232, 235.

Barrett, A.H., Cardello, A.V., Lesher, L.L. and Taub, I.A. (1994) Cellularity, mechanical failure, and textural perception of corn meal extrudates. *Journal of Texture Studies*, **25**, 77–95.

Bartoshuk, L.M. (1974) Taste illusions: Some demonstrations. *Annals of the New York Academy of Sciences*, **237**, 279–285.

Bartoshuk, L.M. (1979) Bitter taste of saccharin related to the genetic ability to taste the bitter substance 6-n-Propylthiouracil. *Science*, **205** (4409), 934–935.

Bartoshuk, L.M. (1980) Sensory analysis of the taste of NaCl, in *Biological and Behavioral Aspects of Salt Intake* (eds M.R. Kare, M.J. Fregley and R.A. Bernard), Academic Press, New York, pp. 83–98.

Bartoshuk, L.M. and Wolfe, J.M. (1990) Conditioned 'taste' aversions in humans: Are they olfactory aversions? Paper presented at the meeting of the Association for Chemoreception Sciences, Sarasota, Fl., April 1990.

Bartoshuk, L.M., Murphy, C. and Cleveland, C. (1978) Sweet taste of dilute NaCl: pyschophysical evidence for a sweet stimulus. *Physiology and Behavior*, **21**, 609–613.

Bartoshuk, L.M., Rennert, K., Rodin, J. and Stevens, J.C. (1982) Effects of temperature on the perceived sweetness of sucrose. *Physiology and Behavior*, **28**, 905–910.

Bartoshuk, L.M., Dateo, G.P., Vandenbelt, D.J., Buttrick, R.L. and Long, L., Jr. (1969) Effects of gymnema sylvestre and synsepalum dulcificum on taste in man, in *Olfaction and Taste* (ed C. Pfaffmann), Rockefeller University Press, New York, pp. 436–444.

Bartoshuk, L.M., McBurney, D.H. and Pfaffmann, C. (1964) Taste of sodium chloride solutions after adaptation to sodium chloride: Implications for the 'water taste'. *Science*, **143** (3609), 967–968.

Bartoshuk, L.M., Riflin, B., Marks, L.E. and Hooper, J.E. (1988) Bitterness of KCI and benzoate: related to genetic status for sensitivity of PTC/PROP. *Chemical Senses*, **13** (4), 517–528.

Bartoshuk, L.M., Fast, K., Karrer, T.A., Marino, S., Price, R.A. and Reed, D.R. (1992) PROP supertasters and the perception of sweetness and bitterness. *Chemical Senses*, **17**, 594.

Bealer, S.L. and Smith, D.V. (1975) Multiple sensitivity to chemical stimuli in single human taste papillae. *Physiology and Behavior*, **14**, 795–799.

Beauchamp, G.K., Cowart, B.J. and Moran, M. (1986) Developmental changes in salt acceptability in human infants. *Developmental Psychology*, **19**, 75–83.

Beebe-Center, J.G. (1932) *Psychology of Pleasantness and Unpleasantness*, Van Nostrand, Princeton, New Jersey.

Beidler, L.M. (1954) A theory of taste stimulation. *Journal of General Physiology*, **38**, 133–139.

Beidler, L.M. (1967) Anion influences on taste receptor response, in *Olfaction and Taste, 2* (ed T. Hayashi), Pergamon Press, Oxford, pp. 509–534.

Beidler, L.M. (1971) Taste receptor stimulation with salts and acids, in *Handbook of Sensory Physiology, 4* (ed L.M. Beidler), Springer-Verlag, New York, p. 200.

Beidler, L.M. (1978) Biophysics and chemistry of taste, in *Handbook of Perception, 2* (eds E.C. Carterette and M.P. Freidman), Academic Press, New York, p. 21.

Bekesy, G. von (1960) *Experiments in Hearing*. McGraw-Hill, New York.

Bekesy, G. von (1964) Sweetness produced electrically on the tongue and its relation to taste theories. *Journal of Applied Physiology*, **19**, 1105–1113.

Bekesy, G. von (1966) Taste theories and the chemical stimulation of single papillae. *Journal of Applied Physiology*, **21**, 1–9.

Bell, R. and Meiselman, H.L. (1995) The role of eating environments in determining food choice, in *Food Choice and the Consumer* (ed D.W. Marshall), Blackie A&P, Glasgow, in press.

Berglund, B., Berglund, T., Lindvall, T. and Svensson, L.T. (1973) A quantitative principle of perceived intensity summation in odor mixtures. *Journal of Experimental Psychology*, **100**, 29–38.

Berlyne, D.E. (1973) The vicissitudes of aplopathematic and thelematoscopic pneumatology (or the hydrography of hedonism), in *Pleasure, Reward, Preference* (eds D.E. Berlyne and K.B. Madsen), Academic Press, New York, pp. 1–33.

Bernstein, I.L. and Webster, M.M. (1980) Learned taste aversions in humans. *Physiology and Behavior*, **25**, 363–366.
Birch, L.L. (1979) Dimensions of preschool children's food preferences. *Journal of Nutrition Education*, **11**, 189–192.
Birch, L.L. and Marlin, D.W. (1982) I don't like it; I never tried it: Effects of exposure on two-year old children's food preferences. *Appetite*, **3**, 353–360.
Birch, L.L., McPhee, L., Pirok and Steinberg, L. (1987) What kind of exposure reduces children's food noephobia? *Appetite*, **9**, 171–178.
Birch, L.L., McPhee, L., Steinberg, L. and Sullivan, S. (1990) Conditioned flavor preferences in young children. *Physiology and Behavior*, **47**, 501–505.
Blaker, G.G., Newcomber, J.L. and Ramsey, E. (1961) Holding temperatures needed to serve hot foods hot. *Journal of American Dietetic Association*, **38**, 455–457.
Blundell, J.E. and Rogers, P.J. (1991) Hunger, hedonics, and the control of satiation and satiety, in *Chemical Senses, 4, Appetite and Nutrition* (eds M.I. Friedman, M.G. Tordoff and M.R. Kare), Dekker, New York, pp. 127–148.
Bolanowski, S.J., Jr. and Gescheider, G.A. (eds) (1991) *Ratio Scaling of Psychological Magnitude*. Lawrence Erlbaum Associates, Hillsdale, New Jersey.
Bonnans, S. and Noble, A.C. (1993) Effect of sweetener type and of sweetener and acid levels on temporal perception of sweetness, sourness and fruitiness. *Chemical Senses*, **18** (3), 273–283.
Booth, D.A. (1972) Conditioned satiety in the rat. *Journal of Comparative Physiology and Psychology*, **81**, 457–471.
Booth, D.A. (1981) The physiology of appetite. *British Medical Bulletin*, **37**, 135–140.
Booth, D.A. (1982) Normal control of omnivore intake by taste and smell., in *The Determination of Behavior by Chemical Stimuli: ECRO Symposium* (eds J. Steiner and J. Ganchrow), Information Retreival, London, pp. 233–243.
Booth, D.A. (1987) Individualized objective measurement of sensory and image factors in product acceptance. *Chemistry and Industry*, **13**, 441–446.
Booth, D.A. (1990) Designing products for individual customers, in *Psychological Basis of Sensory Evaluation* (eds R.L. McBride and H.J.H. MacFie), Elsevier Applied Science, London, pp. 163–193.
Booth, D.A., Thomson, A.L. and Shahedian, B. (1983) A robust, brief measure of an individual's most preferred level of salt in an ordinary foodstuff. *Appetite*, **4**, 301–312.
Booth, D.A., Conner, M.T., Marie, S., Griffiths, R.P., Haddon, A.V. and Land, D.G. (1986) Objective tests of preference amongst foods and drinks, in *Measurement and Determinants of Food Habits and Food Preferences* (eds J.M. Diehl and C. Leitzmann), University Department of Human Nutrition, Wageningen, pp. 87–108.
Booth, D.A., Conner, M.T. and Marie, S. (1987) Sweetness and food selection: Measurement of sweetener's effects on acceptance, in *Sweetness* (ed J. Dobbing), Springer-Verlag, London, pp. 143–160.
Boring, E.G. (1942) *Sensation and Perception in the History of Experimental Psychology*, Appleton-Century-Crofts, Inc., New York.
Boudreau, J.C. (1986) Neurophysiology and human taste sensations. *Journal of Sensory Studies*, **1** (3/4), 185–202.
Bourne, M.C. (1979) Texture of temperate fruits. *Journal of Texture Studies*, **10** (1), 25–44.
Bourne, M.C. (1980) Texture evaluation of horticultural crops. *Hortscience*, **15** (1), 7–13.
Brand, J.G. *et al.* (1989) *Chemical Senses, 1, Receptor Events and Transduction in Taste and Olfaction*, Dekker, New York.
Brandt, M.A., Skinner, E.Z. and Coleman, J.A. (1963) Texture profile method. *Journal of Food Science*, **28**, 404–409.
Breslin, P.A.S., Davidson, T.L. and Grill, H.J. (1990) Conditioned reversal of reactions to normally avoided tastes. *Physiology and Behavior*, **47**, 535–538.
Bruch, R.C. (1990) Signal transduction in olfaction and taste, in *G-Proteins and Calcium Mobilizing Hormones* (eds L. Birnbaumer and R. Iyengar), Academic Press, San Diego, pp. 411–428.
Bruch, R.C. and Teeter, J.H. (1989) Second-messenger signalling mechanisms in olfaction, in *Chemical Senses, 1, Receptor Events and Transduction in Taste and Olfaction* (eds J.G. Brand, J.H. Teeter, R.H. Cagan and M.R. Kare), Dekker, New York, pp. 283–298.

Buck, L. and Axel, R. (1991) A novel multigene family may encode odorant receptors: A molecular basis for odor recognition. *Cell*, **65**, 175–187.
Burdach, K.J., Kroeze, J.H.A. and Koster, E.P. (1984) Nasal, retronasal and gustatory perception: an experimental comparison. *Perception and Psychophysics*, **36**, 205–208.
Burgard, D.R. and Kuznicki, J.T. (1990) *Chemometrics: Chemical and Sensory Data*, CRC Press, Boca Raton, Florida.
Burgess, P.R. and Perl, E.R. (1977) Cutaneous mechanoreceptors and nociceptors, in *Handbook of Sensory Physiology, 2* (ed A. Iggo), Springer-Verlag, New York, pp. 29–78.
Cabanac, M. (1971) Physiological role of pleasure. *Science*, **173**, 1103–1107.
Cain, W.S. (1974) Contribution of the trigeminal nerve to perceived odor magnitude. *Annals of the New York Academy of Science*, **237**, 28–34.
Cain, W.S. (1975) Odor intensity: Mixtures and masking. *Chemical Senses and Flavor*, **1**, 339–352.
Cain, W.S. (1979) Lability of odor pleasantness, in *Preference Behaviour and Chemoreception* (ed J.H.A. Kroeze), IRL, London, pp. 303–315.
Cain, W.S. and Murphy, C. (1980) Interaction between chemoreceptive modalities of odour and irritation. *Nature*, **284**, 255–257.
Cain, W.S. and Johnson, F. (1978) Lability of odor pleasantness: influence of mere exposure. *Perception*, **7**, 459–465.
Cairncross, S.E. and Sjostrom, L.B. (1950) Flavor profiles—a new approach to flavor problems. *Food Technology*, **4**, 308–311.
Calvino, A.M. (1986) Perception of sweetness: the effects of concentration and temperature. *Physiology and Behavior*, **36**, 1021–1028.
Cambell, G.W. (1956) Consumer acceptance of beef. *Arizona University Agricultural Experimentation Station Report*, 145.
Capaldi, E.D. and Powley, T.L. (eds) (1990) *Taste, Experience, and Feeding*, American Psychological Association, Washington, DC.
Capretta, P.J. and Rawls, L.H. (1974) Establishment of a flavor preference in rats: Importance of nursing and weaning experience. *Journal of Comparative and Physiology and Psychology*, **86**, 670.
Cardello, A.V. (1978) Chemical stimulation of single human fungiform taste papillae: sensitivity profiles and locus of stimulation. *Sensory Processes*, **2**, 173–190.
Cardello, A.V. (1979) Taste quality changes as a function of salt concentration in single human papillae. *Chemical Senses and Flavor*, **4**, 1.
Cardello, A.V. (1981) Comparison of taste qualities elicited by tactile, electrical, and chemical stimulation of single human taste papillae. *Perception and Psychophysics*, **29** (2), 163–169.
Cardello, A.V. (1994) Consumer expectations and their role in food acceptance, in *Measurement of Food Preferences* (eds H.J. MacFie and D.M.H. Thomson), Blackie A&P, Glasgow, pp. 253–297
Cardello, A.V. and Maller, O. (1982) Acceptability of water, selected beverages and foods as a function of serving temperature. *Journal of Food Science*, **47**, 1549–1552.
Cardello, A.V. and Maller, O. (1987) Psychophysical bases for the assessment of food quality, in *Objective Methods in Food Quality Assessment* (ed J.G. Kapsalis), CRC Press, Boca Raton, pp. 61–125.
Cardello, A.V. and Murphy, C. (1977) Magnitude estimates of gustatory quality changes as a function of solution concentration of simple salts. *Chemical Senses*, **2**, 327.
Cardello, A.V. and Sawyer, F.M. (1992) Effects of disconfirmed consumer expectations on food acceptability. *Journal of Sensory Studies*, **7**, 253–277.
Cardello, A.V., Maller, O., Bloom-Masor, H., Dubose, C. and Edelman, B. (1985) Role of consumer expectancies in the acceptance of novel foods. *Journal of Food Science*, **50**, 1707–1714, 1718.
Cash, R.M. and Linden, R.W.A. (1982) The distribution of mechanoreceptors in the periodontal ligament of the mandibular canine tooth in the cat. *Journal of Physiology*, **330**, 439–447.
Caul, J.F. (1957) The profile method of flavor analysis, in *Advances in Food Research, 7* (eds E. Mark and G.F. Stewart), Academic Press, New York, pp. 1–40.
Chambers, K.C. and Bernstein, I.L. (1995) Conditioned flavor aversions, in *Handbook of Olfaction and Gustation* (ed R.L. Doty), Dekker, New York, pp. 745–773.
Christensen, C.M. (1977) Texture-taste interactions. *Cereal Foods World*, **22**, 243–256.

Christensen, C.M. (1980a) Effects of solution viscosity on perceived saltiness and sweetness. *Perception and Psychophysics*, **28**, 347–353.
Christensen, C.M. (1980b) Effects of taste quality and intensity on oral perception of viscosity. *Perception and Psychophysics*, **28** (4), 315–320.
Christensen, C.M. (1983) Effects of color on aroma, flavor and texture judgements of foods. *Journal of Food Science*, **48**, 787–790.
Civille, G.V. and Liska, I.H. (1975) Modifications and applications to foods of the General Foods sensory texture profile technique. *Journal of Texture Studies*, **6**, 19–31.
Civille, G.V. and Szczesniak, A.S. (1973) Guidelines to training a texture profile panel. *Journal of Texture Studies*, **4**, 204–223.
Cliff, M. and Heymann, H. (1992) Descriptive analysis of oral pungency. *Journal of Sensory Studies*, **7** (4), 279–290.
Clydesdale, F.M. (1984) Influence of colour on sensory perception and food choices, in *Developments in Food Colours, 2* (ed J. Walford), Elsevier Applied Science, New York, pp. 75–112.
Clydesdale, F.M. (1993) Color as a factor in food choice. *Critical Reviews in Food Science and Nutrition*, **33** (1), 83–101.
Clydesdale, F.M., Gover, R., Philipsen, D. and Fugardi, C. (1992) The effect of color on thirst quenching, sweetness, acceptability and flavor intensity in fruit punch flavored beverages. *Journal of Food Quality*, **15**, 19–38.
Connell, J.J. and Howgate, P.F. (1971) Consumer evaluation of fresh and frozen fish, in *Fish Inspection and Quality Control* (ed R. Kreuzer), Fishing News Books Ltd., London, pp. 155–159.
Conner, M.T. and Booth, D.A. (1992) Combined measurement of food taste and consumer preference in the individual: reliability, precision and stability data. *Journal of Food Quality*, **15**, 1–17.
Conner, M.T., Land, D.G. and Booth, D.A. (1986) Very rapid, precise measurements of effects of constituent variation on product acceptability. *Lebensmittel Wissenschaft und Technologie*, **19**, 486–490.
Conner, M.T., Land, D.G. and Booth, D.A. (1987) Effect of stimulus range on judgements of sweetness intensity in a lime drink. *British Journal of Psychology*, **78**, 357–364.
Conner, M.T., Booth, D.A., Clifton, V.J. and Griffiths, R.P. (1988) Do comparisons of a food characteristic with ideal necessarily involve learning? *Br. J. Psychol.*, **79**, 121–128.
Cooper, H.R. (1987) Texture in dairy products and its sensory evaluation, in *Food Texture* (ed H.R. Moskowitz), Dekker, New York, pp. 251–272.
Cowart, B.J. (1987) Oral chemical irritation, does it reduce perceived taste intensity? *Chemical Senses*, **12**, 467–479.
Cowart, B.J. and Beauchamp, G.K. (1986) Factors affecting acceptance of salt by human infants and children, in *Interaction of the Chemical Senses with Nutrition* (eds M.R. Kare and J.G. Brand, Academic Press, New York, pp. 25–44.
Cowart, B.J. and Beauchamp, G.K. (1990) Early development of taste perception, in *Psychological Basis of Sensory Evaluation* (eds R.L. McBride and H.J.H. MacFie), Elsevier Applied Science, London, pp. 1–17.
Croze, S., Duclaux, R. and Kenshalo, D.R. (1976) The thermal sensitivity of the polymodal nociceptors in the monkey. *Journal of Physiology*, **263**, 539–562.
Cussler, E.L., Kokini, J.L., Weinheimer, R.L. and Moskowitz, H.R. (1979) Food texture in the mouth. *Food Technology*, **33** (10), 89–92.
Davis, L.B. and Porter, R.H. (1991) Persistent effects of early odor exposure on human neonates. *Chemical Senses*, **16** (2), 169–174.
Davis, R.G. and Pangborn, R.M. (1985) Odor pleasantness judgements compared among samples from 20 nations using microfragrances. *Chemical Senses*, **10**, 413.
DeSimone, J.A. and Ferrell, F. (1985) Analysis of amiloride inhibition of chorda tympani taste response of rat to NaCl. *American Journal of Physiology*, **249**, R52–R61.
DeSimone, J.A., Heck, G.L. and DeSimone, S.K. (1981) Active ion transport in dog tongue: A possible role in taste. *Science*, **214**, 1039–1041.
Desor, J.A. and Beauchamp, G.K. (1974) The human capacity to transmit olfactory information. *Perception and Psychophysics*, **16**, 551–556.
Desor, J.A. and Finn, J. (1989) Effects of amiloride on salt taste in humans. *Chemical Senses*, **14** (6), 793–803.

Desor, J.A., Maller, O. and Turner, R.E. (1973) Taste in acceptance of sugars by human infants. *Journal of Comparative and Physiological Psychology*, **84**, 496–501.

Desor, J.A., Maller, O. and Andrews, K. (1975) Ingestive responses of human newborns to salts, sour and bitter stimuli. *Journal of Comparative and Physiological Psychology*, **89**, 966–970.

Dijksterhuis, G.B. (1993) Principal component analysis of time-intensity bitterness curves. *Journal of Sensory Studies*, **8** (4), 317–328.

Domjan, M. (1976) Determinants of the enhancement of flavored-water intake by prior exposure. *Journal of Experimental Psychology, Animal Behavior Processes*, **2** (1), 17–27.

Doty, R.L. (1975) An examination of relationships between the pleasantness, intensity, and concentration of 10 odorous stimuli. *Perception and Psychophysics*, **17** (5), 492–496.

Doty, R.L. (ed.) (1995) *Handbook of Olfaction and Gustation*, Dekker, New York.

Drake, B.K. (1963) Food crushing sounds. An introductory study. *Journal of Food Science*, **28**, 233–241.

Drake, B.K. (1965) Food crushing sounds. Comparisons of subjective and objective data. *Journal of Food Science*, **30**, 556–559.

Drake, B.K. (1974) A comprehensive formula for the acceptance of food texture and its generalization to overall food acceptance. *Journal of Texture Studies*, **5**, 109–113.

Drewnowski, A. (1987) Fats and food texture: Sensory and hedonic evaluations, in *Food Texture* (ed H.R. Moskowitz), Dekker, New York, pp. 251–272.

Drewnowski, A. (1991) Fat and sugar: Sensory and hedonic aspects of sweet, high-fat foods, in *Chemical Senses, 4, Appetite and Nutrition* (eds M.I. Friedman, M.G. Tordoff and M.R. Kare), Dekker, New York.

Drewnowski, A. (1992) Sensory preferences and fat consumption in obesity and eating disorders, in *Dietary Fats* (ed D.J. Mela), Elsevier Applied Science, London, pp. 59–77.

Drewnowski, A. and Greenwood, M.R.C. (1983) Cream and Sugar: human preferences for high-fat foods. *Physiology and Behavior*, **30**, 629–633.

Drewnowski, A., Grinker, J.A. and Hirsh, J. (1982) Obesity and flavor perception: Multidimensional scaling of soft drinks. *Appetite*, **3**, 361–368.

Drewnowski, A., Brunzell, J.D., Sande, K., Iverius, P.H. and Greenwood, M.R.C. (1985) Sweet tooth reconsidered: taste responsiveness in human obesity. *Physiology and Behavior*, **30**, 629–633.

Dubose, C.N., Cardello, A.V. and Maller, O. (1980) Effects of colorants and flavorants on identification, perceived flavor intensity and hedonic quality of fruit flavored beverages and cake. *Journal of Food Science*, **45**, 1393–1415.

Dzendolet, E. (1968) A structure common to sweet-evoking compounds. *Perception and Psychophysics*, **3**, 65–68.

Dzendolet, E. and Meiselman, H. (1967) Gustatory quality changes as a function of solution concentration. *Perception and Psychophysics*, **2**, 29–33.

Ekman, G., Hosman, J. and Lindstrom, B. (1965) Roughness, smoothness and preference: A study of quantitative relations in individual subjects. *Journal of Experimental Psychology*, **70**, 18–26.

Engen, T. (1971a) Psychophysics: discrimination and detection, in *Experimental Psychology, 3rd edn* (eds J.W. Kling and Lorrin A. Riggs), Holt, Rinehart and Winston, Inc., New York, pp. 11–46.

Engen, T. (1971b) Psychophysics: Scaling methods, in *Experimental Psychology, 3rd edn* (eds J.W. Kling and L.A. Riggs), Holt, Rinehart and Winston, Inc., New York, pp. 47–86.

Engen, T. (1982) *The Perception of Odors*, Academic, New York.

Enns, M.P. and Hornung, D.E. (1985) Contributions of smell and taste to overall intensity. *Chemical Senses*, **10**, 357–366.

Enns, M.P. and Hornung, D.E. (1988) Comparisons of the estimates of smell, taste and overall intensity in young and elderly people. *Chemical Senses*, **13**, 131–139.

Enns, M.P., Van Itallie, T.B. and Grinker, J.A. (1979) Contributions of age, sex and degree of fatness on preferences and magnitude estimation for sucrose in humans. *Physiology and Behavior*, **22**, 999–1003.

Erickson, R.P. (1982) Studies in the perception of taste: Do primaries exist? *Physiology and Behavior*, **28**, 57–62.

Erickson, R.P. and Covey, E. (1980) On the singularity of taste sensations: what is a taste primary? *Physiology and Behavior*, **25**, 527–533.

Eysenck, H.J. (1941) A critical and experimental study of colour preferences. *American Journal of Psychology*, **54**, 385–394.

Fanselow, M. and Birk, J. (1982) Flavor-flavor associations induce hedonic shifts in taste preference. *Animal Learning and Behavior*, **10**, 223–228.

Fechner, G.T. (1860) *Elements der Psychophysik*, Brietkopf and Harterl, Leipzig. (English translation by H.E. Adler, Holt, Reinhart and Winston, New York, 1966.)

Firestein, S., Darrow, B. and Shepherd, G.M. (1991) Activation of the sensory current in salamander olfactory receptor neurons depends on a G protein-mediated cAMP second messenger system. *Neuron*, **6**, 825–835.

Fletcher, L., Heymann, H. and Ellersieck, M. (1991) Effects of visual masking techniques on the intensity rating of sweetness of gelatins and lemonades. *Journal of Sensory Studies*, **6** (3), 179–191.

Fox, A.L. (1932) The relationship between chemical constitution and taste. *Proceedings of the National Academy of Science*, NAS Press, Washington, D.C., **18**, 115.

Frank, A.S.T. (1964) Studies on the innervation of the temporomandibular joint and lateral pterygoid muscle in animals. *Journal of Dental Research*, **43**, 947.

Ganchrow, J.R. (1995) Ontogeny of human taste perception, in *Handbook of Olfaction and Gustation* (ed R.L. Doty), Dekker, New York, 715–729.

Ganchrow, J.R., Steiner, J.E. and Munif, D. (1983) Neonatal facial expressions in response to different qualities and intensities of gustatory stimuli. *Infant Behavior and Development*, **6**, 473–484.

Ganzevles, P.G.J. and Kroeze, J.H.A. (1987) The sour taste of acids. The hydrogen ion and the undissociated acid as sour agents. *Chemical Senses*, **12** (4) 563–576.

Garb, J.L. and Stunkard, A. (1974) Taste aversions in man. *American Journal of Psychiatry*, **131**, 1204–1207.

Garcia, J. and Koelling, R.A. (1966) Relation of cue to consequence in avoidance learning. *Psychonomic Science*, **4**, 123–124.

Garcia, J., Ervin, F.R. and Koelling, R.A. (1966) Learning with prolonged delay of reinforcement. *Psychonomic Science*, **5**, 121–122.

Gardner, R.J. (1978) Lipophilicity and bitter taste. *Journal of Pharmacy and Pharmacology*, **30**, 531–532.

Gardner, R.J. (1979) Lipophilicity and the perception of bitterness. *Chemical Senses and Flavor*, **4**, 275–286.

Gardner, R.J. (1980) Lipid solubility and the sourness of acids: implications for models of the acid taste receptor. *Chemical Senses and Flavor*, **5**, 185–192.

Gesell, A., Ilg, F.L. and Ames, L.B. (1956) *Youth, The Years from Ten to Sixteen*, Harper and Bros., New York.

Getchell, T.V. (1986) Functional properties of vertebrate olfactory receptor neurons. *Physiological Review*, **66**, 772–818.

Gibson, J.J. (1966) *The Senses Considered as Perceptual Systems*, Houghton Mifflin, Boston.

Gibson, L. and Ashby, M.F. (1988) *Cellular Solids*, Pergamon Press, Oxford.

Gifford, S.R. and Clydesdale, F.M. (1986) The psychophysical relationship between color and sodium chloride concentrations in model systems. *Journal of Food Protection*, **49**, 977–982.

Gifford, S.R., Clydesdale, F.M. and Damon, R.A., Jr. (1987) The psychophysical relationship between color and salt concentration in chicken flavored broths. *Journal of Sensory Studies*, **2**, 137–145.

Gilbertson, T.A., Avenet, P., Kinnamon, S.C. and Roper, S.D. (1992) Proton currents through amiloride-sensitive Na channels in hamster taste cells: role in acid transduction. *Journal of General Physiology*, **100**, 803–824.

Gill, H.J. (1971) Neuromuscular spindles in human lateral pterygoid muscles. *Journal of Anatomy*, **109**, 157–168.

Gilmore, M.M. and Green, B.G. (1993) Sensory irritation and taste produced by NaCl and citric acid: effects of capsaicin desensitization. *Chemical Senses*, **18** (3), 257–272.

Gold, G.H. *et al.* (1989) A role for cyclic AMP in olfactory transduction, in *Chemical Senses, 1, Receptor Events and Transduction in Taste and Olfaction* (eds J.G. Brand *et al.*), Dekker, New York, pp. 311–317.

Good Housekeeping Institute (1984) *Consumer Food and Nutrition Study*. Good Housekeeping Institute, New York.

Green, B.G. (1986) Sensory interactions between capsaicin and temperature in the oral cavity. *Chemical Senses*, **11**, 371–382.
Green, B.G. (1990) Effects of thermal, mechanical, and chemical stimulation on the perception of oral irritation, in *Chemical Senses, 2, Irritation* (eds B.G. Green, J.R. Mason and M.R. Kare), Dekker, New York, pp. 171–192.
Green, B.G. and Frankmann, S.P. (1987) The effects of cooling the tongue on the perceived intensity of taste. *Chemical Senses*, **12**, 609–619.
Green, D.M. and Swets, J.A. (1966) *Signal Detection Theory and Psychophysics*, John Wiley & Sons, New York.
Green, B.G., Shaffer, G.S. and Gilmore, K.M. (1993) Derivation and evaluation of a schematic scale of oral sensation magnitude with apparent ratio properties. *Chemical Senses*, **18**, 683–702.
Griffin, F.M. (1966) On the interaction of chemical stimuli with taste receptors. Doctoral dissertation, Ohio State University, Ann Arbor, Michigan: University Microfilms, No. 66-15091.
Grossman, R.F. and Hattis B.F. (1967) Oral mucosa sensory innervation and sensory experience: a review, in *Symposium on Oral Sensation and Perception* (ed J.F. Bosma), Charles C. Thomas, Springfield, IL, pp. 5–62.
Guilford, J.P. and Smith, P.C. (1959) A system of color preferences. *American Journal of Psychology*, **72**, 487–502.
Gulick, W.L. (1971) *Hearing: Physiology and Psychophysics*. Oxford University Press, New York.
Hahn, H. and Gunther, H. (1932) Uber die reize und die reizbedingungen des geschmacksinnes. *Pflugers Archiv fuer Gesamte Physiologie*, **231**, 48–67.
Hall, R.L. (1958) Flavor study approach at McCormick and Co. Inc., in *Flavor Research and Food Acceptance*, A.D. Little, Inc., Reinhold, New York, pp. 224–240.
Halpern, B. (1977) Functional anatomy of the tongue and mouth of mammals, in *Drinking Behavior, Oral Stimulation, Reinforcement and Preference* (eds J.A.W.M. Weijner and J. Mendelson), Plenum Press, New York, pp. 1–92.
Harper, R., Bate Smith, E.C. and Land. D.G. (1968) *Odour Description and Odour Classification: A Multidisciplinary Examination*. American Elsevier, New York.
Harries, J.M., Rhodes, D.N. and Chrystall, B.B. (1972) Meat texture. *Journal of Texture Studies*, **3**, 101–114.
Harris, G. and Booth, D.A. (1987) Infants' preference for salt in food: Its dependence upon recent dietary experience. *Journal of Reproductive and Infant Psychology*, **5**, 97–104.
Heath, M.R. and Lucas, P.W. (1987) Mastication: The need for collaborative research. *Journal of Texture Studies*, **18**, 111–123.
Helson, H. (1964) *Adaptation-Level Theory*, Harper and Row, New York.
Henion, E. (1971) Odor pleasantness and intensity: A single dimension? *Journal of Experimental Psychology*, **90**, 275–279.
Hensel, H. (1977) Cutaneous thermoreceptors, in *Handbook of Sensory Physiology*, 2 (ed A. Iggo), Springer-Verlag, New York, p. 79.
Herness, M.S. and Pfaffmann, C. (1986) Iontophoretic application of bitter taste stimuli in hamsters. *Chemical Senses*, **11** (2), 203–211.
Hodgson, R.R., Belk, K.E., Savell, J.W., Cross, H.R. and Williams, F.L. (1992) Development of a quantitative quality grading system for mature cow carcasses. *Journal of Animal Science*, **70** (6), 1840–1847.
Hollingsworth, H.L. and Poffenberger, A.T. (1917) *The Sense of Taste*, Moffat, Yard & Co, New York.
Hornung, D.E. and Enns, M.P. (1986) The contribution of smell and taste to overall intensity: a model. *Perception and Psychophysics*, **39**, 385–391.
Hornung, D.E. and Enns, M.P. (1989) Separating the contributions of smells and tastes in flavor perception, in *Perception of Complex Smells and Tastes* (eds D.G. Laing, W.S. Cain, R.L. McBride and B.W. Ache), Academic Press, Sydney, Australia, pp. 285–296.
Hough, G. and Contarini, A. (1994) Training a texture profile panel and constructing standard rating scales in Argentina. *Journal of Texture Studies,* **25** (1), 45–57.
Howard, A. (1976) Psychometric scaling of sensory texture attributes of meat. *Journal of Texture Studies*, **7**, 95–107.

Hutchinson, S.E., Trantow, L.A. and Vickers, Z.M. (1990) The effectiveness of common foods for reduction of capsaicin burn. *Journal of Sensory Studies*, **4**, 157–164.

Hwang, P.M., Verma, A., Bredt, D.S. and Snyder, S. (1990) Localization of phosphatidylinositol signaling components in rat taste cells: Role in bitter taste transduction. *Proceedings of the National Academy of Sciences*, **87**, 7395–7399.

Iles, B.C. and Elson, C.R. (1972) *Crispness*. BFMIRA Research Report No. 190.

Irwin, R.J., Hautus, M.J. and Stillman, J.A. (1992) Use of the receiver operating characteristic in the study of taste perception. *Journal of Sensory Studies*, **7** (4), 291–314.

Irwin, R.J., Stillman, J.A., Hautus, M.J. and Huddleston, L.M. (1993) The measurement of taste discrimination with the same–different task: A detection-theory analysis. *Journal of Sensory Studies*, **8** (3), 229–240.

Ishii, R. and O'Mahony, M. (1987) Taste sorting and naming: Can taste concepts be misrepresented by traditional psychophysical labelling systems? *Chemical Senses*, **12**, 37–51.

Ishii, R. and O'Mahony, M. (1990) Group taste concept measurement: Verbal and physical definition of the Umami taste concept for Japanese and Americans. *Journal of Sensory Studies*, **4**, 215–227.

Izutsu, T., Taneya, S., Kikuchi, E. and Sone, T. (1981) Effect of viscosity on perceived sweetness intensity of sweetened sodium carboxymethycellulose solutions. *Journal of Texture Studies*, **12**, 259–273.

Jakinovich, W. Jr. and Sugarman, D. (1988) Sugar taste reception in mammals. *Chemical Senses*, **13** (1), 13–31.

Janowitz, H.D., Hollander, F., Orringer, D., Levy, M.H., Winkelstein, A., Kaufman, M.R. and Margolin, S.G. (1950) A quantitative study of the gastric secretory response to sham feeding in a human subject. *Gastroenterology*, **16**, 104–116.

Johnson, J. (1982) *Psychophysical Relationships between Color and Sweetness in Fruit Flavored Solutions*. Ph.D. thesis, University of Massachusetts at Amherst.

Johnson, J.L. and Clydesdale, F.M. (1982) Perceived sweetness and redness in colored sucrose solutions. *Journal of Food Science*, **47**, 747–752.

Johnson, J.L. and Vickers, Z. (1991) Sensory-specific satiety for selected bread products. *Journal of Sensory Studies*, **6** (2), 65–79.

Johnson, J.L., Dzendolet E., Damon E., Sawyer M. and Clydesdale, F.M. (1982) Psychophysical relationship between perceived sweetness and color in cherry flavored beverages. *Journal of Food Protection*, **45**, 601–606.

Johnson, J.L., Dzendolet E. and Clydesdale, F.M. (1983) Psychophysical relationship between sweetness and redness in strawberry flavored drinks. *Journal of Food Protection*, **46**, 21–28.

Jones, F.N. and Woskow, M.H. (1964) On the intensity of odor mixtures. *Annals of New York Academy of Sciences*, **116**, 484–494.

Kauer, J.S. (1991) Contributions of topography and parallel processing to odor coding in the vertebrate olfactory pathway. *TINS*, **14**, 79–85.

Kawamura, Y. and Kare, M.R. (eds) (1987) *Umami: A Basic Taste*, Dekker, New York.

Kalmus, H. (1971) Genetics of taste, in *Handbook of Sensory Physiology*, *4* (ed L. Beidler), Springer-Verlag, New York, pp. 165–179.

Kelly, F.B. and Heymann, H. (1989) Contrasting the effects of ingestion and expectoration in sensory difference tests. *Journal of Sensory Studies*, **3** (4), 249–255.

Kinnamon, S.C. and Roper, S.D. (1988) Membrane properties of isolated mud puppy taste cells. *Journal of General Physiology*, **91**, 351–371.

Kocher, E.C. and Fisher. G.L. (1969) Subjective intensity and taste preference. *Perceptual and Motor Skills*, **28**, 735–740.

Kokini, J.L. (1985) Fluid and semi-solid food texture and texture–taste interactions. *Food Technology*, **39**, 86–94.

Kokini, J.L., Kadane, J.B. and Cussler, E.L. (1977) Liquid texture perceived in the mouth. *Journal of Texture Studies*, **8**, 195–218.

Kostyla, A.R. (1978) *The Psychophysical Relationship between Color and Flavor of Some Fruit Flavored Beverages*. Ph.D. thesis, University of Massachusetts at Amherst.

Kostyla, A.S. and Clydesdale, F.M. (1978) The psychophysical relationships between color and flavor. *Critical Reviews in Food Science and Nutrition*, **10**, 303–378.

Kroeze, J.H.A. (1979) Masking and adaptation of sugar sweetness intensity. *Physiology and Behavior*, **22**, 347–351.

Kroeze, J.H.A. (1990) The perception of complex taste stimuli, in *Psychological Basis of Sensory Evaluation* (eds R.L. McBride and H.J.H. MacFie), Elsevier Applied Science, London, pp. 41–68.

Kroeze, J.H.A. and Bartoshuk, L.M. (1985) Bitterness suppression as revealed by split-tongue taste stimulation in humans. *Physiology and Behavior*, **35**, 779–783.

Kubie, J.L. and Moulton, D.G. (1980) Odorant specific patterns of differential sensitivity inherent in the salamander olfactory epithelium. *The Society of Neuroscience Abstracts*, **6**, 243.

Kubota, T. and Kubo, I. (1969) Bitterness and chemical structure. *Nature*, **223**, 97–99.

Kumazawa, T., Mizumura, K. and Sato, J. (1987) Response properties of polymodal receptors studied using *in vitro* testis superior spermatic nerve preparations of dogs. *Journal of Neurophysiology*, **57**, 702–711.

Kurahashi, T., Lowe, G. and Gold, G.H. (1994) Suppression of odorant responses by odorants in olfactory receptor cells. *Science*, **265**, 118–120.

Kuznicki, J. (1978) Taste profiles from single human taste papillae. *Perceptual and Motor Skills*, **47**, 279–286.

Kuznicki, J.T. and Cardello, A.V. (1986) Psychophysics of single taste papillae, in *Clinical Measurement of Taste and Smell* (eds H.L. Meiselman, Ph.D. and R.S. Rivlin, M.D.), Macmillan, New York, pp. 200–228.

Labows, J.N. and Wysocki, C.J. (1984) Individual differences in odor perception. *Perfumes and Flavors*, **9**, 21–26.

Laing, D.G. (1995) Perception of odor mixtures, in *Handbook of Olfaction and Gustation* (ed R.L. Doty), Dekker, New York, pp. 283–297.

Laing, D.G. and Livermore, B.A. (1992) Perceptual analysis of complex chemical signals by humans, in *Chemical Signals in Vertebrates* (eds R.L. Doty and D. Muller-Schwartz), Plenum Press, New York, pp. 587–593.

Laing, D.G., Panhuber, H., Wilcox, M.E. and Pittman, E.A. (1984) Quality and intensity of binary odor mixtures. *Physiology and Behavior*, **33**, 309–319.

Laird, D.A. and Breen, W.J. (1939) Sex and age alterations in taste preferences. *Journal of the American Dietetic Association*, **15**, 549–550.

Lancet, D. (1986) Vertebrate olfactory reception. *Annual Review of Neuroscience*, **9**, 329–355.

Lancet, D. and Pace, U. (1987) The molecular basis of odor recognition. *Trends in Biochemical Science*, **12**, 63–66.

Lanza, D.C. and Clerico, D.M. (1995) Anatomy of the human nasal passages, in *Handbook of Olfaction and Gustation* (ed R.L. Doty), Dekker, New York, pp. 53–73.

Larmond, E. (1976) Texture measurement in meat by sensory evaluation. *Journal of Texture Studies*, **7**, 87–93.

Laska, M., Hudson, R. and Distel, H. (1990) Olfactory sensitivity to biologically relevant odors may exceed the sum of component thresholds. *Chemoecology*, **1**, 139–141.

Laslett, G.M. and Bremner, H.A. (1979) Evaluating acceptability of fish minces and fish fingers from sensory variables. *Journal of Food Technology*, **14**, 389.

Lawless, H.T. (1978) Evidence for neural inhibition in bittersweet mixtures. Doctoral Dissertation, Brown University. University Microfilms International, Ann Arbor, Michigan, #7906574.

Lawless, H.T. (1979) Evidence for neural inhibition in bittersweet taste mixtures. *Journal of Comparative and Physiological Psychology*, **93**, 538–547.

Lawless, H.T. (1987) Gustatory psychophysics, in *Neurobiology of Taste and Smell* (eds T.E. Finger and W.L. Silver), John Wiley & Sons, Inc., New York, pp. 401–420.

Lawless, H.T. (1988) Odour description and odour classification revisited, in *Food Acceptability* (ed D.M.H. Thomson), Elsevier Applied Science, London, pp. 27–40.

Lawless, H.T. and Cain, W.S. (1975) Recognition memory for odors. *Chemical Senses and Flavor*, **1**, 331–337.

Lawless, H.T. and Malone, G.J. (1986) A comparison of rating scales: Sensitivity, replicates and relative measurement. *Journal of Sensory Studies*, **1** (2), 155–174.

Lawless, H.T. and Stevens, D.A. (1984) Effects of oral chemical irritation on taste. *Physiology and Behavior*, **32**, 995–998.

Lawless, H.T., Rozin, P. and Shenker, J. (1985) Effects of oral capsaicin on gustatory, olfactory and irritant sensations and flavor identification in humans who regularly or rarely consume chilli pepper. *Chemical Senses*, **10**, 579–589.

Lawless, H.T., Glatter, S. and Hohn, C. (1991) Context-dependent changes in the perception of odor quality. *Chemical Senses*, **16** (4), 349–360.

Lawrie, R. (ed.) (1989) *Developments in Meat Science*, Elsevier Science Publishing Company, Inc., New York.

Lawrie, R.A. (5th ed.) (1991) *Meat Science*, Pergamon Press, Oxford.

Lee, W.E. and Pangborn, R.M. (1986) Time-intensity: the temporal aspects of sensory perception. *Food Technology*, **40** (11), 71–82.

Lee, W.E. III, Deibel, A.E., Glembin, C.T. and Munday, E.G. (1988) Analysis of food crushing sounds during mastication: frequency–time studies. *Journal of Texture Studies*, **19**, 27–38.

Lee, W.E. III, Schweitzer, M.A., Morgan, G.M. and Shepherd, D.C. (1990) Analysis of food crushing sounds during mastication: total sound level studies. *Journal of Texture Studies*, **21**, 165–178.

Lester, L.S. and Kramer, F.M. (1991) The effects of heating on food acceptability and consumption. *Journal of Foodservice Systems* **6**, 69–87.

Logue, A.W., Ophir, I. and Strauss, K.E. (1981) The acquisition of taste aversions in humans. *Behavior Research and Therapy*, **19**, 319–333.

Lucas, F. and Bellisle, F. (1987) The measurement of food preferences in humans: Do taste-and-spit tests predict consumption? *Physiology and Behavior*, **39**, 739–743.

Lynch, J., Liu, Y.-H., Mela, D.J. and MacFie, H.J.H. (1993) A time–intensity study of the effect of oil mouthcoatings on taste perception. *Chemical Senses*, **18** (2), 121–129.

MacFie, H.J.H. and Thomson, D.M.H. (eds) (1994) *Measurement of Food Preferences*, Blackie Academic & Professional, Glasgow.

Mackay-Sim, A., Shaman, P. and Moulton, D.G. (1982) Topographic coding of olfactory quality: odorant-specific patterns of epithelial responsivity in the salamander. *Journal of Neurophysiology*, **48**, 584–596.

Mackey, A. (1958) Discernment of taste substances as affected by solvent medium. *Food Research*, **23** (6), 580–583.

Mackey, A. and Valassi, K. (1956) The discernment of primary tastes in the presence of different food textures. *Food Technology*, **10**, 238–240.

Maga, J.A. (1974) Influence of color on taste thresholds. *Chemical Senses and Flavor*, **1**, 115–119.

Makhlouf, G.M. and Blum, A.L. (1972) Kinetics of the taste response to chemical stimulation: a theory of acid taste in man. *Gastroenterology*, **63** (4), 67–75.

Makin, J.W. and Porter, R.H. (1989) Attractiveness of lactating females' breast odors to neonates. *Child Development*, **60**, 803–810.

Maller, O. and Desor, J.A. (1973) Effect of taste on ingestion by human newborns, in *Fourth Symposium on Oral Sensation and Perception: Development in the Fetus and Infant* (ed J.F. Bosma), U.S. Department of Health, Education, and Welfare, Bethesda, Maryland, pp. 279–291.

Maller, O., Cardello, A.V., Sweeney, J. and Shapiro, D. (1982) Psychophysical and cognitive correlates of discretionary usage of table salt and sugar by humans, in *Proceedings of the Fifth European Chemoreception Research Organization Symposium* (eds J.E. Steiner and J.R. Granchrow), IRL Press, London, pp. 205–218.

Marks, W.B., Dobelle, W.H. and MacNichol, E.F. (1964) Visual pigments of single primate cones. *Science*, **143**, 1181.

Marks, L.E., Borg, G. and Westerlund, J. (1992) Differences in taste perception assessed by magnitude matching and by category–ratio scaling. *Chemical Senses*, **17**, 493–506.

Marshall, S.G. and Vaisey, M. (1972) Sweetness perception in relation to some textural characteristics of hydrocolloid gels. *Journal of Texture Studies*, **3**, 173–185.

Martens, H. and Russwurm. H., Jr. (eds) (1983) *Food Research and Data Analysis*, Applied Science Publishers, Essex.

Mattes, R.D. (1987) Sensory influences on food intake and utilization in humans. *Human Nutrition: Applied Nutrition*, **41A**, 77–95.

Mattes, R.D. and Mela, D. (1988) The chemical senses and nutrition: part II. *Nutrition Today*, May/June, 19–25.
McBride, R.L. (1993) Integration psychophysics: The use of functional measurement in the study of mixtures. *Chemical Senses*, **18**, 83–92.
McBride, R.L. and Johnson, R.L. (1987) Perception of sugar–acid mixtures in lemon juice drink. *International Journal of Food Science and Technology*, **22**, 299–408.
McBride, R.L. and MacFie, H.J.H. (eds) (1990) *Psychological Basis of Sensory Evaluation*, Elsevier Applied Science, London.
McBurney, D.H. (1972) Gustatory cross adaptation between sweet-tasting compounds. *Perception and Psychophysics*, **11** (8), 225–227.
McBurney, D.H. (1974) Are there primary tastes for man? *Chemical Senses and Flavors*, **1**, 17–28.
McBurney, D.H. and Gent, J.F. (1979) On the nature of taste qualities. *Psychological Bulletin*, **36**, 151–167.
McBurney, D.H. and Shick, T.R. (1971) Taste and water taste of twenty-six compounds for man. *Perception and Psychophysics*, **10**, 249–252.
McBurney, D.H., Smith, D.V. and Schick, T.R. (1972) Gustatory cross adaptation: Sourness and bitterness, *Perception and Psychophysics*, **11**, 228–232.
McBurney, D.H., Collings, V.B. and Glanz, L.M. (1973) Temperature dependence of human taste responses. *Physiology and Behavior*, **11**, 89–94.
McCutcheon, N.B. and Saunders, J. (1972) Human taste papilla stimulation: stability of quality judgements over time. *Science*, **175**, 214–216.
McLaughlin, S. and Margolskee, R.F. (1994) The sense of taste. *American Scientist*, **82**, 538–545.
Meheil, R. and Bolles, R.C. (1988) Learned flavor preferences based on calories are independent of initial hedonic value. *Animal Learning and Behavior*, **16**, 383–387.
Meilgaard, M., Civille, G.V. and Carr, B.T. (eds) (1991) *Sensory Evaluation Techniques*, CRC Press, Boca Raton.
Meiselman, H.L. (1968) Adaptation and cross-adaptation of the four gustatory qualities. *Perception and Psychophysics*, **4**, 368–372.
Meiselman, H.L. and Halpern, B.P. (1970a) Human judgments of *Gymnema sylvestre* and sucrose mixtures. *Physiology and Behavior*, **5**, 945–948.
Meiselman, H.L. and Halpern, B.P. (1970b) Effects of *Gymnema sylvestre* on complex tastes elicited by amino acids and sucrose. *Physiology and Behavior*, **5**, 1379–1384.
Meiselman, H.L., Halpern, B.P. and Dateo, G.P. (1976) Reduction of sweetness judgments by extracts from the leaves of *Ziziphus jujuba*. *Physiology and Behavior*, **17**, 313–317.
Mela, D.J. (1988) Sensory assessment of fat content in fluid dairy products. *Appetite*, **10**, 37–44.
Mela, D.J. (1989) Bitter taste intensity: the effect of tastant and thiourea taster status. *Chemical Senses*, **14** (1), 131–135.
Mela, D.J. and Christensen, C.M. (1988) Sensory assessment of oiliness in a low moisture food. *Journal of Sensory Studies*, **2**, 273–281.
Mela, D.J. and Sacchetti, A. (1991) Sensory preferences for fats: relationship with diet and body composition. *American Journal of Clinical Nutrition*, **53** (4), 908–915.
Midkiff, E.E. and Bernstein, I.L. (1985) Targets of learned food aversions in humans. *Physiology and Behavior*, **34**, 839–841.
Miller, I.J., Jr. (1995) Anatomy of the peripheral taste system, in *Handbook of Olfaction and Gustation* (ed R.L. Doty), Dekker, New York, pp. 521–547.
Moncrieff, R.W. (1966) *Odour Preferences*, John Wiley, New York.
Moncreiff, R.W. (1967) *The Chemical Senses*, Leonard Hill, London.
Moskowitz, H.R. (1971) The sweetness and pleasantness of sugars. *American Journal of Psychology*, **84** (3), 387–405.
Moskowitz, H.R. (1973) Effects of solution temperature on taste intensity in humans. *Physiology and Behavior*, **10**, 289–292.
Moskowitz, H.R. (1976) The nature of acceptability functions in texture. *Journal of Texture Studies*, **7**, 235–242.
Moskowitz, H.R. (1977) Intensity and hedonic functions for chemosensory stimuli, in *The Chemical Senses and Nutrition* (eds M.R. Kare and O. Maller), Academic Press, New York, pp. 71–101.

Moskowitz, H.R. (1980) Psychometric evaluation of food preferences. *Journal of Foodservice Systems*, **1**, 149–167.
Moskowitz, H.R. (1981) Sensory intensity versus hedonic functions: classical psychophysical approaches. *Journal of Food Quality*, **5**, 109–137.
Moskowitz, H.R. (1984) Relative importance of sensory factors to acceptance: Theoretical and empirical analyses. *Journal of Food Quality*, **7**, 75–90.
Moskowitz, H.R. (ed.) (1987) *Food Texture*. Dekker, New York.
Moskowtiz, H.R. (1991) Optimizing consumer product acceptance and perceived sensory quality, in *Food Product Development* (eds E. Graf and I.S. Saguy), Van Nostrand Reinhold, New York, pp. 157–187.
Moskowitz, H.R. (1992) Importance of sensory factors in processed seafood: methods and results. *Journal of Sensory Studies*, **7** (2), 147–156.
Moskowitz, H.R. (1994) *Food Concepts and Products: Just-In-Time-Development*, Food and Nutrition Press, Connecticut.
Moskowitz, H.R. and Arabie, P. (1970) Taste intensity as a function of stimulus concentration and solvent viscosity. *Journal of Texture Studies*, **1**, 502–510.
Moskowitz, H.R. and Barbe, C.D. (1977) Profiling of odor componens and their mixtures. *Sensory Processes*, **1**, 212–226.
Moskowitz, H.R. and Dubose, C. (1977) Taste intensity, pleasantness and quality of aspartame, sugars, and their mixtures. *Journal de l'Institut Canadien Science de Technologie Alimentaire*, **10** (2), 126–131.
Moskowitz, H.R. and Krieger, B. (1993) What sensory characteristics drive product quality? An assessment of individual differences. *Journal of Sensory Studies*, **8** (4), 271–282.
Moskowitz, H.R. and Krieger, B. (1995) The contribution of sensory liking to overall liking: An analysis of 6 food categories. *Food Quality and Preference*, **6**, 83–90.
Moskowitz, H.R., Kluter, R.A., Westerling, J. and Jacobs, H.L. (1974) Sugar sweetness and pleasantness: Evidence for different psychological laws. *Science*, **184**, 583–585.
Moskowitz, H.R., Sharma, K., Kumariah, V.N., Jacobs, H.L. and Sharma, S.D. (1975) Cross cultural differences in simple taste preferences. *Science*, **190**, 1217–1218.
Moskowitz, H.R., Kumraiah, V., Sharma, K.N., Jacobs, H.L. and Sharma, S.D. (1976a) Effects of hunger, satiety and glucose load upon taste intensity and taste hedonics. *Physiology and Behavior*, **16**, 471–475.
Moskowitz, H.R., Dravnieks, A. and Klarman, L.A. (1976b) Odor intensity and pleasantness for a diverse set of odorants. *Perception and Psychophysics*, **19** (2), 122–128.
Moskowitz, H.R., Dubose, C.N. and Reuben, M. (1977) Flavor chemical mixtures—A psychophysical analysis, in *Flavor Quality: Objective Measurement* (ed R.A. Scanlan), ACS Symposium Series, American Chemical Society, Washington D.C., pp. 29–44.
Moulton, R. (1955) Oral and dental manifestations of anxiety. *Psychiatry*, **18** (3), 1.
Mozell, M.M. and Hornung, D.E. (1985) Peripheral mechanisms in the olfactory process, in *Taste, Olfaction and the Central Nervous System* (ed D.W. Pfaff), Rockefeller University Press, New York, pp. 253–279.
Munoz, A.M. and Chambers, E. (1993) Relating sensory measurements to consumer acceptance of meat products. *Food Technology*, **47** (11), 128–131, 134.
Murphy, C. (1982) Effects of exposure and context on hedonics of olfactory-taste mixtures. *Sensory Processes*, **1**, 212–226.
Murphy, C. (1983) Age-related effects on the threshold, psychophysical function, and pleasantness of menthol. *Journal of Gerontology*, **38**, 217–222.
Murphy, C. (1986) Taste and smell in the elderly, in *Clinical Measurement of Taste and Smell* (eds H.L. Meiselman and R.S. Rivlin), Macmillan, New York, pp. 343–371.
Murphy, C. and Cain, W.S. (1980) Taste and olfaction: independence versus interaction. *Physiology and Behavior*, **24**, 601–605.
Murphy, C. and Cain, W.S. (1986) Odor identification: The blind are better. *Physiology and Behavior*, **37**, 177–180.
Murphy, C. and Gilmore, M.M. (1989) Quality-specific effects of aging on the human taste system. *Perception and Psychophysics*, **45**, 121–128.
Murphy, C. and Gilmore, M.M. (1990) Effects of aging on sensory functioning: Implications for dietary selection, in *Psychological Basis of Sensory Evaluation* (eds R.L. McBride and H.J.H. MacFie), Elsevier Applied Science, London, pp. 19–39.

Murphy, C. and Withee, J. (1986) Age-related differences in the pleasantness of chemosensory stimuli. *Psych. Aging*, **1**, 312–318.
Murphy, C., Cain, W.S. and Bartoshuk, L.M. (1977) Mutual action of taste and olfaction. *Sensory Processes*, **1** (3), 204–211.
Murphy, C., Cardello, A.V. and Brand, J.G. (1981) Tastes of fifteen halide salts following water and NaCl: anion and cation effects. *Physiology and Behavior*, **26**, 1083–1095.
Murphy, C., Cain, W.S., Gilmore, M.M. and Skinner, R.B. (1991) Sensory and semantic factors in recognition memory for odors and graphic stimuli: Elderly versus young persons. *American Journal of Psychology*, **104**, 161–192.
Murphy, C., Nordin, S., Wijk, R.A. de, Cain, W.S. and Polich, J. (1994) Olfactory-evoked potentials: assessment of young and elderly, and comparison to psychophysical threshold. *Chemical Senses*, **19** (1), 47–56.
Nasrawi, C.W. and Pangborn, R.M. (1989) The influence of tastants on oral irritation by capsaicin. *Journal of Sensory Studies*, **3**, 287–294.
Neitz, M. and Neitz, J. (1995) Numbers and ratios of visual pigment genes for normal red-green color vision. *Science*, **267**, 1013–1021.
Normand, M.D. and Peleg, M. (1988) Evaluation of the 'Blanket Algorithm' for ruggedness assessment. *Powder Technology*, **54**, 255–259.
O'Brien-Nabors, L. and Gelardi, R.C. (eds) (1986) *Alternative Sweeteners*, Dekker, New York.
O'Mahony, M. (1983) Adapting short-cut signal detection measures to the problem of multiple difference testing: The R-index, in *Sensory Quality in Foods and Beverages. Definition, Measurement and Control* (eds A.A. Williams and R.K. Atkin), Ellis Horwood, Chichester, pp. 69–81.
O'Mahony, M. (1991) Descriptive analysis and concept alignment, in *Sensory Science Theory and Applications in Foods* (eds H.T. Lawless and B.P. Klein), Dekker, New York, pp. 223–267.
O'Mahony, M. (1992) Understanding discrimination tests: A user friendly treatment of response bias, rating and ranking R-index tests and their relationship to signal detection. *Journal of Sensory Studies*, **7** (1), 1–47.
O'Mahoney, M., Atassi-Sheldon, S., Rothman, L. and Murphy-Ellison, T. (1983) Relative singularity/mixedness judgements for selected taste stimuli. *Physiology and Behavior*, **31**, 749–755.
O'Mahony, M., Rothman, L., Ellison, T., Shaw D. and Buteau, L. (1990) Taste descriptive analysis: Concept formation, alignment and appropriateness. *Journal of Sensory Studies*, **5**, 71–103.
Okabe, M. (1979) Texture measurement of cooked rice and its relationship to the eating quality. *Journal of Texture Studies*, **10**, 131–152.
Overbosch, P., Enden, J.C. Van Den and Keur, B.M. (1986) An improved method for measuring perceived intensity/time relationships in human taste and smell. *Chemical Senses*, **11** (3), 331–338.
Pak, W.L. and Shortridge, R.D. (1991) Inositol phospholipid and invertebrate photoreceptors. *Photochemistry and Photobiology*, **53**, 871–875.
Pangborn, R.M. (1960) Influence of color on the discrimination of sweetness. *American Journal of Psychology*, **73**, 229–238.
Pangborn, R.M. (1981) Individuality in responses to sensory stimuli, in *Criteria of Food Acceptance* (eds J. Solms and R.L. Hall), Forster Verlag, Zurich, pp. 177–219.
Pangborn, R.M. and Dunkley, W.L. (1964) Sensory discrimination of fat and solids-not-fat in milk. *Journal of Dairy Science*, **47**, 719–725.
Pangborn, R.M. and Hansen, B. (1963) The influence of color on discrimination of sweetness and sourness in pear nectar. *American Journal of Psychology*, **26**, 315–317.
Pangborn, R.M. and Szczesniak, A.S. (1974) Effect of hydrocolloids and viscosity on flavor and odor intensities of aromatic flavor compounds. *Journal of Texture Studies*, **4**, 467–482.
Pangborn, R.M., Chrisp, R.B. and Bertolero, L. (1970) Gustatory, salivary and oral thermal responses to solutions of sodium chloride at four temperatures. *Perception and Psychophysics*, **8**, 69–75.
Pangborn, R.M., Trabue, I.M. and Szczesniak, A.S. (1973) Effect of hydrocolloids on oral viscosity and basic taste intensities. *Journal of Texture Studies*, **4**, 224–241.
Pangborn, R.M., Gibbs, Z.M. and Tassan, C. (1978) Effect of hydrocolloids on apparent

viscosity and sensory properties of selected beverages. *Journal of Texture Studies*, **9**, 415–436.

Pangborn, R.M., Braddock, K.S. and Stone, L.J. (1983) *Ad Lib Mixing to Preference* vs. *Hedonic Scaling: Salts in Broths and Sucrose in Lemonade*. American Chemoreception Society V Poster Presentation, Sarasota, FL.

Pantone (1986) Fun foods, in *Color News*, Pantone Color Institute, Woodland Hills, CA, **1** (2), 3.

Pelchat, M.L. and Rozin, P. (1982) The special role of nausea in the acquisition of food dislikes by humans. *Appetite*, **3**, 341–351.

Peleg, M. (1993) Fractals and food. *Critical Reviews in Food Science and Nutrition*, **33** (2), 149–165.

Perng, C.M. and McDaniel, M.R. (1989) Optimization of a blackberry juice drink using response surface methodology. Presented at the 49th Annual Meeting of the Institute of Food Technologists, Chicago, IL.

Peryam, D.R. and Girardot, N.F. (1952) Advanced taste-test method. *Food Engineering*, **24**, 58–61.

Peryam, D.R. and Pilgrim, F.J. (1957) Hedonic scale method of measuring food preferences. *Food Technology*, **11**, 9–14.

Pfaffmann, C. (1939) Afferent impulses from the teeth due to pressure and noxious stimulation. *Journal of Physiology*, **97**, 207.

Piggot, J. and Harper, R. (1976) Variations of odour quality and pleasantness with intensity. *Journal of the Science of Food and Agriculture*, **27** (8), 787–788.

Porter, R.H. and Schaal, B. (1995) Olfaction and development of social preferences in neonatal organisms, in *Handbook of Olfaction and Gustation* (ed R.L. Doty), Dekker, New York, pp. 299–321.

Porter, R.H., Makin, J.W., Davis, L.B. and Christensen, K.M. (1992) Breast-fed infants respond to olfactory cues from their own mother and unfamiliar lactating females. *Infant Behavior and Development*, **15**, 85–93.

Powers, J.J., Godwin, D.R. and Bargmann, R.E. (1977) Relations between sensory and objective measurements for quality evaluations of green beans, in *Flavor Quality-Objective Measurement* (ed R.A. Scanlon), American Chemical Society, Washington, D.C., pp. 51–70.

Prescott, J., Allen, S. and Stephens, L. (1993) Interactions between oral chemical irritation, taste and temperature. *Chemical Senses*, **18** (4), 389–404.

Price, S. and DeSimone, J.A. (1977) Models of taste receptor cell stimulation. *Chemical Senses and Flavor*, **2**, 427–456.

Rasekh, J. and Kramer, A. (1970) Objective evaluation of canned tuna sensory quality. *Journal of Food Science*, **35** (4), 417–423.

Reid, T. (1785) *Essays on the Intellectual Powers of Man*, Macmillan and Co., London.

Renqvist, Y. (1919) Ueber den Geschmack. *Skand. Arch. Physiologie*, **38**, 97–201.

Rhodes, V.J., Kiehl, E.R. and Brody, D.E. (1955) Visual preferences for grades of retail beef cuts. Missouri University Agricultural Experimentation Station Research Bulletin, 583.

Richardson, J.T.E. and Zucco, G.M. (1989) Cognition and olfaction: A review. *Psychological Bulletin*, **105**, 352–360.

Ringel, R.L. (1970) Oral region two-point discrimination in normal and myopathic subjects, in *Second Symposium on Oral Sensation and Perception* (ed J.F. Bosma), Charles C. Thomas, Springfield, IL, pp. 309–322.

Ringo, L. (1982) M.S. Thesis, University of California, Davis.

Riskey, D.R. (1986) Use and abuse of category scales in sensory measurement. *Journal of Sensory Studies*, **1** (3/4), 217–236.

Rogers, P.J. (1990) Dietary fat, satiety and obesity. *Food Quality and Preference*, **2**, 103–110.

Rolls, B.J., Rolls, E.T., Rowe, E.A. and Sweeney, K. (1981) Sensory specific satiety in man. *Physiology and Behavior*, **27**, 137–142.

Rolls, B.J., Rowe, E.A. and Rolls, E.T. (1982) How sensory properties of foods affect human feeding behavior. *Physiology and Behavior*, **29**, 409–417.

Rolls, B.J., Hetherington, M. and Burley, V.J. (1988a) The specificity of satiety: the influence of foods of different macronutrient content on the development of satiety. *Physiology and Behavior*, **43**, 145–153.

Rolls, B.J., Hetherington, M. and Burley, V.J. (1988b) Sensory stimulation and energy density in the development of satiety. *Physiology and Behavior*, **44**, 727–733.

Roper, S.D. (1989) Ion channels and taste transduction, in *Chemical Senses, 1, Receptor Events and Transduction in Taste and Olfaction* (eds J.G. Brand, J.H. Teeter, R.H. Cagan and M.R. Kare), Dekker, New York, pp. 137–149.

Rosenstein, D. and Oster, H. (1988) Differential facial responses to four tastes in newborns. *Child Development*, **59**, 1555–1568.

Roth, H.A., Radle, L., Gifford, S.R. and Clydesdale, F.M. (1988) Psychophysical relationships between perceived sweetness and color in lemon and lime flavored beverages. *Journal of Food Science*, **53**, 1116, 1162.

Roy, G.M. (1990) The applications and future implications of bitterness reduction and inhibition in food products. *Critical Reviews in Food Science and Nutrition*, **29** (2), 59–71.

Rozin, P. (1989) The role of learning in the acquisition of food preferences by humans, in *Handbook of the Psychophysiology of Human Eating* (ed R. Shepherd), John Wiley & Sons, Chichester, pp. 205–227.

Rozin, P. and Schiller, D. (1980) The nature and acquisition of a preference for chili pepper by humans. *Motivation Emotion*, **4**, 77–100.

Rozin, P. and Tuorila, H. (1993) Simultaneous and temporal contextual influences on food acceptance. *Journal of Food Quality and Preference*, **4**, 11–20.

Rushton, W.A.H. (1958) Visual pigments in the colour blind. *Nature*, **182**, 690–692.

Sawyer, F.M., Cardello, A.V. and Prell, P.A. (1988) Consumer evaluation of the sensory properties of fish. *Journal of Food Science*, **53** (1), 12–18, 24.

Scheuplein, R.J. (1976) Permeability of the skin: A review of major concepts and some new developments. *Journal of Investigative Dermatology*, **67**, 672–676.

Schifferstein, H.N.J. and Frijters, J.E.R. (1991) The perception of the taste of KCl, NaCl and quinine-HCl is not related to PROP-sensitivity. *Chemical Senses*, **16** (4), 303–317.

Schiffman, S.S. (1974) Contributions to the physicochemical dimensions of odor: a psychophysical approach. *Annals of the New York Academy of Sciences*, **137**, 164–83.

Schiffman, S.S. (1986) Age-related changes in taste and smell and their possible causes, in *Clinical Measurement of Taste and Smell* (eds H.L. Meiselman and R.S. Rivlin), Macmillan, New York, pp. 326–342.

Schiffman, S.S. (1987) Recent development in taste enhancement. *Food Technology*, **41** (6), 72–73, 124.

Schiffman, S.S. (1991) Taste and smell losses with age. *Contemporary Nutrition*, **16** (2), 1–2.

Schiffman, S.S. and Erickson, R.P. (1980) The issue of primary tastes versus a taste continuum. *Neuroscience and Behavioral Reviews*, **4**, 109–117.

Schiffman, S.S. and Warwick, Z.S. (1988) Flavor enhancement of foods for the elderly can reverse anorexia. *Neurobiology of Aging*, **9**, 24–26.

Schiffman, S.S. and Warwick, Z.S. (1991) Changes in taste and smell over the life span: effects on appetite and nutrition in the elderly, in *Chemical Senses, 4, Appetite and Nutrition* (eds M. Friedman, M. Tordoff and M. Kare), Dekker, New York, pp. 341–365.

Schiffman, S.S., Orlandi, M. and Erickson, R.P. (1979) Changes in taste and smell with age: biological aspects, in *Sensory Systems and Communication in the Elderly*, Vol. 10 (eds J.M. Ordy and K. Brizzee), Raven Press, New York, pp. 247–268.

Schiffman, S.S., Reynolds, M.L. and Young, F.W. (1981) *Introduction to Mulitdimensional Scaling: Theory, Methods, and Applications*, Academic Press, New York.

Schiffman, S.S., Lockhead, E. and Maes, F.W. (1983) Amiloride reduces the taste intensity of Na^+ and Li^+ salts and sweeteners. *Proceedings of the National Academy of Sciences*, **80**, 6136–6140.

Schiffman, S.S., Simon, S.A., Gill, J.M. and Beeker, T.G. (1986) Bretylium tosylate enhances salt taste. *Physiology and Behavior*, **36**, 1129–1137.

Schmidt, H.J. and Beauchamp, G.K. (1988) Adult-like preferences and aversions in three-year-old children. *Child Development*, **59**, 1136–1143.

Schutz, H.G. and Wahl, O.L. (1981) Consumer perception of the relative importance of food appearance, flavor and texture to food acceptance, in *Criteria of Food Acceptance: How Man Chooses What He Eats* (eds J. Solms and R.L. Hall), Forster Verlag, Zurich, pp. 97–116.

Scott, T.R. and Giza, B.K. (1995) Theories of gustatory neural coding, in *Handbook of Olfaction and Gustation* (ed R.L. Doty), Dekker, New York, pp. 611–633.
Shallenberger, R.S. and Acree, T.E. (1967) Molecular theory of sweet taste. *Nature*, **216**, 480–482.
Shepherd, R. (1989) *Handbook of the Psychophysiology of Human Eating* (ed R. Shepherd), John Wiley & Sons Ltd., Chichester.
Shepherd, R. and Stockley, L. (1985) Fat consumption and attitudes towards food with a high fat content. *Human Nutrition: Applied Nutrition*, **39A**, 431–442.
Sherrington, C.S. (1906) *The Integrative Action of the Nervous System*, Constable, London.
Siegel, P.S. and Pilgrim, F.J. (1958) The effect of monotony on the acceptance of food. *American Journal of Psychology*, **71**, 756–759.
Sizer, F. and Harris, N. (1985) The influence of common food additives and temperature on threshold perception of capsaicin. *Chemical Senses*, **10**, 279–286.
Stang, D.J. (1975) When familiarity breeds contempt, absence makes the heart grow fonder: Effects of exposure and delay on taste pleasantness ratings. *Bulletin of the Psychonomic Society*, **6** (3), 273–275.
Steenkamp, J.B.E.M. and van Trijp, H.C.M. (1988) Free-choice profiling in cognitive food acceptance research, in *Food Acceptability* (ed D.M.H. Thomson), Elsevier Appled Science, London, pp. 363–376.
Steiner, J.E. (1973) The gustofacial response: Observation on normal and anencephalic newborn infants, in *Fourth Symposium on Oral Sensation and Perception* (ed J.F. Bosma), Superintendent of Documents, U.S. Government Printing Office, Washington, pp. 254–278.
Steiner, J.E. (1977) Facial expressions of the neonate infant indicating the hedonics of food-related chemical stimuli, in *Taste and Development: The Genesis of Sweet Preference* (ed J.M. Weiffenbach), DHEW Publication No. (NIH) 77-1068, U.S. Department of Health, Education, and Welfare, Bethesda, MD, pp. 173–189.
Steiner, J.E. (1979) Human facial expression in response to taste and smell stimulation, in *Advances in Child Development and Behavior, 1* (eds H.W. Reese and L.P. Lipsitt), Academic Press, New York, pp. 257–295.
Stevens, D.A. and Lawless, H.T. (1986) Putting out the fire, Effects of tastants on oral chemical irritation. *Perception and Psychophysics*, **39**, 346–350.
Stevens, D.A. and Lawless, H.T. (1988) Responses by humans to oral chemical irritants as a function of locus of stimulation. *Perception and Psychophysics*, **43**, 72–78.
Stevens, S.S. (1957) On the psychophysical law. *Psychological Review*, **64**, 153–181.
Stevens, S.S. (1961) To honor Fechner and repeal his law. *Science*, **133**, 80–86.
Stevens, S.S. (1975) *Psychophysics: Introduction to its Perceptual, Neural and Social Prospects*, John Wiley & Sons, New York.
Stevenson, R.J. and Yeomans, M.R. (1993) Differences in ratings of intensity and pleasantness for the capsaicin burn between chili likers and non-likers; implications for liking development. *Chemical Senses*, **18** (5), 471–482.
Stillman, J.A. (1993) Color influences flavor identification in fruit-flavored beverages. *Journal of Food Science*, **58** (4), 810–812.
Stone, H. and Oliver, S. (1966) Effect of viscosity on the detection of relative sweetness intensity of sucrose solutions. *Journal of Food Science*, **31**, 129–134.
Stone, H., Sidel, J.L., Oliver, S., Woolsey, A. and Singleton, R.C. (1974) Sensory evaluation by qualitative descriptive analysis. *Food Technology*, **28**, 24–34.
Striem, B.J., Pace, U., Zehavi, U., Naim, M. and Lancet, D. (1989) Sweet tastants stimulate adenylate cyclase coupled to GTP binding protein in rat tongue membranes. *Biochemistry Journal*, **260**, 121–126.
Stryer, L. (1988) Molecular basis of visual excitation. *Cold Spring Harbor Symposia on Quantitative Biology*, **53**, 283–294.
Swithers, S.E. and Hall, W.G. (1994) Does oral experience terminate ingestion? *Appetite*, **23** (2), 113–138.
Szczesniak, A.S. (1963) Classification of textural characteristics. *Journal of Food Science*, **28**, 385–389.
Szczesniak, A.S. (1971) Consumer awareness of texture and of other food attributes II. *Journal of Texture Studies*, **2**, 196–206.

Szczesniak, A.S. (1972) Consumer awareness of and attitudes to food texture, 2, Children and teenagers. *Journal of Texture Studies*, **3**, 206–217.
Szczesniak, A.S. (1987) Correlating sensory with instrumental texture measurements—An overview of recent developments. *Journal of Texture Studies*, **18**, 1–15.
Szczesniak, A.S. (1991) Textural perceptions and food quality. *Journal of Food Quality*, **14**, 75–78.
Szczesniak, A.S. and Kahn, E.L. (1971) Consumer awareness of and attitudes to food texture I: adults. *Journal of Texture Studies*, **2**, 280–295.
Szczesniak, A.S. and Kahn, E.L. (1984) Texture contrasts and combinations: a valued consumer attribute. *Journal of Texture Studies*, **15** (3), 285–302.
Szczesniak, A.S. and Kleyn, D.H. (1963) Consumer awareness of texture and other food attributes. *Food Technology*, **27**, 74–77.
Szczesniak, A.S. and Ilker, R. (1988) The meaning of textural characteristics-juiciness in plant foodstuffs. *Journal of Texture Studies*, 61–78.
Szczesniak, A.S., Brandt, M.A. and Friedman, H. (1963) Development of standard rating scales for mechanical parameters of texture and correlation between the objective and the sensory methods of texture evaluation. *Journal of Food Science*, **28**, 397–403.
Szolcsanyi, J. (1977) A pharmacological approach to elucidation of the role of different nerve fibres and receptor endings in mediation of pain. *Journal of Physiology* (Paris), **73**, 251–259.
Tancredi, T., Lelj, F. and Temussi, P.A. (1979) Three dimensional mapping of the bitter taste receptor site. *Chemical Senses and Flavor*, **4**, 259–265.
Tanimura, S. and Mattes, R.D. (1993) Relationships between bitter taste sensitivity and consumption of bitter substances. *Journal of Sensory Studies*, **8** (1), 31–41.
Tanner, W.P. and Swets, J.A. (1954) A decision-making theory of visual detection. *Psychological Review*, **61**, 401–409.
Tansley, K. (1965) *Vision in Vertebrates*, Chapman & Hall Ltd., London.
Taylor, D.E and Pangborn, R.M. (1990) Temporal aspects of hedonic responses. *Journal of Sensory Studies*, **4**, 214–247.
Temussi, P.A., Lelj, F. and Tancredi, T. (1978) Three dimensional mapping of the sweet taste receptor site. *Journal of Medicinal Chemistry*, **21**, 1154–1158.
Thomson, D.M.H. (1988) *Food Acceptability*, Elsevier Applied Science, London.
Thompson, J.D. and Johnson, D.J. (1963) Food temperature preferences of surgical patients. *Journal of the American Dietetic Association*, **43**, 209–211.
Titchener, E.B. (1909) *Textbook of Psychology*, Macmillan, New York.
Tonosaki, K. and Funakoshi, M. (1988) Cyclic nucleotides may mediate taste transduction. *Nature*, **331**, 354–356.
Torrance, E.P. (1958) Sensitization versus adaption in preparation for emergencies: Prior experience with an emergency ration and its acceptability in a stimulated survival situation. *Journal of Applied Psychology*, **42**, 63–67.
Trant, A.S. and Pangborn, R.M. (1983) Discrimination, intensity, and hedonic responses to color, aroma, viscosity, and sweetness of beverages. *Lebensmittel Wissenschaft und Technologie*, **16**, 147–152.
Troland, L.T. (1928) *Fundamentals of Human Motivation*, Van Nostrand-Reinhold, Princeton, New Jersey.
Tuorila, H., Cardello A.V. and Leshner, L. (1994) Antecedents and consequences of expectations related to fat-free regular-fat foods. *Appetite*, **23**, 247–263.
Tuorila-Ollikainen, H., Mahlamaki-Kultanen, S. and Kurkela, R. (1984) Relative importance of color, fruity flavor and sweetness in the overall liking of soft drinks. *Journal of Food Science*, **49**, 1598–1600.
Valkenburgh, B.V. and Hertel, F. (1993) Tough times at La Brea: Tooth breakage in large carnivores of the Late Pleistocene. *Science*, **261**, 456–459.
Vickers, Z.M. (1975) Development of a psychoacoustical theory of crispness. Ph.D. Thesis, Cornell University, Ithaca, New York.
Vickers, Z.M. (1979) Crispness and crunchiness of food, in *Food Texture and Rheology* (ed P. Sherman), Academic Press, London, pp. 33–41.
Vickers, Z.M. (1983) Pleasantness of food sounds. *Journal of Food Science*, **48**, 783–786.
Vickers, Z.M. (1988) Instrumental measures of crispness and their correlation with sensory assessment. *Journal of Texture Studies*, **19**, 1–14.

Vickers, Z.M. (1991) Sound perceptions and food quality. *Journal of Food Quality*, **14** (1), 87–96.
Vickers, Z.M. and Bourne, M.C. (1976) A psychoacoustical theory of crispness. *Journal of Food Science*, **41**, 1158–1164.
Vie, A., Gulli, D. and O'Mahoney, M. (1991) Alternative hedonic measures. *Journal of Food Science*, **56**, 1–5.
Von Sydow, E., Moskowitz, H., Jacobs, H. and Meiselman, H. (1974) Odor-taste interaction in fruit juices. *Lebensmittel Wissenschaft und Technologie*, **7**, 9–16.
Wald, G. (1964) The receptors of human color vision. *Science*, **145**, 1007–1016.
Wald, G. (1968) Molecular basis of visual excitation. *Science*, **162**, 230–239.
Walker, J.C., Reynolds, J.H., Warren, D.W. and Sidman, J. (1990) Responses of normal and anosmic subjects to odorants, in *Chemical Senses, 2, Irritation* (eds B.G. Green, J.R. Mason and M.R. Kare), Dekker, New York, pp. 95–121.
Weiffenbach, J.M., Cowart, B.J. and Baum, B.J. (1986) Taste intensity perception in aging. *Journal of Gerontology*, **41**, 460–468.
Wesson, J.B., Lindsay, R.C. and Stuiber, D.A. (1979) Discrimination of fish and seafood quality by consumer populations. *Journal of Food Science*, **44**, 878–882.
Westenhoefer, J. and Pudel, V. (1993) Pleasure from food: Importance for food choice and consequences of deliberate restriction. *Appetite*, **20**, 246–249.
Wever, E.G. (1962) Development of traveling-wave theories. *Journal of Acoustical Society of America*, **34**, 1319–1324.
Wheatley, J. (1973) Putting color into marketing. *Marketing* (Oct. 23–29), 67.
Williams, A.A. and Langron, S.P. (1984) The use of free-choice profiling for the evaluation of commercial ports. *Journal of the Science of Food and Agriculture*, **35**, 558–568.
Wiseman, J.J. and McDaniel, M.R. (1989) Modification of fruit flavors by aspartame and sucrose. Presented at the 49th Annual Meeting of the Institute of Food Technologists, Chicago, IL.
Wisniewski, L., Epstein, L. and Caggiula, A.R. (1992) Effect of food change on consumption, hedonics and salivation. *Physiology and Behavior*, **52**, 21–26.
Wright, R.W. (1982) *The Sense of Smell*. CRC Press, Boca Raton, FL.
Wysocki, C.J. and Beauchamp, G.K. (1984) Ability to smell androsterone is genetically determined. *Proceedings of National Academy of Sciences*, **81**, 4899–4902.
Wysocki, C.J. and Beauchamp, G.K. (1991) Individual differences in human olfaction, in *Chemical Senses, 3, Genetics of Perception and Communication* (eds C.J. Wysocki and M.R. Kare), Dekker, New York, pp. 353–373.
Yamaguchi, S. (1979) The Umami taste, in *Food Taste Chemistry* (ed J.C. Boudreau), American Chemistry Society Symposium Series, No. 115, Washington, D.C., pp. 33–51.
Yamaguchi, S. (1987) Fundamental properties of Umami in human taste sensation, in *Umami: A Basic Taste* (eds Y. Kawamura and M.R. Kare), Dekker, New York, pp. 41–74.
Yokomukai, Y., Cowart, B.J. and Beauchamp, G.K. (1993) Individual differences in sensitivity to bitter-tasting substances. *Chemical Senses*, **18** (6), 669–681.
Yoshida, M. (1964) Studies in the psychometric classification of odor. *Japanese Psychological Research*, **6**, 111, 124–155.
Zellner, D.A., Rozin, P., Aron, M. and Kulish, C. (1983) Conditioned enhancement of humans' liking for flavor by pairing with sweetness. *Learning and Motivation*, **14**, 338–350.
Zellner, D.A., Stewart, W.F., Rozin, P. and Brown, J.M. (1988) Effect of temperature and expectations on liking for beverages. *Physiology and Behavior*, **44**, 61–68.
Zellner, D.A., Bartoli, A.M. and Eckard, R. (1991) Influence of color on odor identification and liking ratings. *American Journal of Psychology*, **104** (4), 547–561.
Zuniga, J.R., Davis, S.H., Englehardt, R.A., Miller, I.J., Jr, Schiffman, S.S. and Phillips, C. (1993) Taste performance on the anterior human tongue varies with fungiform taste bud density. *Chemical Senses*, **18** (5), 449–460.

2 The socio-cultural context of eating and food choice

PAUL ROZIN

For convenience, we can assume that the 5 billion people in the world eat an average of 3 meals a day, or 15 billion meals in total. It is virtually certain (though no one has counted) that the great majority of these meals are eaten with someone else. Let us generously estimate that 25% of all meals are taken alone. This may be literally true, but virtually none of these meals could have happened without others: the people who raised or hunted or gathered the food, and those who prepared it. And, the particular form of the food eaten is determined by recipes which are social constructions, and social communications, as are the food beliefs and attitudes of the supposedly solitary eater. So the study of the socio-cultural context of eating and food selection is almost the same as the study of eating and food selection.

In most, if not all cultures, food is laden with meaning, and constitutes a major form of social exchange. For our particular order of animals, the mammals, initial feeding events involve a tight social linkage with maternal interaction; ordinarily, it is impossible to separate the social and feeding aspects of nursing. And in the Judeo-Christian tradition, a socially embedded food transaction over an apple is a foundation event for the story of humans on earth.

It is widely believed in traditional cultures that 'you are what you eat' (Frazer, 1890/1959), that is, that people take on some of the properties of the foods they eat. There is also evidence for an unacknowledged belief in the same in western, developed cultures (Nemeroff and Rozin, 1989). This belief is very sensible; why shouldn't the things that enter the body impart their characteristics to the body? And the mouth is the principal and most salient route of entry of material things into the body.

The 'you are what you eat' principle, by itself, does not bring the social world into eating. But it does so when coupled with another principle, the law of contagion (Rozin, 1990). This law, originally proposed by Tylor (1871/1974) and elaborated by Frazer (1890/1959) and Mauss (1902/1972), was a description of beliefs of 'primitive' people, under the heading of sympathetic magic. The law of contagion holds that: once in contact, always in contact (see Rozin and Nemeroff, 1990, for a review). That is, when two objects touch, they pass their properties into the touched object, and this transfer is permanent.

If you are what you touch, a generous rephrasing of the law of contagion holds that all those who gather, prepare and serve a food are likely to have passed their properties to that food. And, through 'you are what you eat' (which can be taken to be a special case of contagion), these social 'essences' become part of the eater. Hence, every bite of a food, every sip of a drink, is a form of social incorporation. This is explicitly acknowledged in some cultures, including the Hua of Papua, New Guinea (Meigs, 1984) and the some 650000000 Hindu Indians (Appadurai, 1981).

One possible domain for non-social aspects of intake and food choice is genetic. Hard-wired systems, resistant to change, would set out a domain moderately independent of social forces. In the area of intake, there does seem to be a complex homeostatic system in the brain and other organs, which controls the tendency to eat in relation to various body-state variables. However, even this system is very much influenced by cultural norms, and is easily over-ridden in the appropriate contexts. In the area of food choice, there are few genetic constraints. Because humans are food generalists genetic specification of edibles and inedibles is extraordinarily difficult, and hence there are very few genetic constraints. There are genetically based positive biases to sweet and negative biases to bitter and irritant 'tastes'. However, these innate positives and negatives encompass the properties of a minority of actual human foods. Furthermore, they are easily reversed, as in distaste for high levels of sweetness in certain parts of cuisines (e.g. American main courses), and the extraordinary popularity of innately negative substances such as chili pepper and tobacco.

In short, the act of eating is usually overtly social, and the context of eating is invariably social, in many ways. Thus, a chapter on social influences on eating must, of necessity, overlap heavily with chapters on 'other' aspects of eating. In this volume, the chapters on attitudes and beliefs, information/communication, development in children, economic influences, dietary change, contextual factors, and changing dietary patterns all have a substantial, if not almost exclusive, social basis.

Under the circumstances, in this chapter, I propose to organize and summarize many of those socio-cultural influences that will also be covered in other chapters, concentrating more on those specific social influences and contexts that might escape discussion in other chapters.

The range of social influences/contexts that influence eating, and an outline for this chapter, is presented in Table 2.1. One major distinction that divides these influences is whether the social influence is explicity present, in the form of other persons (which I will call direct social influence) or not (which I will call indirect social influence). (For related discussions of types of social influence, see Galef (1985), Birch (1986, 1987) and Rozin (1988, 1990a).)

There are two aspects of eating that we are trying to explain. How much is eaten, and what is eaten? Almost the entire literature in psychology that

Table 2.1 Ways in which sociocultural contexts and influences affect food intake and food choice

1. Indirect environmental/cultural (culturally prescribed, with no necessary social agent intervention during the lifetime)
   - availability
   - price
   - convenience (technology)

2. Indirect personal (carried in the head of the individual, and introduced almost invariably by socioculturally influenced past experiences)
   - norms, beliefs, knowledge, attitudes

food intake	food choice
(i) body image and ideals	(i) role of food in health and life
(ii) norms about meal size	(ii) likes and dislikes
(iii) role of food in health and life	(iii) cuisine and appropriateness

3. Direct on-line influences

a. inadvertent (unintentional)	b. advertent (intentional/teaching)
(i) meal size	(i) explanations
(ii) food choice	(ii) norm identification
exposure	(iii) social pressure
imitation	
mood contagion	
(iii) sense of self, self-esteem and attributions of others	

4. Sociocultural aspects of the acquisition of norms, beliefs, knowledge and attitudes.
   - (1) programming the environment
   - (2) parental influence and the family paradox
   - (3) other social influences
   - (4) mechanisms
     - mere exposure
     - hedonic shift
     - identification
     - addiction
     - self-related meanings
       - disgust
       - self-image
     - values and morals
       - moralization

deals with eating concerns how much is eaten: regulation of food intake, and 'disorders' related to food intake, such as obesity, anorexia nervosa and bulimia. This may well be, in part, because it is easier to imagine a reasonable non-social context for studying amount ingested, allowing for laboratory settings, and in part because of the salient pathologies of amount eaten (see Rozin, 1981, 1995a). I do not plan to contribute to this overemphasis in this chapter, and will devote most of it to what is eaten. I believe that most students of amount eaten have seriously underestimated the importance of social factors. I also believe that the 'pathologies' of food choice are greatly underestimated in the USA, that a great deal of money is wasted on what are believed to be healthy foods, and a great deal of time and energy is wasted in worrying about healthy diets.

2.1 Indirect socio-cultural effects

2.1.1 Availability, price and convenience/technology

These influences do not require the presence of a social agent (person) at the time of eating or choice, or at any particular point in development. They are framings of the possible established by the joint operation of culture and the environment. They manifest themselves principally in three forms: determination of availability, price and convenience.

2.1.2 Food intake

We can only eat what is there to eat. If the environment/culture limit the amount available, there is no escaping this constraint. Cultural factors limit availability in a number of ways. They co-determine, with environmental/economic factors what is grown or raised, and what is imported. Traditions often determine appropriate portion sizes, and hence appropriate packaging. Palatability (along with visual attractiveness) is a powerful determinant of how much is eaten on any given occasion, and this is a function, in good part, of availability. Price is a major practical determinant of what is effectively available, and hence intake. And convenience is regularly cited as an important factor in determining food choice and presumably amount eaten. The convenience of fast foods is one of their appeals, as is the convenience of prepared frozen full dinners, or other microwavable foods. As a result of cultural changes, including the developments of technology, the average American now has available in one supermarket, modestly priced, more food choices, and more foods that can be consumed with minimal or no preparation, than have ever been available to anyone. Indeed, just 100 years ago even the wealthiest people anywhere in the world had a range of choices orders of magnitude less than what any American has now.

2.1.3 Food choice

In the domain of food choice, this availability and technology opens wide vistas, and brings the raw foods and prepared dishes of the world to the average hearth. Through restaurants, cookbooks and technologies such as microwave ovens and food processors, enormous opportunities are opened. All of these influences are independent of the direct social experience of an individual, though that social experience undoubtedly influences the extent to which these opportunities are explored and engaged.

Of course, availability is ultimately the product of individuals and their preferences and abilities. Widespread interest in a food within a culture spurs attempts to obtain more, by local means or by importation, and to develop technologies to increase availability and lower price. High levels of desire in Europe over the last hundreds of years have caused coffee, chocolate and sugar to move from prohibitively expensive, luxury items, to commonplace parts of the daily diet (see, e.g. Mintz, 1985).

The brevity of this section on availability, price, and convenience is no indication of the importance of these factors. Rather, the understanding of the effects of availability and price falls largely outside of the domain of the psychology of food choice.

2.2 Indirect personal effects

Norms, beliefs, knowledge and attitudes are prime determinants of both food intake and food choice. They exist in the heads of individuals, but they got there primarily via sociocultural influences in the past.

2.2.1 Food intake

2.2.1.1 Body image and ideals. Given the close relation between amount ingested and body weight (as a result, essentially, of Newton's Laws), culturally introduced ideals of body size are modulating forces in amount eaten. In the USA and probably most other Western cultures, women internalize an ideal shape that is considerably thinner than most women's actual shape (Fallon and Rozin, 1985). Furthermore, body shape and body image are of primary importance for these women (Rozin and Fallon, 1988). As a result there is a preoccupation with amount eaten and the possibility of becoming 'fat' in American women that has major implications for amount eaten, and concerns about eating (Rodin *et al.*, 1985). This preoccupation, in turn, may be linked to the relatively high incidence of depression, anorexia and bulimia in women in cultures which promote the value and importance of slim bodies (McCarthy, 1990). Interestingly, bulimia and anorexia are virtually absent in cultures, like Hindu India, in which women do not seek a slimmer body (Fallon *et al.*, 1995).

2.2.1.2 Norms about meal size. Every culture has traditions relating to the size, content, duration and context of meals. For example, traditionally (less today), it is customary in American culture to finish what is on one's plate. In America (as in most other cultures) there are three daily meals, each served at a typical time, with supper, for example, served earlier in America than in most of Europe. There are also standards about a normal size for a meal: for lunch, a sandwich, perhaps a side dish (e.g. potato chips or salad) and a beverage is appropriate. Two sandwiches seems excessive for many.

These traditions and norms may have a surprising amount of control over the amount eaten, in opposition, as it were, to the physiological signals that promote a link between meal size and energy balance. People tend to eat a meal if it is presented to them at an appropriate time, and if they know that they have not eaten recently. This strong influence on food intake was recently demonstrated in a study on amnesic patients, who cannot remember that they have just eaten. Two patients were presented on three occasions with three consecutive full lunches, each separated by ten or so minutes, so that memory for the prior meal eating episode was gone. On all six occasions, the second meal was consumed without question, and in most cases, the third meal was begun (Rozin *et al.*, 1996b).

2.2.1.3 Food intake, health and body image. Beliefs about the importance of body weight for both health and appearance can be strong influences on amount eaten. In general, Americans are very concerned about the health risks of obesity. American women are greatly concerned about their appearance with respect to body shape (Rodin *et al.*, 1985; Rozin and Fallon, 1988).

2.2.2 Food choice

2.2.2.1 The role of food in health and life. Food and eating occupy particular niches, and have particular connections with other activities in each culture. In some cultures, such as the Hua of Papua New Guinea (Meigs, 1984) or Hindu Indians (Appadurai, 1981; Marriott, 1968), food is at the center of life; what one eats and who one eats with has deep social and moral implications. Among Hindu Indians, the maintenance of bodily purity is a moral principle, and food constitutes a basic vehicle for establishing or violating purity. In all cultures, food has its principal nutritional function, but this function varies in salience, and even in valence; in some cultures, getting enough food (nutrition) is a central concern, whereas in others, the problem is to avoid eating too much. Food is also a source of basic pleasure, as well as of aesthetic experiences, and this varies as well in different cultural settings. And finally, food can be thought of on the medicine-poison dimension, and its location on that dimension varies across cultures.

In a recent survey of food attitudes in four different cultures (Japan, France, Flemish Belgium and the USA), distinct differences appeared in food attitudes, with the French and Americans anchoring the extremes (Rozin *et al.*, 1996c). On average, the French were more likely to think of food in terms of cuisine and pleasure, and Americans to think of food in terms of nutritional values and health risks. Food had more positive associations for the French, and was more central to their life. In all four cultures, women showed greater concerns about the food–health link, and relatively more interest in nutrition than in cuisine, compared to men.

Traditions and religions establish a framework within which food is experienced. For a Hua or a Hindu Indian, the most important thing about a food is who prepared it, whereas for an American, the most important things are what's in it (nutrition, toxins) and how good it tastes. Similarly, Americans are inclined to think foods are better if they are natural, an opinion probably not shared by most other human beings.

2.2.2.2 Preferences, likes and dislikes. Traditions have a powerful influence on what we eat, what we prefer and that we like. Humans often eat what is available and cheap (these factors, themselves, the product of culture), even if they don't prefer the most widely consumed foods. Surely, it would be a mistake to estimate world food preferences by looking at amount consumed of different foods. But, even restricting consideration to situations where a price-neutral choice is available, there is still a fundamental distinction between preference and liking (Rozin, 1979). Liking is one reason for preferring A to B, but it is not the only reason. One might like B more, but think it is unhealthy or impolite to eat it. Generally, the foods we like or dislike are those we consume primarily because we like or dislike their sensory properties; we call these good tastes or distastes. They contrast with foods we eat because we believe that they will be good for us (beneficial foods) or that they will harm us (dangerous foods) (Rozin and Fallon, 1981).

This distinction is neatly illustrated by the taste aversion phenomenon. When a person gets nauseous after eating a food, the food becomes disliked, that is, it now tastes bad. On the other hand, when lower gut cramps, pains in other organs, respiratory distress, or skin rashes develop after eating a food, it is typically avoided as dangerous, but does not become disliked (Pelchat and Rozin, 1982).

As we move from consumption through preference to liking, the type of cultural influence changes. Availability and price dominate consumption, and have some influence on preference, whereas these indirect social influences have much less effect on what is liked (indeed, liking, and to some extent, preference, may be enhanced by high price and low availability). Preference, especially in the USA, is heavily influenced by views about the healthfulness of individual foods, whereas liking is less susceptible to such influences. With the exception of the taste aversion learning paradigm, we do not have clear indications of the determinants of liking (see later section on acquisition, and chapter by Birch), but it is clear that the behavior of respected others, in terms of choice, attitude, and expressions, has a powerful influence on liking. These social factors operate directly, both at the time of choice (see section on direct effects) and in development (see section on acquisition).

2.2.2.3 Cuisine and appropriateness. Cultural and historical factors are primary influences on availability and price, but they also permeate, and determine thinking about food, the pattern of ingestion, and the meanings of food. That portion of this cultural influence that relates directly to food and its presentation may be called cuisine. There is not a full explication of the range and taxonomy of cuisines, but a number of aspects have been elaborated. With respect to individual dishes, Elisabeth Rozin (1982, 1983) has provided a useful taxonomy. She isolates three basic dimensions or factors in the composition/preparation of dishes within any particular cuisine. These are:

(i) staple foods (e.g. rice for China; corn for Mexico);
(ii) recurrent flavorings, which she calls flavor principles (e.g. soy sauce, ginger root and rice wine for China; chili pepper and either tomato or lime for Mexico); and
(iii) processing techniques (e.g. stir frying for China: stewing for Mexico).

Most of the world's cuisines employ a characteristic family of flavorings (flavor principles) on virtually all main course foods. In addition to dish components, one can consider meal structure (e.g. Douglas and Nicod, 1974): the order of dishes, the acceptable sequences and combinations of dishes. There are also many traditions having to do with foods for particular occasions or times of day, or for people in particular roles in life (adolescents, infants, old people, head of household). Schutz (1989) has described many of these factors under the appropriate heading of appropriateness. Finally, there are less well defined, but extremely important aspects of food (discussed above) having to do with the importance of food, the relation of food to social relations (food sharing, deference at the table, table manners), etc. All of these culinary traditions are operative in older children and adult natives of a culture, and modulate experience, attitudes and beliefs.

2.3 Direct on-line influences

We divide direct influences into those that are inadvertent, that is, resulting from a direct social presence without any intent on the part of the social agent, and advertent, in which the social agent assumes an intentional role to influence the food intake or choice of another person.

2.3.1 Inadvertent (unintentional) influences

2.3.1.1 Food intake. A social presence, in both animals and humans, seems to increase food intake. This has been demonstrated most directly in a series of diary studies on Americans by John deCastro and his colleagues (e.g. deCastro, 1990). Overall, intake increases with the number of people involved in an eating episode. Recently, this work has been extended to other cultures (France and Holland), with similar results (deCastro *et al.*, 1995). There has also been one laboratory demonstration that more food is eaten by individuals in groups than by individuals alone (Clendenen *et al.*, 1995). The mechanism of social facilitation of intake is not yet fully analyzed. Recent work (Feunekes *et al.*, 1995) suggests that the principal cause of social facilitation of human food intake is that social agents increase duration of meals, and hence the amount eaten. We do not know the effects of chronic differences in amount of social contact on eating. That is, is there adaptation to a person's typical meal companions, or is there a continued increased intake in social eaters? The work by deCastro and Feunekes, since it is based on diary keeping in normal life, suggests that the effects are long-term. Presumably, it follows from this that those who eat in more social situations will be heavier.

The presence of particular others may influence intake, e.g. if that other is a spouse who is intent on the slimming of him/herself and his/her partner, or a parent concerned about an underweight child. The presence of others may make cultural ideals salient, and engage issues of self-esteem and self-presentation. For example, in the USA and Canada, the feminine ideal includes eating sparingly (Pliner and Chaiken, 1990). This is explicitly realized by both males and females (Pliner and Chaiken, 1990). Furthermore, specific social effects have been demonstrated in the laboratory. Females eat less, that is, behave in a more feminine way, in the presence of a desirable male companion, as opposed to a female companion or a less desirable male companion (Mori *et al.*, 1987). In short, in Canada and the USA, eating lightly is part of the self-presentation of femininity.

2.3.1.2 Food choice. Social agents are usually responsible for the presentation of foods, and hence by controlling exposure, influence choice. This is particularly true in a home environment in which one person prepares food for a whole family. Furthermore, there is abundant literature in developmental psychology (not focused on food) indicating that children tend to imitate respected others (see chapter by Birch). Thus, at least for children, the presence of respected others enjoying a particular food probably promotes that choice. There are also reasons to believe that mood influences food intake and choice (see chapter by Rogers in this volume), and mood typically has social determinants. There is abundant evidence for mood or emotional contagion (Hatfield *et al.*, 1994); the mood of others can influence the mood of an individual, and hence food choice.

The particular foods consumed, at least in public, have significance in terms of self-esteem and public esteem in many cultural settings. For example, in a recent survey of American college undergraduates, 12% of female students said they would feel embarrassed to walk up to the check-out counter of a supermarket with a purchase of chocolate bars or ice cream (Rozin, unpublished). Being seen as an eater of indulgent, fattening foods, is upsetting to some females. This relates to general concerns about eating heartily in females, as illustrated by the work of Pliner *et al.* described above (Pliner and Chaiken, 1990; Mori *et al.*, 1987).

The social and moral implications of food choice, among American college students, are strikingly illustrated by a recent impression formation study by Stein and Nemeroff (1995). Students read a brief description of a male or female student. There were two versions of the description, identical except for what was listed as most regularly eaten foods. In one case, these were 'fruit (especially oranges), salad, homemade whole wheat bread, chicken, and potatoes' and in the other description it was 'steak, hamburgers, French fries, doughnuts, and double-fudge ice cream sundaes'. After reading one or the other description, subjects rated the person portrayed on a number of dimensions. These dimensions included moral judgments, such as immoral/virtuous, concerned/unconcerned, or cruel/kind-hearted. The high calorie 'junk food' eater was consistently rated as less moral! One might have expected such results in a culture in which food is laden with moral meaning, such as a case of meat-eating by a Brahmin in India, but the robustness of this effect in the United States is *surprising*, and opens the door to other studies of moral and character implications of food choice in America.

2.3.1.3 Advertent (intentional/teaching) influences. Humans seem to spend a fair amount of time teaching other humans facts about food and giving advice about what is healthy, tasty and appropriate. This substantial human acitivity may be unique; there is not good evidence for explicit teaching about food in any other animal species (Galef, 1990).

2.3.1.4 Information and explanations. Attitudes to foods, like other attitudes, are influenced by relevant beliefs and knowledge. This is a major feature of the dominant Ajzen–Fishbein model of attitude formation in social psychology, and has been articulated in the domain of food choice by Richard Shepherd and his colleagues (Shepherd, 1989; and chapter by Shepherd in this volume). Whether or not it is the learner's intention, in the USA, one is bombarded by information about food and health, in the media and in conversation. Information about the health consequences of consuming different foods, and the preparation of foods, is intentionally provided and popularized for the public by professionals, providers and parents. Some of this information is highly personalized, as in attempts to match a person's preferences with what is considered a healthy diet. Some of this information influences food intake and choice on-line, as when the healthiness of a particular food is discussed while someone present is eating it.

2.3.1.5 Social pressure. There is a large literature that demonstrates, outside of the domain of food, substantial effects of social influence on choice. Leann Birch, among others (reviewed in Birch, 1987), has shown that food choices by peers or teachers can influence, on-line, the choices of children. Motivations such as doing what older and other admired people are doing, in their presence, prompt children's interest in adult foods like coffee, alcohol and tobacco. Although such influences, especially from peers and older children are powerful, social influence appears to be often quite ineffective in the hands of parents. Problems with narrow food choice, inappropriate food choices (from the parents' point of view) or inadequate food intake are high on the agenda of parents speaking to their pediatricians (Bakwin and Bakwin, 1972). Parents often respond to these problems by cajoling, persuading, threatening and bribing. These social influences are sometimes successful (especially bribing) in the short-term, but are generally not successful strategies in the long run. Many parents are aware of this fact, as they are of the difficulty in changing their children's food preferences by direct intervention (Casey and Rozin, 1989), but they often cannot restrain themselves from intervening.

2.4 Socio-cultural aspects of the acquisition of norms, beliefs, knowledge and attitudes

In referring to the acquisition of food habits and preferences, we naturally think of children. Although it is probably true that a disproportionate amount of acquisition occurs in the first ten years of life, it is probably less than the total amount acquired in the remainder of life. Adults frequently get to like or dislike foods and beverages, such as coffee, tobacco and particular ethnic cuisines. Information about foods, such as health effects, is probably more influential with adults than children, and the physical limitations that occur with old age probably produce substantial changes in exposure and preference. In this section, we review what is known about social influences in acquisition of attitudes and beliefs about food. The focus will be on attitudes (liking) and on children, because much of the literature deals with children (see chapter by Birch in this volume, and Birch, 1987).

2.4.1 Parental influence and the family paradox

It has been an assumption for many decades, by both lay people and psychologists in the Western world, that early experience shapes and strongly determines later preferences, attitudes and personality. Freud promoted and popularized this view. Recently, psychologists have come to realize that this bit of common sense probably is not true. Most strikingly, studies of twins, especially identical twins reared apart (e.g. Bouchard *et al.*, 1990) have shown rather convincingly that for many ability and personality variables, *all* of the considerable similarity shown by siblings can be accounted for by their genetic similarity.

Along the same lines, it has been generally assumed that well known similarities in spouses in preferences and attitudes were substantially due to their common experience together. However, this popular wisdom has been challenged by the finding that on most variables, the similarity between recently married spouses is about as great as that between spouses married for decades (Price and Vandenberg, 1980). That is, most spousal similarity results from assortative mating, rather than common experience. The Price and Vandenberg (1980) study that reports these important facts also notes that food preferences are one of the few areas in which clear effects of living together are manifested.

In the framework of skepticism about conventional wisdom, we will now consider three commonly held views about the role of parents in shaping their children's food preferences:

(i) parents have a strong influence on their children's food preferences, presumably by virtue of controlling food availability and choice, imparting food beliefs and modelling for food preferences.
(ii) given her traditional role as caretaker, procurer and preparer of foods, the mother's food preferences should be more influential with children than the father's.
(iii) in line with the extensive literature on modelling, children should be inclined to resemble their same sex parent in food preferences more than their opposite sex parent.

All three of these suppositions turn out to be false, and have been labelled as the family paradox (Rozin, 1991). We will consider the evidence for each claim.

The correlation between food preferences of parents and their children, across a number of studies, has varied from zero to marginally significant; a reasonable estimate of this value is a Pearson correlation in the 0.1 to 0.2 range (reviewed in Rozin, 1991, including Birch, 1980b; Pliner, 1983). These results derive from studies of young children and their parents or college students and their parents; they include actual tests of food choice and survey responses. The same samples have generated substantial parent–child correlations in other domains, as for example in value-related matters such as attitude to abortion, or in disgust sensitivity (Rozin, 1991; Rozin *et al.*, 1984).

The low mid-parent child correlation might result from discordant parental preferences; it does not make sense to predict that a child with one parent who loves broccoli and one who hates it would end up neutral. However, if the parental pairs used in generating these correlations are restricted to parents who are concordant for the particular preference in question, the parent–child correlations rise slightly, but remain very low (mean correlation of 0.18 across 12 foods) (Rozin, 1991).

These same studies (also reviewed in Rozin, 1991) indicate that mother–child correlations are not consistently higher than are father–child correlations, nor are same sex parent–child correlations consistently higher than opposite sex parent–child correlations.

One reason for the lack of a mother effect may be that in selection of foods to serve the family, mothers may be more influenced by their husbands' preferences than by their own (Burt and Hertzler, 1987; Weidner *et al.*, 1985). This may account for some of the failure to find a better match between mother's preferences and those of their children.

The minimal influence of parents on their children's food preferences is in accord with the fact that parents, in the USA, frequently complain to their pediatricians about their children's food choice, and recognize their own lack of success in modifying their children's food habits (Casey and Rozin, 1989).

The lack of evidence for all three of the reasonable suppositions about parent–child food preference linkages is collectively a paradox. It is a paradox for two reasons: first, parents should be the primary influence, and second, there are no other powerful candidates to account for the extensive variability in children's food preferences.

However, from an adaptive point of view, these low correlations may make sense. As mammals, we spend the first few years of life (in traditional societies) on a diet dominated by mother's milk. It is experienced in association with warmth, touch, a nurturing mother and relief of hunger. Before the origin of dairying, in accordance with the situation for all other mammals, milk was not available as a food after the first few years of life. It would therefore be maladaptive for young children to develop strong milk preferences. The gradual development of lactose intolerance after the first few years of life, the relative unpalatability of milk sugar, and a resistance to food imprinting, all present in mammals including humans, act to reduce the ultimate importance of early food experiences (Rozin and Pelchat, 1988).

2.4.2 Other social influences

The fact is that there is wide within-culture variation in food preferences; many people like broccoli, but many do not. What is the source of this difference, if it is neither a result of interaction of the child with its parents nor genetic (note that the low parent–child correlation precludes any substantial genetic effect on food preferences)? We do not know, but we can be confident that the answer has much to do with social influences. Even if chance encounters with foods, or chance pairing of foods with significant events turn out to be important, the probability of these chance occurrences would be largely determined by socially influenced factors, including availability. Other more immediate social influences include effects of the media, teachers, peers and siblings. There is evidence that sibling influences are more powerful (in terms of sibling correlations in preferences) than parental influences (Pliner and Pelchat, 1986).

Culture-wide food preferences (e.g. for the native cuisine) are transmitted rather well to children. One would be inclined to point to parental influence in this domain as well, but this may not be so. Perhaps, culture-wide preferences are transmitted so faithfully because all sources of influence, including availability, the media, and other individuals, all provide a common message/example to children.

2.4.3 Mechanisms of acquisition

The acquisition of information about foods (whether correct or not) presumably derives from typical sources of information, such as parents, friends, the media and health professionals. There are interesting sociological issues involved in this information transfer, but they do not represent a puzzle from the point of view of a food-oriented psychologist. Of course, in representative approaches to food choice in humans (e.g. Shepherd, 1989, and Shepherd chapter in this volume), in the framework of the Ajzen–Fishbein model of attitude formation, beliefs play a major role. Our focus in this section will be on the determinants of liking, or more generally, the process of internalization of food preferences.

Internalization of a preference means that it originates from the self, or is desired by the self. This contrasts with compliance-based preferences, which are expressed for extrinsic reasons (see McCauley *et al.*, 1995; Kelman, 1958). Thus, a person who prefers cottage cheese to ice cream because she likes cottage cheese better has an internalized preference for cottage cheese, whereas a person who chooses the cottage cheese, contrary to desire, in an effort to lose weight, would be demonstrating a compliant preference. A third way of describing or processing a preference is to essentially remove the act of choice. Very well-practiced tasks or choices may become automatized, that is, carried out without thought, and almost reflexively. Certain food-related acts, like managing the use of a fork and knife, or adding sugar to coffee may become automated.

While preferences based on compliance are important, they are neither interesting nor durable. They are not interesting because there is no problem in explaining why someone does or does not choose a particular option under social constraint. They are not durable because the behavior or choice is unlikely to remain when the social constraint is removed. When internalization occurs, on the other hand, control of the choice or behavior changes, so that it is not dependent on any external contingencies or presence. The person who likes cottage cheese will continue to choose it after hearing that it actually may be as fattening as ice cream. The compliant cottage cheese chooser will gladly abandon this preference on hearing the same news. Internalization is interesting, psychologically, because we do not know how it happens. The remainder of this paper consists of a review of what we know about internalization, with an emphasis on social influences.

There is no doubt that social influences can produce internalization. The studies that demonstrate this influence, however, do not indicate the process through which internalization occurs. The majority of studies concentrate on the creation of likes (as opposed to dislikes) (see Birch, 1986; Rozin, 1988; and the Birch chapter in this volume).

Early studies by Duncker (1938) showed that children preferred a food if it was chosen by admired others, and Duncker (1938) and Marinho (1942) showed that children would prefer a food that was preferred by a fictional hero. More systematic investigations by Leann Birch and her colleagues followed on this early work. Preschoolers show enduring preferences (presumably likes) for foods that their peers select (Birch, 1980a), and also for foods favored by an admired adult (their teacher) (Birch *et al.*, 1980). Birch interprets these results in terms of the action of a positive social-affective context (with controls for mere exposure). However, the precise mode of operation of this context is not identified in these important studies of social influence.

A research program on the acquisition of liking for the innately unpalatable burn of chili pepper by young Mexican children, also suggests the importance of a positive social-affective context (the apparent enjoyment of this food, in the family eating situation, by older siblings and parents) (Rozin, 1990b).

A major influence of social context is demonstrated by research, in the sociological tradition, which accounts for the disparity between high levels of alcohol intake but low levels of alcohol abuse in Italy (Lolli *et al.*, 1958). Lolli *et al.* trace this to deeply rooted Italian attitudes to alcohol, as a food, as an integral part of meals, and as a family tradition. It is introduced to children early, in the positive context of a meal and a family event. This role for alcohol (wine in this case) places it in a situation where its absorption will be slower, and where it becomes a part of normal life, rather than a focus for rebellion from family values.

2.4.4 *Mechanisms of inadvertent social effects*

We (McCauley *et al.*, 1995) have attempted to review processes through which internalization is accomplished, and to enumerate possible and demonstrated mechanisms. Our list includes: mere exposure, addiction, evaluative conditioning, identification and self-relationship. Mere exposure and addiction both have indirect social components, but do not properly belong in center stage in an article on social influences. The other possible mechanisms will be elaborated below.

2.4.4.1 Evaluative conditioning. It has been demonstrated in both animals and humans that evaluations of objects can change, as a result of contingent pairing of an event (e.g. a flavor) with an already positive or negative event. This Pavlovian process has been termed evaluative conditioning (Martin and Levey, 1978; Baeyens *et al.*, 1990; Rozin and Zellner, 1985). Conditioned taste aversions are an example of evaluative conditioning. Although these pairings are not necessarily socially mediated, they are often arranged in a social context and often with intent. For example, a possible route to the liking of unsweetened black coffee is earlier experiences of coffee with cream and sugar, followed by a gradual reduction in cream and sugar.

It is likely that direct social mediation is involved in much of the evaluative conditioning that goes on with respect to food. The 'unconditioned stimulus' typically associated with a new food or beverage is another person, who is expressing enjoyment or displeasure, or approval or disapproval, of the food in question. This contingency between food and a relevant positive or negative social event may produce conditioning in a number of ways. For example, the positive expression of another may directly serve as an unconditioned stimulus, or by a process of mimicry or social contagion, it may induce a comparable feeling in the observer, which then serves as an unconditioned stimulus. In either event, through evaluative conditioning, social signals with affective content may influence likes and dislikes for foods and other things.

The first laboratory demonstration of such a social evaluative conditioning with respect to food was recently published by Baeyens and his colleagues (Baeyens *et al.*, 1995). Subjects watched videotapes in which a person sampled a particular beverage and expressed (facially) either pleasure or displeasure. The critical (conditioned) stimulus was the shape of the glass. Independent of the color of the beverage, if the glass had a stem, the drinker showed a positive expression, and if the glass did not have a stem, a negative expression was shown on sampling the drink. Subjects, unaware of the contingency for the most part, demonstrated increased liking for the type of glass contingently paired with the positive expression.

2.4.4.2 Identification. Identification is a poorly understood process through which a person becomes 'invested' in another person, social entity or object. This investment is some combination of admiration, wanting to be like, and deep caring. It seems to happen quite readily in human beings (see McCauley *et al.*, 1995). One result of identification is that the values and preferences of the object of identification become those of the identifier. This is presumably one pathway through which the food likes and dislikes of an admired other can become those of the self. Although this scenario seems very reasonable, and is widely used in advertising, there is little known about the properties of identification or the process through which it occurs.

2.4.4.3 Self-consistency. The theme of self-consistency has emerged many times in the history, within psychology, of the study of attitudes (see, e.g. Eagly and Chaiken, 1993). Cognitive dissonance and self-perception theory, among others, stress the importance to the self of holding consistent attitudes, such that the perception of inconsistency motivates some sort of change. In particular, when a person finds him- or herself behaving in a way that is not consistent with basic attitudes, beliefs or values, three options arise. The person can ignore the disparity, rationalize it, or change attitudes to be in line with behaviors. Much of the literature suggests that the presence of a perceptible social constraint, which essentially forces the behavior, provides a rationalization, and prevents attitude change (Lepper, 1983; Deci and Ryan, 1985). When behavior is perceived as originating, unconstrained from the self, it becomes a force for change of attitudes, with the motive of self-consistency (Bem, 1967). A consequence of this view is that intrinsic value is promoted if one can get a target person to perform a behavior in the absence of an obvious extrinsic force (the 'minimal sufficiency' principle).

Birch and her colleagues (Birch *et al.*, 1982) have demonstrated the operation of these principles in the food domain. Explicitly rewarding selection of a particular food by a child has the long-term effect of reducing the preference for the food. Ironically, using the target food as a reward, under appropriate conditions, may be a more effective way of inducing a liking for the food. A parent survey indicates that, in contrast to these findings, parents are more optimistic about the value of rewarding ingestion of a target food than about the efficacy of using the target food as a reward (Casey and Rozin, 1989).

2.4.4.4 Self-relevance. Many preferences are just that; nothing more significant than a particular choice or liking with no further implications. Liking broccoli or not has that quality, for most people. However, some choices are more central to the self, and say something about what a person stands for, what is valued. Political and moral choices have this quality. Value-based choices tend to be internalized, because they derive from basic, meaningful personal principles.

Two studies (Cavalli-Sforza *et al.*, 1982; Rozin, 1991) suggest that values are transmitted much more effectively from parent to child than are preferences. This finding is of particular significance in the food domain, because, even in the USA, food selection can engage value/moral issues. The previously cited study by Stein and Nemeroff (1995) indicated how healthy *vs.* high fat/high calorie food preferences had moral implications for Americans.

The value/preference distinction is important in internalization, because preferences can become values, or vice versa. For example, cigarette smoking used to be a preference among Americans, a neutral act in the moral domain. It is now much more of a value, and a common reason, for example, for a person to not date another person. This process has been described as 'moralization' (Rozin, 1995b). It can occur at the level of society, as is the case with cigarettes in the last few decades in the USA, or in individuals, as in the frequent development of meat rejection based on moral issues in many young Americans (Amato and Partridge, 1989).

2.4.5 A summary of basic issues: The example of disgust

A European or American, faced with a putrid piece of meat, swarming with maggots, will experience disgust. This is a basic emotion. This disgust experience is acquired; little children do not show such a response to insects or putrefaction. This is a uniquely human response, a rejection based on the idea of what the food is and/or its origin. It is probably the strongest emotional response people have to foods. By our analysis (Rozin and Fallon, 1987; Rozin *et al.*, 1993), this ideational rejection of foods originates in the common animal rejection response to distasteful foods. The sense of rejection and its facial and nausea manifestations are extended, through human cultural evolution and a culturally programmed developmental trajectory, to certain potential foods. Almost all of these 'foods', cross-culturally, are of animal origin. Thus, disgust is an inherently socio-culturally based response/emotion. Our analysis also indicates that, depending on the culture, the domain of elicitors and meanings expands from the food core to include reminders of our animal origin (such as blood and gore, certain aspects of sex, poor hygiene, and death/decay; what we call animal origin disgust), situations involving contact with strangers (interpersonal disgust), and certain moral violations in some cultures (moral disgust). In general, we believe that food core and animal origin disgust occupy a central position in disgust in Western-developed countries, while the interpersonal and moral domains assume more prominence in Hindu India. And of course, the specific potential foods that elicit disgust vary considerably from one culture to another. Most cultures consume at least a few decayed foods that many other cultures find disgusting.

According to this analysis, there are some universals about disgust, e.g. the centrality of feces as an elicitor, the origin in the food system, the basic facial expression, and the contagious properties of disgusting entities (they spoil a good food if they touch it). But there are also substantial cultural differences.

Disgust has emerged as the major emotion of negative socialization or negative internalization. If a 'culture' desires that something be avoided, in both thought and action, it can do nothing better than make this something an object of disgust. It then assumes value properties, and the rejection is supported by a strong feeling of revulsion.

Disgust serves, in this way, as a major component of moralization. We (Rozin *et al.*, 1995) have recently shown that American moral vegetarians, people who reject meat on moral (value-based/internalized) grounds, are more likely to find meat disgusting than health vegetarians (people who reject meat because they feel it is unhealthy, an instrumentally based/ non-internalized response).

Although it has not been documented, it is highly likely that disgust is communicated and acquired in social situations, with verbalizations and the disgust face as critical parts of the social context. Thus we can see that this basic, food-related emotion plays all the roles we have set out for social factors. It influences availability, by eliminating certain animal products from the domain of choice. Socialized individuals carry in their heads a set of negative attitudes to a range of animal products and decayed foods, along with many non-food disgust elicitors. These do not require a social presence to be elicited, but are based on culture-historical and developmental social events. Expressions of disgust by others, on line, have major influences on an individual's food choices. And the acquisition of disgust, in a social context, is both a major feature of socialization and a major mechanism through which further socialization is accomplished.

2.5 Conclusion

Suppose you wish to know as much as you can about the foods a person likes and eats, and can ask only one question. What should that question be? Without doubt, the question should be a distinctly social one: what is your culture or ethnic group? There is no other question that is nearly as informative. We don't really know much about how these powerful socio-cultural factors operate in the present, or in development. Although culture/ethnic group is the best predictor of food habits and preferences, there is a great deal of within-culture variance that this predictor cannot account for. We know surprisingly little about the causes of this within-culture variation.

Acknowledgements

This paper was prepared with the assistance of funding from the University of Pennsylvania Research Fund.

References

Amato, P.R. and Partridge, S.A. (1989) *The New Vegetarians: Promoting Health and Protecting Life*. New York: Plenum Press.

Appadurai, A. (1981) Gastro-politics in Hindu South Asia. *American Ethonolgist*, **8**, 494–511.

Baeyens, F., Eelen, P., van den Bergh, O. and Crombez, G. (1990) Flavor–flavor and color–flavor conditioning in humans. *Learning & Motivation*, **21**, 434–455.

Baeyens, F., Kaes, B., Eelen, P. and Silverans, P. (1995) Observational evaluative conditioning of an embedded stimulus element (submitted manuscript).

Bakwin, H. and Bakwin, R.M. (1972) *Behavior Disorders in Children*. Philadelphia: Saunders.

Bem, D.J. (1967) Self-perception: An alternative interpretation of cognitive dissonance phenomena. *Psychological Review*, **74**, 183–200.

Birch, L.L. (1980a) Effect of peer model's food choices and eating behaviors on pre-schoolers' food preferences. *Child Development*, **51**, 489–496.

Birch, L.L. (1980b) The relationship between children's food preferences and those of their parents. *Journal of Nutrition Education*, **12**, 14–18.

Birch, L.L. (1986) Children's food preferences: Developmental patterns and environmental influences. In G. Whitehurst and R. Vasta (eds), *Annals of Child Development, Vol. 4.*, Greenwich, Connecticut: JAI.

Birch, L.L. (1987) The acquisition of food acceptance patterns in children. In R. Boakes, D. Popplewell and M. Burton (eds), *Eating Habits*. (pp. 107–130). Chichester, England: Wiley.

Birch, L.L., Birch, D., Marlin, D.W. and Kramer, L. (1982) Effects of instrumental eating on children's food preferences. *Appetite*, **3**, 125–134.

Birch, L.L., Zimmerman, S.I. and Hind, H. (1980) The influence of social-affective context on the formation of children's food preferences. *Child Development*, **51**, 856–861.

Bouchard, T.J., Lykken, D.T., McGue, M., Segal, N.L. and Tellegen, A. (1990) Sources of human psychological differences: The Minnesota study of twins reared apart. *Science*, **250**, 223–228.

Burt, J.V. and Hertzler, A.A. (1978) Parental influence on the child's food preference. *Journal of Nutrition Education*, **10**, 127–128.

Casey, R. and Rozin, P. (1989) Changing children's food preferences: Parent opinions. *Appetite*, **12**, 171–182.

Cavalli-Sforza, L.L., Feldman, M.W., Chen, K.H. and Dornbusch, S.M. (1982) Theory and observation in cultural transmission. *Science*, **218**, 19–27.

Clendenen, V.I., Herman, C.P. and Polivy, J. (1995) Social facilitation of eating among friends and strangers. (submitted manuscript).

de Castro, J.M. (1990) Social facilitation of duration and size but not rate of the spontaneous meal intake of humans. *Physiology & Behavior*, **47**, 1129–1135.

deCastro J.M., Bellisle, F., Feunekes, G.I.J., Dalix, A.-M. and deGraaf, C. (1995) Culture and meal patterns: A comparison of food intake of free-living American, Dutch and French students (submitted manuscript).

Deci, E.L. and Ryan, R.M. (1985) *Intrinsic Motivation and Self-determination in Human Behavior*. New York: Plenum Press.

Douglas, M. and Nicod, M. (1974) Taking the biscuit: The structure of British meals. *New Society* (December), 744–747.

Duncker, K. (1938) Experimental modifications of children's food preferences through social suggestion. *Journal of Abnormal & Social Psychology*, **33**, 489–507.

Eagly, A.H. and Chaiken, S. (1993) *The Psychology of Attitudes*. Fort Worth, Texas: Harcourt, Brace, Jovanovich.

Elias, N. (1978) *The History of Manners. The Civilizing Process: Vol. I* (E. Jephcott, Trans.). New York: Pantheon Books. Original work published 1939.

Fallon, A.E. and Rozin, P. (1985) Sex differences in perception of desirable body shape. *Journal of Abnormal Psychology,* **94**, 102–105.

Fallon, A.E., Rozin, P., Byrnes, D., Gogeneni, R. and Desai, K. Body image and eating attitudes: A cross cultural comparison between Asian Indians and Americans. (manuscript).

Feunekes, G.I.J., de Graaf, C. and van Stavaren, W.A. (1995) Social facilitation of food intake is mediated by meal duration. *Physiology & Behavior* (in press).

Frazer, J. G. (1890/1959) *The Golden Bough: A Study in Magic and Religion*. New York: Macmillan (reprint of 1922 abridged edition, edited by T.H. Gaster; original work published, 1890).
Galef, B.G. Jr. (1985) Direct and indirect pathways to the social transmission of food avoidance. *Annals of the New York Academcy of Sciences*, **443**, 203–215.
Galef, B.G. Jr. (1990) Tradition in animals: Field observations and laboratory analysis. In M. Bekoff and D. Jamieson (eds), *Methods, Inference, Interpretation and Explanation in the Study of Behavior*. Boulder, Colorado: Westview Press.
Hatfield, E., Cacioppo, J.T. and Rapson, R.L. (1994) *Emotional Contagion*. Cambridge, England: Cambridge University Press.
Kelman, H. (1958) Compliance, identification, and internalization: Three processes of opinion change. *Journal of Conflict Resolution*, **2**, 51–60.
Lepper, M.R. (1983) Social control processes and the internalization of social values: An attributional perspective. In E.T. Higgins, D.N. Ruble and W.W. Hartup (eds). *Social Cognition and Social Development* (pp. 294–330). New York: Cambridge University Press.
Lolli, G., Serianni, E., Golder, G.M. and Luzzatto-Fegiz, P. (1958) *Alcohol in Italian Culture*. Glencoe, Illinois: The Free Press.
Marinho, H. (1942) Social influence in the formation of enduring preferences. *Journal of Abnormal and Social Psychology*, **37**, 448–468.
Marriott, M. (1968) Caste ranking and food transactions: A matrix analysis. In M. Singer and B.S. Cohn (eds), *Structure and Change in Indian Society* (pp. 133–171). Chicago: Aldine.
Martin, I. and Levey, A.B. (1978) Evaluative conditioning. *Advances in Behavior Research & Therapy*, **1**, 57–102.
Mauss, M. (1902/1972) *A general theory of magic* (R. Brain, Trans.). New York: W.W. Norton (original work published 1902).
McCarthy, M. (1990) The thin ideal, depression and eating disorders in women. *Behaviour Research & Therapy*, **28**, 205–215.
McCauley, C.R., Rozin, P. and Schwartz, B. (1995) *The Origin and Nature of Preferences and Values*. (Book manuscript).
Meigs, A.S. (1984) *Food, Sex, and Pollution: A New Guinea Religion*. New Brunswick, New Jersey: Rutgers University Press.
Mintz, S.W. (1985) *Sweetness and Power*. New York: Viking.
Mori, D., Chaiken, S. and Pliner, P. (1987) 'Eating lightly' and self-presentation of femininity. *Journal of Personality & Social Psychology*, **53**, 693–702.
Nemeroff, C. and Rozin, P. (1989) An unacknowledged belief in 'you are what you eat' among college students in the United States: An application of the demand-free 'impressions' technique. *Ethos. The Journal of Psychological Anthropology*, **17**, 50–69.
Pelchat, M.L. and Rozin, P. (1982) The special role of nausea in the acquisition of food dislikes by humans. *Appetite*, **3**, 341–351.
Pliner, P. (1983) Family resemblance in food preferences. *Journal of Nutrition Education*, **15**, 137–140.
Pliner, P. and Chaiken, S. (1990) Eating, social motives, and self-presentation in women and men. *Journal of Experimental Social Psychology*, **26**, 240–254.
Pliner, P. and Pelchat, M.L. (1986) Similarities in food preferences between children and their siblings and parents. *Appetite*, **7**, 333–342.
Price, R.A. and Vandenberg, S.G. (1980) Spouse similarity in American and Swedish couples. *Behavior Genetics*, **10**, 59–71.
Rodin, J., Silberstein, L. and Striegel-Moore, R. (1985) Women and weight: A normative discontent. In T.B. Sonderegger (ed.) *Psychology and gender: Nebraska Symposium on Motivation*. Lincoln, Nebraska: University of Nebraska Press.
Rozin, E. (1982) The structure of cuisine. In L.M. Barker (ed.) *The Psychobiology of Human Food Selection* (pp. 189–202). Westport, Connecticut: AVI.
Rozin, E. (1983) *Ethnic Cuisine: The Flavor Principle Cookbook*. Brattleboro, VT.: Stephen Greene Press.
Rozin, P. (1979) Preference and affect in food selection. In J.H.A. Kroeze (ed.), *Preference Behavior and Chemoreception* (pp. 289–302). London: Information Retrieval.
Rozin, P. (1981) The study of human food selection and the problem of 'Stage One Science.' In S. Miller (ed.), *Nutrition and Behavior* (pp. 9–18). Philadelphia: Franklin Institute Press.

Rozin, P. (1988) Social learning about foods by humans. In T. Zentall and B.G. Galef, Jr. (eds), *Social Learning: A Comparative Approach* (pp. 165–187). Hillsdale, New Jersey: Erlbaum.
Rozin, P. (1990a) Social and moral aspects of eating. In I. Rock (ed.), *The Legacy of Solomon Asch: Essays in Cognition and Social Psychology*. Potomac, Maryland: Lawrence Erlbaum, (pp. 97–110).
Rozin, P. (1990b) Getting to like the burn of chili pepper: Biological, psychological and cultural perspectives. In B.G. Green, J.R. Mason and M.L. Kare (eds), *Chemical Irritation in the Nose and Mouth*, New York: Marcel Dekker.
Rozin, P. (1991) Family resemblance in food and other domains: The family paradox and the role of parental congruence. *Appetite*, **16**, 93–102.
Rozin, P. (1995a) Towards a psychology of food and eating: From motivation to model to morality to metaphor. *Current Directions in Psychological Science* (in press).
Rozin, P. (1995b) Psychological perspectives on the process of moralization. In A. Brandt and P. Rozin (eds), *Morality and Health*. New York: Routledge, in press.
Rozin, P. and Fallon, A.E. (1981) The acquisition of likes and dislikes for foods. In J. Solms and R.L. Hall (eds), *Criteria of Food Acceptance: How Man Chooses What He Eats. A Symposium* (pp. 35–48). Zurich: Forster.
Rozin, P. and Fallon, A.E. (1987) A perspective on disgust. *Psychological Review*, **94**, 23–41.
Rozin, P. and Fallon, A.E. (1988) Body image, attitudes to weight, and misperceptions of figure preferences of the opposite sex: A comparison of males and females in two generations. *Journal of Abnormal Psychology*, **97**, 342–345.
Rozin, P. and Nemeroff, C.J. (1990) The laws of sympathetic magic: A psychological analysis of similarity and contagion. In J. Stigler, G. Herdt and R.A. Shweder (eds), *Cultural Psychology: Essays on Comparative Human Development* (pp. 205–232). Cambridge, England: Cambridge University Press.
Rozin, P. and Pelchat, M.L. (1988) Memories of mammaries: Adaptations to weaning from milk in mammals. In A.N. Epstein and A. Morrison (eds), *Advances in Psychobiology, Volume 13* (pp. 1–29). New York, Academic Press.
Rozin, P. and Zellner, D.A. (1985) The role of Pavlovian conditioning in the acquisition of food likes and dislikes. *Annals of the New York Academy of Sciences*, **443**, 189–202.
Rozin, P., Fallon, A.E. and Mandell, R. (1984) Family resemblance in attitudes to food. *Developmental Psychology*, **20**, 309–314.
Rozin, P., Haidt, J. and McCauley, C.R. (1993) Disgust. In M. Lewis and J. Haviland (eds), *Handbook of Emotions* (pp. 575–594). New York: Guilford.
Rozin, P., Markwith, M. and Stoess, C. (1996a) Becoming a vegetarian: Moralization, the conversion of preferences into values and the recruitment of disgust. *Psychological Science* (in press).
Rozin, P., Dow, S., Moscovitch, M., and Rajaram, S. (1996b) The role of memory for recent eating experiences in onset and cessation of meals. Evidence from the amnesic syndrome (submitted manuscript).
Rozin, P., Fischler, C., Imada, S., Sarubin, A. and Wrzesniewski, A. (1996c) Attitudes to food and the role of food in life: Cultural comparisons that enlighten the diet-health debate (submitted manuscript).
Schutz, H.G. (1989) Beyond preference: Appropriateness as a measure of contextual acceptance of food. In D.M.H. Thomson (ed.), *Food Acceptability* (pp. 115–134). Essex, England: Elsevier Applied Science Publishers.
Shepherd, R. (1989) Factors influencing food preferences and choice. In R. Shepherd (ed.) *Handbook of the Psychophysiology of Human Eating* (pp. 3–24). Chichester, England: Wiley.
Stein, R.I. and Nemeroff, C.J. (1995) Moral overtones of food: Judgments of others based on what they eat. *Personality & Social Psychology Bulletin*, **21**, 480–490.
Tylor, E.B. (1871/1974) *Primitive Culture: Researches into the Development of Mythology, Philosophy, Religion, Art and Custom.* New York: Gordon Press. (Original work published, 1871).
Weidner, G., Archer, S., Healy, B. and Matarazzo, J.D. (1985) Family consumption of low fat foods: Stated preference versus actual consumption. *Journal of Applied Social Psychology*, **15**, 773–779.

3 What animal research tells us about human eating

HARRY R. KISSILEFF, JANET L. GUSS and
LAURENCE J. NOLAN

3.1 Introduction

3.1.1 Goals and strategies

The thesis of this chapter is that understanding of the control of eating in animals is essential for understanding control of eating in humans. Animal research has provided four tools for the study of human eating, each of which will be a focus of this paper:

(a) the methods, both strategic and practical
(b) the observed phenomena
(c) the theoretical framework and experimental analysis
(d) models for the study of eating disturbances and their treatments.

Our approach will be to review a few key questions within each focus, and to provide examples that illustrate their answers. The guiding questions will be: 1. What do we know about human eating? 2. What contributions has animal research made to what we know? We will therefore omit from serious discussion animal research that has no application to humans and human research which has no relation to animals. While others have reviewed research from their own laboratories (e.g. Schachter, 1971; Stellar, 1967) which bears on our theme, and some specialized reviews have appeared which utilized animal models of human eating and its disorders (Smith, 1989; Harris, 1993; Grinker, 1981; McHugh *et al.*, 1989; Coscina and Garfinkel, 1991), we could find no other review designed specifically to identify general findings in animal research which bear on human eating.

3.1.2 Human eating has a biological aspect

Whereas other chapters in this book have focused primarily on human aspects of eating, this chapter will concentrate on concerns about human eating that are shared with other members of the animal kingdom. Biological controls of eating have been demonstrated for such widely diverse phenomena as choice among alternatives, quantity consumed over time, and the influence of conspecifics, i.e. social effects (Galef and Beck, 1990). While much of this book may give the impression that human eating is largely under the control of perceptions within the mind of the eater which are strongly influenced by factors outside the person (who prepares the food, where it is presented, ease of availability of food, appearance and composition of the food, habits ingrained by socialization during childhood), there are also many other factors which have a physiological underpinning. Furthermore, this field has also traditionally ignored the context of eating until recently (see chapter by Meiselman). Studies in animals can help us to unravel the biological controls of choice of food and amount of any particular food consumed in an eating episode. Because eating episodes (Smith, 1982) determine the energy intake side of the energy balance equation (Brobeck, 1955), understanding their control is essential to an understanding of energy balance and its major disturbances, obesity and anorexia.

3.1.3 *Quantification of eating*

The amount of food consumed in a given period of time is referred to as 'food intake'. A key underlying principle in understanding the control of food intake is that during the course of an eating event, regardless of whether it is a meal, a snack, a party, or a banquet, an individual's decision to stop eating a particular item of food or to stop eating for that event, is influenced by physiological processes which have cognitive representation in the brain and which the individual may choose to ignore or to follow. Therefore education about the role these processes play in the control of food intake is as important for the development of healthy dietary habits as is education about any other aspect of health, such as the importance of exercise, cleanliness, and sexual practices. It is important for leaders of government, industry and education to understand how this research can be applied to help people everywhere develop eating patterns that will enable them to lead healthier lives.

3.1.4 *Need for integration of the biological and socio-cultural aspects of eating*

A major goal of this chapter is to provide evidence for the proposal that without consideration of the biological controls of human eating, along with the socio-cultural, it will be difficult to effect changes in human dietary practices. It may be well and good for a practitioner or panel of experts to recommend that people consume less fat or less meat (e.g. Thomas, 1991), but unless they understand the biological underpinnings of preferences, hunger, satiety, and controls of quantities consumed, it may be impossible to attain the goals that are sought. Conversely armed with knowledge of how particular foods lead to changes in preference or changes in the effectiveness with which they induce satiety, to which there are certainly biological limits, it may be possible to redesign foods and to re-educate the public. A key question is, if it were possible to produce a meat or fat substitute which was indistinguishable from meat or fat in its sensory properties, would the post-ingestional effects be sufficiently similar for subjects to continue consuming it as if it were the item substituted? However, a casual perusal of the literature on the design of new foods (see Frier, 1993; Booth, 1994: ch. 7) and their lists of references suggests that purely sensory tests are employed during product development but post-ingestional effects are not assessed until the finished product is ready for consumer testing (e.g. Rolls *et al.*, 1992; Burley *et al.*, 1994; Hulshoff *et al.*, 1993). At that point in product generation emphasis on liking during brief exposure sensory tests overrides considerations of long term commitment to dietary change, which may require re-education of the consumer for ultimate acceptance. Studies of the relationship between sensory stimuli and post-ingestional effects in animals (e.g. Sclafani, 1991), which have not yet been rigorously applied to humans (see section 3.4.2), suggest that post-ingestive testing along with sensory testing during product development could reduce the rate of product failure.

3.2 Methods for studying eating—both strategic and practical

The primary methods of studying human eating (Hetherington and Rolls, 1987; Wilson, 1993), except those which involve clinical interview, utilize procedures that have parallels in animal studies. These animal studies established both the strategic and practical methods for studying human eating, although certain aspects of consumer testing methodology, such as responsiveness to advertising, have few discernible roots in animal psychology. In this section emphasis will be placed on describing the methods and factors involved in their choice. Discussion of how these techniques have been used in analytical studies on theories and mechanisms, where available, will be found in section 3.4.

3.2.1 Importance of selecting a method

The choice of a method frequently determines the interpretation of a product's effectiveness, such as the influence of the presentation of foods or food components on choice, acceptability, or consumption. For example, if brief exposure tests are used exclusively, long term effects, which might be entirely opposite, will be missed. An excellent case in point is the preference shown by rats for glucose, which is sweet, over Polycose (a maltodextrin), which tastes bland to humans, yet both are identical after digestion. Short term tests show that glucose is consistently preferred, but when rats are exposed to chronic consumption of both substances, their preference switches to Polycose (Sclafani, 1991: p. 78). Similar findings apply to selection of dilute *vs.* concentrated solutions, starch *vs.* Polycose, and sucrose *vs.* Polycose (see Sclafani, 1991 for details and references). The advantages and limitations of any procedure for the scientific, not just economic, considerations need to be carefully considered before a study is undertaken. For example, the relative costs of developing new products could be weighed against the costs of educating the public about the benefits of traditional or well-accepted products, which could include research on the possible physiological reasons that some products, such as bread or fruit juice tend to become stable items in the diet. The choice of a method in conducting such tests is a critical decision. Comparisons of brief and long term effects of food choices for humans are rare (Meiselman, 1992), yet the need for such basic studies is urgent if we are to begin understanding how dietary choices and intakes are actually composed.

3.2.2 *Types of methods*

Many methods for studying human eating are based on animal models. They vary along several dimensions, for example: goal of study (strategic *vs.* practical), length of study or test (short term *vs.* long term), amount of interference with subject (observation *vs.* intervention), number of items available (choice *vs.* intake), and location (laboratory *vs.* field).

Strategic methods refer to adoption of decision rules about the types of studies to do and methods used to do them, the problems selected and the approach of the investigator. For example, investigators may probe the short term choice or the long term intake, depending on the intended use of the research. The brain or gut might be emphasized in a mechanistic study. Food diaries collected by subjects themselves may or may not be favored over studies in metabolic wards to obtain data on long term food intake. These are strategic considerations.

Practical methods refer to decision rules regarding what can actually be accomplished with a given procedure and have to do with considerations such as the physical properties of foods, the capacity of the individual to endure testing, and expense of conducting the tests.

Observational methods refer to collecting data by watching what subjects do without interfering with their behavior. They can be conducted in naturalistic settings (Stunkard and Kaplan, 1977; Warner and Balagura, 1975) or laboratory, either on an individual eating alone (Green and Tapp, 1986) or in a group (Foltin *et al.*, 1988), by a trained observer equipped with a device for recording behavior emitted by the subject. Data can also be recorded and analyzed later with the help of a computer program (Foltin *et al.*, 1988). The issue of introduction of artificiality into laboratory settings has sparked controversy over what constitutes a 'naturalistic' or 'real' setting, and consequently over whether data obtained in a laboratory setting are applicable outside the laboratory. This influence of setting in relation to other variables can, however, be assessed by testing any variable, from number of people present to administration of drugs, or changes in food composition, simply by measuring the effect of the variable in multiple settings. Although it has been claimed that laboratories may be artificial environments (Meiselman, 1992), like zoos, estimates of amounts eaten in laboratory and field settings for humans agree. For example, in the field meal sizes averaged 410 kcal when subjects ate alone and 591 when they ate with others (de Castro and de Castro, 1989), whereas subjects eating in a laboratory setting consumed 537 kcal when eating alone and 592 when eating with others (Kim and Kissileff, 1996). These differences in means are within the typical range of within subject variability [±15% (Kissileff *et al.*, 1982)]. Obviously these comparisons are not clean, because the numbers and genders of subjects, types of meals and times of eating were different and variable, but they serve to illustrate that the two settings do not produce widely divergent differences.

Intervention methods refer to deliberate perturbations in either the subject or the ingesta in order to test a hypothesis about control of eating or amount consumed. These methods include restriction of food availability or setting in which it is eaten, change in food composition, administration of food before a free access test, or change in physiology of the subject by administration of a drug or manipulation of an internal organ. Although studies in section 3.2.4 are categorized for ease in presentation, observational studies are not necessarily without some interventional component, even if that component is merely the social setting in which the observations take place. Studies utilizing intervention require some means of observation, although usually the observation involves measurement with more than the unaided eye.

3.2.3 *Animal research established the strategies for studying human eating behavior*

An important lesson derived from animal research is the critical distinction between causation and confounding variables. Causation has been used in two completely different ways by different schools of thought. On the one hand, causation may refer to mechanism (Booth, 1989a), that is, to the sequence of events by which an output variable, for example, rating of a disposition to eat, is generated by an input, such as the state of distention of the stomach or number of chemoreceptors activated in the intestine (Booth, 1989b: p.129).

On the other hand, causation may refer to the goal or the function of behavior (Tinbergen, 1969: p. 4). For example, one could say that the heart and arteries pulsate in order to keep a continuous flow of aerated blood to the brain. Causation can also refer to subjective sensations such as hunger. Tinbergen rejects both of these uses for explanations of behavior in favor of mechanism, much as Booth uses it. He is particularly vehement in rejecting the use of subjective variables with functional attributes, such as hunger, as causal explanations for objectively observed behavior, such as eating. He points out that 'the conclusion that an animal hunts because it is hungry will satisfy many people at first glance. Yet the use of the word "because" is ambiguous, since "hungry" may be used as a convenient description of the state of the animal, based on subjective as well as objective criteria'. The use of the word in this way does not satisfy the scientist who wants to know what is happening inside the animal when it is in this state. Tinbergen claims that hunger, like anger, fear and so forth, is a phenomenon that can only be known by introspection. We believe that it may be time to reconcile these previously conflicting points of view, by showing that human report of hunger can be related to causal input, namely how much food has *just been* consumed (Wentzlaff *et al.*, 1995), and under certain conditions, how much food *will be* consumed. Booth might object that input and output are confounded by such usage, but it will become clear that where human eating is concerned, the feedback from the act of eating, to the behavioral tendency to continue, and its cognitive representation as 'how hungry' one feels can be made into objective and reproducible phenomena.

Another important strategic lesson from animal research is the necessity for a 'complete inventory of the behavioral patterns of a species. It is a natural tendency of the experimental worker to select a special problem, as for instance color vision, or homing or the delayed response. This specialization is often accompanied by a narrow point of view and a neglect of other aspects of behavior' (Tinbergen, 1969: p. 7). Eating behavior is certainly an example of a specialized behavior which has received extensive attention both from naturalistic observers and experimentalists. However, without recognizing the context in which the eating is set, i.e. the other aspects of the individual's behavioral repertoire, understanding of its control will be difficult if not impossible (*cf.* Meiselman, this book).

For the study of eating behavior in humans, interpretation of context for the control of an eating episode is no less important, and perhaps even more so, than it is in animals. For example, eating while viewing an athletic activity at a stadium with thousands of other people may share some of the controls as eating when viewing the same activity on TV at one's home, alone, but may also introduce additional factors. Eating at parties may be controlled differently than eating with the same people in a structured meal setting. Eating episodes in animals can be both qualitatively and quantitatively different depending on whether the animals appear threatened or relaxed, whether they are migrating or stationary, whether they have just risen from sleep or are about to retire (Goodall, 1986). Indeed examination of primate behavior in the wild (Wrangham, 1974) may be more relevant for understanding some aspects of human eating than the study of primate eating in laboratories. However, it is far more convenient to study rats than primates. On the other hand, for analytical work on the factors that control intake, studies on primates in the laboratory (e.g. Hansen *et al.*, 1981; Foltin and Moran, 1989; Foltin and Fischman, 1990) will be indispensable (Kissileff, 1992; Stellar, 1992).

However, just as important as understanding the behavioral repertoire, is the complete description of the muscular movements involved in different aspects of eating as well as in activities leading to eating. Studies of microbehavioral responses to tastes (Grill and Berridge, 1985) are now giving way to studies of muscular contractions (Grill *et al.*, 1995). This information will be useful in preparing the way for a neurological analysis of the control of eating. The most fruitful approach for general understanding of the control of food intake will be to bring those aspects of eating which occur most frequently in the field into the laboratory for more physiological study. Methods for this type of study are available for humans (Bellisle and Le Magnen, 1980; Stellar and Schrager, 1985), but to date they have been utilized mainly in a descriptive rather than analytic fashion.

Both the advantages and limitations of purely observational studies must be recognized, and observational studies must be integrated with analytical studies in order for progress to be made. As in animals, any eating event in humans must be considered in relation to its role in the overall behavioral repertoire. This consideration leads to questions about social function *vs.* physiological function. For example, is the eating event under study a party, or an opportunity to get rid of an unpleasant sensation so that one can get on with one's activities? The relative influence of the stomach and intestine, as well as metabolism, may be different than the influence of cognitive function—i.e., what the individual is thinking, fearing, liking, etc.—depending on the role of the eating event in the overall behavioral repertoire of the individual. This consideration makes generalization about eating in any species highly suspect, unless a particular influence has actually been tested in a variety of behavioral situations. A start on this type of integration of social setting and physiological control has been made in humans (Kim and Kissileff, 1996). Dilute (1%) and concentrated (15%) glucose solutions (500 ml) were given to subjects in either a social or isolated setting. Differences in intake of a test meal after the two solutions were nearly identical (*c.* 140 g) in both settings. The physiological effects on consumption may be conserved across social settings, but this work is only a beginning.

In summary, the strategic design of studies of human eating must be guided by the use to which the outcomes will be put. Studies needed to establish mechanisms of action for drugs will require different methods than studies to determine whether a new product will ultimately become a dietary staple.

3.2.4 Observational methods in eating (Hetherington and Rolls, 1987)

3.2.4.1 Field studies. Field studies are those made in an animal's natural habitat or in the case of humans, places where people eat, as opposed to studies in which the animal or person is observed in a setting designed by the investigator. Animal observational studies preceded systematic study of the factors that govern human food intake, which was an outgrowth of the standard medical history which assesses appetite, food habits and gastrointestinal symptoms. Long before medical science in general and clinical nutrition in particular, developed dietary recall (Burke, 1947) and the food diary (Beaudoin and Mayer, 1953), naturalists and animal husbandmen had used the observational method of determining on what, where, when, and how much animals ate (see Eibl-Eibesfeldt, 1970, for further discussion and references). However, systematic recording and quantification of eating in primates most closely related to man, in the wild, is relatively recent (e.g. Mackinnon, 1974; Goodall, 1986).

The observational method has been employed in studies of eating by humans in various settings (Stunkard and Kaplan, 1977), i.e. cafeterias, in homes, schools, at social events, or in military installations (Hirsch, 1995) (see also Meiselman, this book). In these types of studies simple recording is made of when, where, with whom, what, and under what circumstances eating takes place. When possible, the amount of each item consumed is also recorded.

The method of observation requires a minimum of equipment. In studies of primates in the field, for example (Goodall, 1986), the observer carries paper and pen and wears a reliable watch. A second observer carries charts on which he records association and travel patterns. The observers must also understand the various behaviors well enough to make an acceptably detailed report of complex behaviors. For records of eating, it is usually sufficient to record the species of plant or animal eaten, and the time of onset and termination of the episode. This information can then be transcribed to charts showing temporal patterns of consumption and selection (Wrangham, 1974; Goodall, 1986). This method is completely analogous to a human recording someone else's or one's own eating pattern and associated activity.

An important lesson from animal observational research is that careful measurement of food actually consumed is crucial to obtaining accurate assessment of food intake. Merely observing the duration of a meal, or noting what the individual animal or person was eating at various points in time is not as effective in advancing understanding of the factors that control eating, as is careful measurement of amount eaten. On the other hand, the fact that the amount consumed can be influenced by the eater's awareness of the presence (Herman *et al.*, 1979), or the activities (Polivy *et al.*, 1979), of an observer, in both human and animal studies (Mackinnon, 1974) as well as by the food that was eaten, informs us that accuracy must sometimes be sacrificed for reality. There is therefore a behavioral analog to the Heisenberg uncertainty principle, that the true position of a particle cannot be known with certainty because in order to measure its position the particle must be disturbed, which disturbance changes its position. Recording of undisturbed eating behavior in humans may only be approximated because whenever it is studied, by almost any means, its study leads to some distortion. The only possible exception would be in covert observations made without the knowledge of the observed, a situation which invites the ethical problem of the invasion of privacy, even when individuals are not identified.

Keeping food diaries is a method of self-observation which has the advantage of obtaining data on spontaneous eating, but the disadvantage of being biased, particularly by underreporting (Hetherington and Rolls, 1987). However, others (de Castro, 1994; Obarzanek and Levitsky, 1985) have reported that the diary method is reliable when subjects are carefully trained and the observations are verified. This method has also been combined with laboratory manipulations (e.g. Hulshoff *et al.*, 1993), and its potential accuracy has been improved by having subjects actually weigh, rather than estimate, items consumed. Dietary recall is even less accurate than diaries (see Hetherington and Rolls, 1987), although it is the only method available for large scale surveys (e.g. National Center for Health Statistics, 1979; US Dept. of Agriculture, 1984). To the best of our knowledge, attempts have not even been made to validate food frequency questionnaires.

In an effort to combine the precision of the laboratory with the naturalness of the field, the preweighed portion method has been employed, first with formula diets (Spiegel, 1973), and more recently with conventional foods (Stubbs *et al.*, 1995). Subjects in this type of procedure obtain all their food from the laboratory and return any uneaten portion, but they are not confined to the laboratory. Studies on military rations (Hirsch, 1995) fit this category as well.

3.2.4.2 Laboratory observation. Laboratory observations may be classified into two types: patterns of eating across time and patterns of eating within an eating episode. Laboratory observations may also be classified by the type of device used to record eating.

Although laboratory observations allow more detailed and accurate investigation of patterns of consumption over time than field observations, they should be considered starting points not ends for understanding control of food intake in humans as well as animals, because they restrict the range of other activities in which the individual might engage. The normal activities of people, just as those of animals, influence the pattern of eating.

Laboratory observations of either gross food intake or microstructural variables, without experimental intervention and theoretical guidance, may therefore, in themselves be sterile (Kissileff, 1991).

The paradigm for animal studies of feeding behavior in the laboratory is Richter's classic (Richter, 1927), in which he followed almost instinctively the best ethological principles (Tinbergen, 1969) even before they had been enunciated. Richter recorded a variety of activities in a relatively naturalistic situation in the laboratory. The housing ranged from a single cage to a complex set of interconnecting cages where 7 different activities could be performed and monitored. The main finding, which has since been replicated (Nicolaidis *et al.*, 1979), was that the frequency of eating was greatly reduced when other outlets for behavior were available. Unfortunately, the technical difficulties which could easily be overcome with modern equipment have never been surmounted, and the conclusions drawn about the relationships between meal size and intermeal interval in animals living in restricted quarters (Le Magnen, 1966), have never been reinvestigated in a more naturalistic situation. Subsequent studies of meal patterns utilized more sterile environments (Fitzsimons and Le Magnen, 1969; Kissileff, 1969), but their goals were different. Richter was searching for the basis of the periodicity in the rats' activity cycles, while later investigators were more concerned with the physiological mechanisms driving food and fluid consumption.

Richter's paradigm served as an inspiration to many other investigators of meal patterns in both animals (see Clifton, 1987, for a thorough and critical review) and humans (Green *et al.*, 1987; Bernstein *et al.*, 1981; Campbell *et al.*, 1971). Several theories have been proposed to account for them, including mathematical physiological (Booth, 1978a), metabolic (Le Magnen, 1981), and ecological (Collier, 1987).

3.2.4.3 Patterns of eating within an episode. Patterns of eating over a day or several days have parallels to patterns of eating within an episode. Both were studied extensively in animals and then in humans. McCleery (1977) has critically reviewed mathematical models of 'satiation' curves, as he terms them, and Davis and Levine (1977) have proposed an elaborate engineering model of a physiological mechanism that could underlie one mathematical function which these curves can fit, the negative exponential. The model has been updated for the 1990s to include the effects of learning to sham feed by the addition of another parameter (Davis and Smith, 1990). Kissileff and Thornton (1982), building on earlier work of Pudel (1971), proposed that a quadratic equation provided a simpler theoretical framework than the exponential and accounted satisfactorily (Kissileff *et al.*, 1982) for intake data generated by humans, a fact which has been confirmed by others (Spiegel *et al.*, 1989). All these models have separate parameters for estimating excitation, associated with initial rate of eating, and inhibition, associated with rate of deceleration of eating. However, neither the exponential, nor the quadratic models, explain the abrupt cessation of ingestion before either reaches asymptote.

3.2.4.4 Devices/procedures for observing eating. At least seven types of devices/procedures patterned on some aspect of animal behavior, have been used to study human eating. (1) The earliest was what could be called a 'suckometer' for measuring sucks in infants (Kron *et al.*, 1963), which was analogous to the 'drinkometer' used to measure licking in rats (Hill and Stellar, 1951). (2) Next came the 'feeding machine' of Hashim and Van Itallie (1964) patterned on the operant techniques pioneered by Skinner (1932). (3) Jordan *et al.* (1966) developed a liquid dispenser which was also patterned on the famous 'Richter tube' used for measuring consumption of liquid foods in rats simply by the experimenter's recording the change in level in the tube over time. (4) The automated food dispenser (Silverstone *et al.*, 1980) was similar in principle to the feeding machine for solid foods, but allowed a variety of foods to be offered simultaneously. (5) The universal eating monitor (UEM) (Kissileff *et al.*, 1980) continuously weighed the food, like its animal predecessors (Baile, 1974), while the subject was eating. The UEM is particularly versatile because it can be used with both solid and liquid foods (Kissileff *et al.*, 1980), and with the food visible or not (Kissileff *et al.*, 1982). In addition, subjects can be interrupted after fixed amounts are eaten to determine their reactions on sensitive psychological scales (Silverstone, 1975) to feelings or moods (Wentzlaff *et al.*, 1995). (6) Solid food units (Stellar and Schrager, 1985), little canapes, taken from a box one at a time, while an experimenter monitors each removal, have also been used. This procedure is analogous to animal feeding procedures in which consumption of single food pellets is automatically monitored (Kissileff, 1970). (7) Finally, a technique has been developed for measuring impulsiveness (Logue and King, 1991) toward food which is patterned on an identical procedure used in animals (Ainslie, 1974). Subjects are required to press a rod for an aliquot of juice. They can choose to obtain a large amount with a long delay or small amount with a short delay. The procedure could conceivably be valuable in the investigation of factors which control food intake in subjects with eating disorders.

3.2.5 Animal research established the basis for study of human eating by interventions

In this section we review three broad classes of interventions in which animal research has either preceded or supplemented corresponding human research: (1) administering food or food ingredients, (2) manipulating the eating situation, and (3) pharmacological manipulation and measurement.

3.2.5.1 Administration of food. The administering of food or food components to various organs or allowing the subject to consume food to determine its effect on subsequent food intake has a rich history in the ingestive behavior literature (see Kissileff and Van Itallie, 1982; Kissileff, 1984 for review). The earliest of these types of studies appears to be that of Janowitz and Grossman (1949) who placed varying amounts of food into the stomachs of esophagostomized (esophagus divided so that when disconnected from the stomach, swallowed food drains before entering the stomach) dogs before meals to determine whether nutrients or distention was the critical factor in the progressive inhibition of food intake that eventually terminates the meal. Their work was based on an earlier study of the inhibition of thirst in dogs with a similar procedure (Bellows and Van Wagenen, 1938). Food given to a subject before a meal either eaten voluntarily, or injected or infused into an organ is known as a 'preload', and the subsequent meal or course is called the 'test meal'. In some cases in humans, a subjective report is substituted for a test meal. The main objective of these types of studies is to determine the effects of food components and/or organ systems on the control of food intake, food choice, or desire to eat. Studies in rats (Berkun *et al.*, 1952; Epstein and Teitelbaum, 1962) and dogs (Grossman, 1955) became models for human studies (Stellar, 1967; Jordan, 1969). In recent years the techniques have been extended from intravenous (Gil *et al.*, 1990) and gastric preloads (Spiegel and Jordan, 1978; Shide *et al.*, 1995) to preloads to the liver via the portal vein (Pruvost *et al.*, 1973), the jejunum, and the ileum (Welch *et al.*, 1985, 1988).

3.2.5.2 Manipulation of the eating situation. Eating situations which have analogs in the animal literature can be partitioned into those where subjects eat alone or with others (de Castro, 1995; Galef and Beck, 1990), where the food is familiar or novel (Rozin, 1977) and where associated activities can range from none to complicated tasks to be performed either before, after or during the eating episode (Schachter and Rodin, 1974). In most cases the influence of situational variables has not been fully explored either in animals or humans when a particular product or physiological mechanism is tested and, therefore, generalization about the effects of particular sensory or physiological post-ingestive manipulations should be withheld until the possible interactions of such situational variables with other variables of interest are measured. For example, Collier (1989) has repeatedly explored the effects of forcing animals to press levers on various reinforcement schedules which he considers analogs to natural foraging situations, but his laboratory has not reported the effects of combined changes in reinforcement schedule with physiological manipulation, such as a preload, to determine whether the effects of changing reinforcement schedule are mediated by cognitive/economic or physiological variables. Scheduling manipulations might induce different results on food intake if the scheduling manipulations operated through a cognitive control which was independent of other physiological controls, than if the scheduling manipulations operated on physiological controls, such as those induced by food deprivation. The consequences of this lack of information about the possible interaction of situational variables in the control of human eating, for advertising claims or the introduction of new foods, is staggering and is a research problem which desperately needs attention. (See Meiselman and other chapters in this book for other points of view.)

3.2.5.3 Pharmacological intervention. Pharmacological manipulations have been developed in animals and then applied to humans for both increasing and decreasing food intake. These agents have been used for both therapeutic and investigational purposes. They have been administered by peripheral injection usually into the peritoneal cavity, but also intravenously, intragastrically, and intrahepatically. They have also been placed in food or drink or injected directly into brain tissue or ventricles. Virtually every conceivable outcome from licking to long term (months) food intake has been studied. Animal research has taught us all that we currently know about how brain function in humans may control food intake. Pharmacological agents affect food intake because they act upon sites within the nervous system that normally transmit neural activity from one cell to another at synapses. These receptive sites are classified by the type of neurotransmitter involved. Beginning in the early 1960s with the identification of norepinephrine (Grossman, 1960, 1962) as the first neurotransmitter known to affect eating behavior, at least six different additional neurotransmitters have been identified that have direct effects on food intake. These include serotonin (Blundell, 1977), GABA (Cooper, 1980), endogenous opioids (Levine *et al.*, 1989), neuropeptide Y (NPY) (Stanley and Leibowitz, 1984; Clark *et al.*, 1984; Morley *et al.*, 1985; Leibowitz, 1994), galanin (Leibowitz, 1994; Akabayashi *et al.*, 1994), and dopamine (Hoebel *et al.*, 1989). However, claim for neurotransmitter involvement in the control of eating requires, in addition to demonstration that its administration affects eating, demonstration that blockade of its action has the opposite effect and that its level or turnover changes in a temporal pattern that coincides with a change in eating as has recently been shown for NPY (Leibowitz, 1994). Because norepinephrine injections into the paraventricular nucleus of the rat hypothalamus have been shown to affect learned, not merely unlearned, eating responses in animals (Matthews *et al.*, 1985; Booth *et al.*, 1986), Booth (1994) has suggested that activation of these systems has produced a ‘minor binge’ and that ‘if people have the same kind of neural hook-up some forms of emotional excitation might simply interrupt habitual loss of appetite and make it harder to stop eating and perhaps easier to start’ (Booth, 1994: p. 167). Additional information about the role of animal studies in the understanding of the pharmacology of human food intake control may be found in several other reviews (Cooper, 1987; Hoebel *et al.*, 1989; Levin, 1986) and below in section 3.5.3.

3.3 Phenomena of eating in humans uncovered by studies in animals

What might now be recognized as the classical phenomena of human eating were initially elucidated by studies in animals. The phenomena will be described in this section, and in section 3.4 we will describe the theoretical basis for these phenomena and the analytical experiments undertaken to test them.

3.3.1 Food intake is relatively stable

3.3.1.1 Stability and variability in energy intake. Total daily energy and weight of food consumed by several mammalian species measured in the laboratory has been reported to vary by as little as 10.7% (i.e. coefficient of variation of an individual rat over 91 days of observation) (Adolph, 1947) and as much 28% (Hamilton, 1972) (see Kissileff and Van Itallie, 1982, for specific species and additional references). Accurate data on intake stability for free ranging animals are virtually impossible to obtain, and, given the difficulty of accurately measuring human food intake, comparable data for humans has only recently been reported (Stubbs *et al.*, 1996). A study of 6 men who took all of their food in conventional form, from a metabolic ward to which they were not confined, showed daily coefficients of variation over 14-day periods ranging from 7.1% to 19.8%. There was no relationship among these coefficients of variation to subjects or diets. Another way of assessing variability is by determining the number of days needed to achieve a given level of precision. The fewest number of days of food intake records needed to predict 'intake within 10% of usual' of an individual was 31 days (Basiotis *et al.*, 1987).

3.3.1.2 Stability and variability of macronutrient intakes. Animals are capable of selecting a diet that appears to meet their needs for macronutrients (Evvard, 1915; Stern *et al.*, 1988; Le Magnen, 1992), but this ability is limited by nutrient density and palatability (Kissileff and Van Itallie, 1982; Galef, 1991). More specifically, rats are capable of selecting a diet that contains adequate protein, on average 13% (Rozin, 1968), from a mixture of foods (Richter, 1942; Osborne and Mendel, 1918) or from an array of macronutrients. Carbohydrate and fat intakes, as percentages of total energy, varied widely, ranging from 1 to 59% for fats and from 51 to 84% for carbohydrates (Richter *et al.*, 1938; Rozin, 1968). Manipulation of nutrient density in the diets offered revealed that protein intake resisted attempts to perturb it, whereas fat and carbohydrate intakes did not resist such attempts

(Rozin, 1968). The conclusion that there is tremendous variability in fat and carbohydrate intakes but more stability in protein intakes of individuals is further supported by studies (Shor-Posner *et al.*, 1991) in which individual animals have been characterized by their percentage of intake for three macronutrients in purified form. A few rats selected up to 40% of their daily energy from protein. These dietary intakes were subsequently attributed to meals consumed at specific times of the day. However, it is not clear whether these reported effects on selection (Shor-Posner *et al.*, 1991, 1994) are simply random fluctuations in choice, or whether they are controlled by biological processes, including learning (see section 3.3.7 for further discussion). In the case of protein, there is solid evidence that protein deficient animals can select diets containing protein (Deutsch *et al.*, 1989) but it is not clear whether learning is necessary for maintaining adequate intake, or whether the learning that has been demonstrated (Simson and Booth, 1973, 1974) ensures that a particular level of protein intake will be maintained.

Humans, like animals, can also select an adequate diet from an array (Davis, 1928, 1939), although once again the ability may depend on the choices offered (Story and Brown, 1987; Galef and Beck, 1990). The protein content of the human diet averages 10% to 15% of daily energy consumption in humans over a wide range of cultural and economic circumstances (Drewnowski, 1992), while fat and carbohydrate intakes are more highly variable. For example carbohydrates can range from 50 to 80% of the diet, and fats from 12 to 40%. The hypothesis that protein intake is more tightly controlled by biological processes than fat or carbohydrate intake is further supported, but certainly not proven, by the stability of protein intake in the US food supply at 11% since 1909, while carbohydrate intake has decreased from 57 to 46% and fat has risen from 32 to 43%. These data are critical for proponents of dietary change, because they indicate that there is no inherent biological (e.g. genetic) reason that fat intake cannot be reduced even as low as 20%, as long as the diet contains sufficient essential fatty acids. For prevention of atherosclerosis, an equal mix of the various saturated fatty acids and lineoleic acid is recommended. This amount results in 10 to 13% of metabolizable energy in an American diet (Linscheer and Vergroesen, 1988).

Thus, animal and human research suggests that the reason for resistance to such low fat intake is likely to be primarily behavioral, since food habits are extremely difficult to change (Rozin, 1976). Because preferences for food are relatively stable and 'people select and consume foods, rather than macronutrients' (Drewnowski, 1994: p. 145), it is unlikely that any particular mechanism for macronutrient selection demonstrated in animal studies will apply to human feeding. Pure fat, for example, is rarely consumed unmixed in the human diet (Block *et al.*, 1985). None the less, the fact that fat intake in rats is stimulated in part by the neurotransmitter galanin (Shor-Posner *et al.*, 1994) and inhibited by enterostatin, vasopressin, and corticotropin releasing factor (Bray, 1994), might, under the right circumstances, be used to develop therapeutic agents that would reduce preferences in humans for sources of pure fat, like butter, or some ingredients in salad dressings. Furthermore, Drewnowski (1992) has proposed, and obtained evidence for, the hypothesis that excessive intake of sweet high fat foods is driven by a biological mechanism involving secretion of endogenous opioids (Drewnowski *et al.*, 1995).

3.3.1.3 Patterns of intake across time. Eating behavior occurs in cycles with smaller more frequent fluctuations embedded within larger and less frequent cycles. In humans there are typically three to five bursts of eating activity per day (Green and Tapp, 1986), whereas in other animals eating is typically most concentrated at the time when light intensity is changing, i.e. morning and evening (Siegel and Stuckey, 1947; Goodall, 1986). Within the diurnal cycle in humans, an ultradian rhythm with a 75 min period has also been identified (Green and Tapp, 1986). There are weekly (Taggart, 1962), as well as monthly (Taggart, 1962) and annual cycles (Sargent, 1954), although more recent studies have failed to confirm the changes with time of year (Van Staveren *et al.*, 1986). Rhythms extend to specific nutrients as well (Krauchi and Wirz-Justice, 1992). The picture of eating over longer periods is completed by the well known patterns of migrators (Odum, 1960) and hibernators (Mrosovsky and Faust, 1985) which have cycles of hyperphagia followed by reduction or cessation of eating before, and during, migration and hibernation. The latter has a parallel in humans in developing countries who are subjected to periodic fluctuations in food supply (Prentice *et al.*, 1981). In this latter group body weight fluctuated by a mean of 1.5 kg, whereas in a comparable group of Dutch women, the fluctuation was only 0.5 kg (Westerterp, 1994). The importance of these findings is that they are more consistent with a hypothesis that body energy stores resist change (Van Itallie *et al.*, 1978) rather than the hypothesis that body weight in general is a highly regulated variable with a set point mechanism controlling it as some have argued (e.g. Levin, 1986; Keesey, 1986).

3.3.2 Episodic meal patterns

3.3.2.1 How often do animals and humans eat? Intake over smaller units of time, e.g. a single day, is discrete, rather than continuous like metabolism, in many species. Nevertheless, there is tremendous variability in the pattern of eating, ranging from animals which eat almost continuously, through grazing animals (Hafez and Schein, 1962) to meal takers like ourselves (Green and Tapp, 1986), the rat (Richter, 1927; Le Magnen and Tallon, 1963; Kissileff, 1970; Strubbe, 1994) and other primates (Goodall, 1986).

However, some animals voluntarily eat less than once a day, such as certain reptiles (Myer and Kowell, 1971). This pattern of eating discrete meals persists even when the macronutrients are presented in purified form (Shor-Posner *et al.*, 1994).

3.3.2.2 Are these patterns of eating biological rhythms? Eating in animals is partly under the control of biological rhythms (Richter, 1965; Strubbe, 1994). These rhythms are controlled by neural oscillators, deep within the brain (Rusak and Zucker, 1979), in the case of mammals. The physiological basis of rhythms independent of external signals known as 'Zeitgebers' (literally time givers) is established by placing the animal in a cue-free environment and noting whether the rhythm in question persists. Not only do rhythms of eating persist in so-called 'time isolation', but responsiveness to food loads is also modulated by phase in the circadian cycle. Rats were least responsive to food infusion between 1800 and 2400 (end of dark cycle) and most responsive between 1200 and 1500 (Strubbe *et al.*, 1986). Time isolation studies (Green *et al.*, 1987) in humans indicate that, like rats, they also possess an internal rhythm. The neural origin of the eating rhythm in humans is not known. However, with so many people flying back and forth around the globe, the impact of disruption of circadian rhythms induced by this travel is an important area needing further study.

3.3.2.3 Why the meal is the considered unit of study for human eating. A crucial issue in the advancing knowledge on the control of eating in humans and its relation to obesity is the dichotomy between so-called long-term and so-called short-term controls (Mayer, 1955). The functional unit of feeding is the meal (Smith, 1982), mainly because it occurs in a discrete period of time and is easily amenable to experimental study. Considering the variety of eating styles and patterns both in animals and humans, investigators should be aware that the meal may be merely a convenient benchmark, and not lose site of other possible functional units such as a 24-hour period. It has been popular (Kissileff, 1991) to consider that long-term effects of manipulations, such as exercise, are mediated by short-term controls of what initiates and terminates a spontaneous bout of eating, despite the fact that such a link has never been, and need not be (Kissileff, 1991), formally demonstrated. The fundamental difference between the way energy enters the body, in discrete packets, and leaves, continuously, makes it unlikely that there would be any mechanism to monitor either system in relation to the other. Yet the idea persists that energy balance and its control is the key to understanding obesity. Undoubtedly there are detectors of energy metabolism which are linked to the control of food intake (Friedman, 1991), but these need not be responsible for initiating every feeding event in order to ensure that over a period of several days energy balance is maintained. Like many other constructs in the field of ingestive behavior, energy balance is a useful computation, but not necessarily one which has a discrete physiological mechanism.

3.3.3 Food intake responds to excesses and deficits in energy stores

Changes in food intake after manipulations that change energy stores [e.g. deprivation of food (Bjorntorp and Yang, 1982; Bjorntorp *et al.*, 1982), overfeeding (Van Itallie and Kissileff, 1985; Cohn and Joseph, 1962), change in ambient temperature (Hamilton, 1967) or exercise (Woo and Pi-Sunyer, 1985)] have been taken as evidence that the body detects these deficits and excesses and mobilizes the brain to activate food seeking and ingestive behavior (e.g. Bray, 1994; Van Itallie and Kissileff, 1990). Both animals (Adolph, 1947; Castonguay, 1981; Janowitz and Grossman, 1949; Hansen *et al.*, 1981) and humans (Spiegel, 1973; Porikos *et al.*, 1982; Hill *et al.*, 1987; Birch and Deysher, 1985; Campbell *et al.*, 1971; Louis-Sylvestre *et al.*, 1989) maintain energy intake over days within limits when the energy density of a monotonous or a mixed diet is varied. However, it would be premature to draw the conclusion, that such results prove that the body regulates energy balance (Booth, 1978b), in the sense described by Brobeck (1965), where there is a detector for the variable being regulated and feedback systems reporting its perturbation. Obesity is, from an abstract perspective, as Bray (1994) says, 'a failure of this homeostatic model for nutrient balance' (p. 29), but the suggestion that because adrenalectomy reverses obesity, obesity is caused by high levels of adrenal steroids (Bray, 1994) does not necessarily follow. There are many ways that obesity can be caused, and they do not all involve excessive adrenal steroids.

3.3.4 Food intake is self-limiting

During the course of individual meals in deprived rats, the rate of eating is faster at the beginning of the meal than at the end (Skinner, 1932). However, deceleration of the rate of intake in a meal does not appear in rats that are feeding *ad libitum* either on solids (Kissileff, 1970) or liquids (Becker and Kissileff, 1974). Studies in animals have shown that both the initial rate of eating and rate of deceleration can be influenced by variables which affect underlying physiological processes such as deprivation, taste, and manipulation of the GI tract (Davis and Levine, 1977). As mentioned in section 3.2.4.3, a quadratic model fits cumulative intake curves in humans well (Kissileff *et al.*, 1982). Although the parameters of cumulative intake curves have been investigated thoroughly in animals (Davis, 1989), it is not completely clear what they may mean for humans (Kissileff and Thornton, 1982). For example, men but not women increased the initial rate of eating after deprivation. The initial rate of eating was correlated with the rate of deceleration. The rate of deceleration did not change when subjects were given the putative satiety inducing hormone cholecystokinin (Kissileff *et al.*, 1981).

Until cumulative intake curves are examined under a variety of experimental conditions, it will not be possible for any general picture to emerge. Cumulative intake curves and the rate of chewing and swallowing can be affected by many factors, particularly the percentage of solid, chewable food in whatever is being consumed, and its palatability (Bellisle and Le Magnen, 1981; Bellisle *et al.*, 1984; Bobroff and Kissileff, 1986; Spiegel *et al.*, 1989). For example, subjects showed reduced initial rates of eating when the food was adulterated with cumin which made it less liked (Bobroff and Kissileff, 1986). The animal work (Davis, 1989) and human studies on the cumulative intake curve have been mutually reinforcing in their attempts to illuminate underlying physiological mechanisms and this aspect of the control of eating behavior is expected to engender even more research in the future (Westerterp-Plantenga *et al.*, 1992).

3.3.5 Availability and orosensory factors affect food intake

Without getting bogged down in terminology and theory which we shall discuss in the next main section on concepts, we note humans and animals are capable of overcoming impediments to adequate food intake. This idea is commonly expressed in the notion that eating is 'motivated' behavior. Evidence for motivation is that, as food becomes easier or more difficult to obtain, meal size and frequency change (Kanarek, 1976; Collier, 1989) in rats, and meal size changes in people in both laboratory (Schachter, 1968) and cafeteria (Meiselman *et al.*, 1994) settings. Intake is also changed when orosensory factors make the food more or less attractive (Rolls *et al.*, 1981; Bellisle, 1995; Mook, 1989; Sclafani, 1989; Cabanac, 1989). Animal studies have been particularly illuminating in suggesting brain areas where such effects could be mediated (Schneider *et al.*, 1989; Grill *et al.*, 1987), and in providing model systems in which these kinds of effects can be physiologically explored.

3.3.6 Food intake is controlled both by genetics and environment

While this statement may appear to be a truism, attempts to partition variability have been fraught with difficulty. It is clear that certain obese strains of rats (Barry and Bray, 1969; Cruce *et al.*, 1974) and mice (Fuller, 1972) exhibit overeating even before marked obesity is present. It is not clear whether these effects on food intake are primary genetic alterations in neurotransmitter systems (Leibowitz, 1994; Akabayashi *et al.*, 1994) or whether they are secondary to genetic effects on metabolism, possibly by increasing lipoprotein lipase (Greenwood and Vasselli, 1981; Gruen and Greenwood, 1981) or cortisol (Bray, 1994), which in turn drive food intake

(Fuller, 1981; Grinker, 1981). Although it has been shown that genetically differing strains of rats (Schemmel *et al.*, 1970) and mice (West *et al.*, 1995) fatten differently on high fat diets, neither the critical genetic locus in animals, nor the extent of genetic mediation of this fattening in humans on high fat diets is known (Bouchard, 1992; Roberts *et al.*, 1994; Heitmann *et al.*, 1995). Furthermore it is unclear what the appropriate phenotype for food intake control or eating behavior would be, although such aspects as rate of consumption, meal size, total daily intake, response to a physiological manipulation, or response to change in taste, among others, could conceivably serve, provided standardized testing conditions were provided. Additionally, it is not clear to what extent the fattening effect of a high fat diet is mediated by oral and by postingestional effects (Warwick, 1995), thus making it even more difficult to unravel the pathway from gene to behavior.

The foregoing notwithstanding, an important breakthrough has been made in the cloning and sequencing of the ob gene in the mouse and its human homologue (Zhang *et al.*, 1994). Fat cells from genetically obese mice with a mutation of this gene fail to make a particular protein as the cells enlarge. A biologically active form of the recombinant mouse ob protein, called leptin (Halaas *et al.*, 1995), has been purified from a bacterial expression system. This protein reduced food intake and body weight in mice (Campfield *et al.*, 1995; Pelleymounter *et al.*, 1995; Halaas *et al.*, 1995). This work supports an earlier theory that food intake was partly controlled by a feedback signal from adipose tissue (Kennedy, 1953; Hervey, 1959; Faust *et al.*, 1977). It can be expected that this work from an animal model will have a big impact on how scientists think about the control of human eating for at least the next decade. Thus, animal research has developed powerful strategies for answering these questions which can be applied to the genetic study of eating behavior and help to suggest gene loci homologous to those in humans where effects on intake might be controlled.

3.3.7 Feeding over the lifespan

Animal studies have helped to elucidate changes in intake over the lifespan. Developmental studies in which various internal organs, including the brain, have been manipulated in animals have revealed how various processes and organs contributed to the development and decline of adult feeding behavior. As parallel studies become available in humans, it is becoming clear that humans share many of the same controls with animals. For example, the suckling rat, like the suckling child (Fomon *et al.*, 1964, 1969), at a very young age seems entirely dependent upon gastric filling (Grinker, 1981) to terminate intake (Houpt and Houpt, 1975). The competency of rats to respond to various physiological challenges appears partially dependent on the effort needed to respond. When foods or fluids are injected directly into the mouth, regulatory responses appear at an earlier age than when the animal is required to perform even as simple a response as moving to the food source (Phifer and Hall, 1987). Animal studies in which food is injected into the mouth of infant animals are more analogous to human infants whose initiation of eating is controlled entirely externally. At the other end of the developmental spectrum are studies in aging animals and humans. As an example of the kinds of work that can be done, Roberts *et al.* (1994) have shown for the first time in aging men, unlike younger men, a deficiency in both the reduction of food intake after overfeeding and the increase after underfeeding. While it is not clear which candidate systems for eating control (see below) were responsible for the failure of these older men to compensate behaviorally for previously imposed changes in food intake, the results are consistent with studies in animal models which have shown alterations in either response to, or levels of, neurotransmitters with aging (Gosnell *et al.*, 1983; Silver and Flood, 1988).

3.4 Theoretical framework and experimental evidence

Animal research established the theoretical framework and experimental evidence for a physiological control of food intake in humans. Having reviewed the procedures for studying eating and the phenomena which they have revealed, we come now to the heart of the matter, the link between biologically programmed behavior and human eating activity. Although we shall not pursue the issue here, it is important to recognize that there is a tremendous philosophical problem in proposing that human behavior may be programmed. We are beyond the realm of science and into metaphysics. As we implied in the introduction, signals from the internal environment of the body are only one source of control, the other two being those from the environment and those from the mind (will). Although we shall describe experiments that show that signals from within the body are able to affect food intake in humans as they do in animals, the relative strengths of the three types of influences (body, mind, environment) on food intake have yet to be ascertained.

3.4.1 Constructs and the use of animal experiments to explain human feeding

3.4.1.1 The hunger construct. The study of eating behavior in animals has had a long and rich history in the field of psychology (Bolles, 1975) where theories have led to experiments in the animal laboratory. However, a more eclectic approach to this history and a better insight into the origin of many of the current problems was reviewed by Stellar, in two papers

(Stellar, 1976, 1989) thirteen years apart. The framework for the study of human eating could be said to begin with Walter Cannon's demonstration on his student Washburn (Cannon and Washburn, 1912), that the origin of the hunger sensation was the contractions of the stomach, demonstrated by the concomitance of gastric contractions and reports of hunger by Washburn, who had swallowed a balloon connected to a pressure sensor. Richter's (1927) demonstration that the two-hour rhythm of contractions in the rat stomach paralleled its meal pattern seemed to confirm the hypothesis. Bard (1928) however, was not totally convinced of Cannon's hypothesis and believed that emotion to which hunger was related was a central brain state and that the sensations were the peripheral effects of central output. Richter also later changed his view to a more global conception that the internal state of the animal's periphery had representation in the brain somewhere which led to 'self-regulation' (Richter, 1942).

Lashley (1938), in whose lab Stellar first began working, proposed that hunger, and other behaviors associated with what we commonly call motivation, were aroused by a central motive state which produced the different sorts of motivated behavior. Hormones and the internal environment along with external and internal sensory stimuli all had to be integrated by the central nervous system to yield specific organized goal directed behavior. But he had no idea where that integrative mechanism might be. By utilizing the lesion work of Brobeck *et al.* (1943) and Anand and Brobeck (1951) showing reciprocal hypothalamic control of food intake, Stellar was able to conceptualize the system that has served us well for 40 years, namely the idea that 'the amount of motivated behavior was a function of the amount of activity in certain excitatory centers of the hypothalamus' (Stellar, 1954: p. 6). That theory led to a host of studies in animals and humans culminating in appetite enhancing and suppressing drugs which are believed to work on certain structures in the human brain. Were it not for the theory and testing in animal work, the current pharmacological treatments for eating disturbances would probably not be available.

3.4.1.2 Behaviorism. Another powerful strain of animal research which has influenced human behavioral studies is the behavioristic approach of Skinner and his school (see Bolles, 1975: pp. 304*ff*, for review). Where Stellar's theory required a change in internal state and neural signal to detect it, Skinner's theory required only that a response be made contingent on presentation of a food item that could be consumed. Hunger was never an issue and the changes in responsiveness that were seen upon eating were attributable to changes in 'reinforcement'. What made food 'reinforcing' was not particularly important to this group. According to Bolles 'If a theorist is willing to attach sufficient importance to the principles of learning, then he has little need for a theory of motivation (e.g. Skinner, 1953); . . . the explanation of behavior can proceed better without any motivational concepts, without involving needs or drives or wants or anything else of the kind either in data, language or among the theoretical concepts' (Bolles, 1975: p. 304). Learning explains all. However, the theory that reinforcement can explain everything begs the question because there is no explanation for changes in reinforcement that account for stopping of behavior. Nevertheless, this line of reasoning has not died, but it is retained in the current notions that rewarding value of stimuli are modulated by internal state (e.g. Cabanac, 1971; Sclafani, 1989; Grill *et al.*, 1987). For studies of food intake, oropharyngeal stimuli are the most powerful of such stimuli.

3.4.1.3 Reward and palatability. Given that eating behavior is not a reflex which is stimulated at each presentation of food, but rather waxes and wanes with internal state, as a well as with food offered, one hypothesis that has been proposed to explain this variability in behavior is that the central motive state changes the reward value of stimuli (Grill and Berridge, 1985; Sclafani, 1989). The consequence of this theory for human eating is that food choice or the choice to eat or not may be conditioned by previous contacts with a particular stimulus and its effects (see section 3.2).

This view has been apposed to the alternate that there may be more than one sort of palatability (Kissileff, 1990) and that under standardized conditions, foods vary in the effects they have on reports of liking, which variation has been termed 'intrinsic' palatability in contrast to 'learned' palatability which depends on post-ingestive effects. According to this latter view:

> 'The desire for food is a relation between two terms, the subject and the food. Differences between foods, taken against some assumed state of the subject, are incentive variables or in ordinary language, palatability. Differences between states of the subject, implicitly with respect to some food presentation condition, are degrees or types of hunger/appetite or satiety' (Booth, 1976: p. 419).

In following up this line of reasoning Kissileff argued that 'it is easier for me to conceive of an ingestive response as being under the control of at least two components, the item to be ingested and the individual doing the ingesting, than for it to be considered some kind of nebulous interaction between two components (see Booth *et al.*, 1972)' (Kissileff, 1990: p. 163). Regardless of the issue of terminology, it is clear that the internal state changes responsiveness to stimuli both in humans and animals, and it is the mechanism of this response, not the words we choose to describe it that needs explanation. The best way out of this dilemma is by what Booth has described as 'mechanistic analysis' (Booth and Weststrate, 1994), in which a behavioral outcome (e.g. food intake, rate of eating, report of pleasantness) is measured as a function of physiological or cognitive input, such as a preload, change in the food stimulus, administration of a drug, or instruction to a subject.

3.4.2 Learning

3.4.2.1 Sensory cue to physiological consequence learning. Animal research on the influence of learning on food intake has told us much about this aspect of human eating. Of particular interest for this chapter is research concerning the phenomenon known as 'taste-to-postingestive consequence' learning. In this paradigm, when the postingestive effects of an ingestant are associated after several trials with an oropharyngeal cue, when next encountered, the cue alone will elicit behavior similar to that elicited by the food. Depending on the nature of the postingestive consequences of the food (i.e. induction of malaise, provision of nutrients), and the metabolic state (fasting or recently fed) of the subject, this behavior may be one of a rejection of the cue (e.g. conditioned taste aversion or CTA, or conditioned satiety), or a preference for the taste over another when presented with a choice (conditioned preference, conditioned desatiation). This learning may be measured either by change in intake when a choice is offered or other acute behavioral sign of change such as an orofacial response (Grill and Berridge, 1985).

The classic *CTA procedure* involves the induction of a state of malaise following the presentation of a taste stimulus (Garcia *et al.*, 1955). It is known that there can be a considerable amount of time between the two stimuli (up to 24 hours!) when the taste stimulus is novel (Etscorn and Stephens, 1973). This work has been useful for a variety of reasons (e.g. pest control, testing of whether pharmacological agents produce malaise in laboratory animals), and the prevalence and role of these aversions in human eating has been investigated (de Silva and Rachman, 1987; Pelchat and Rozin, 1982). Primary among these has been the attempt to use it therapeutically to reduce the ingestion of certain items (e.g. alcohol (Elkins, 1980)) and to avoid its formation in, and therefore improve the nutritional status of, cancer patients undergoing chemo- and/or radiotherapy (Mattes *et al.*, 1991). Little research has been conducted to determine the role of naturally formed aversions to nutritionally valuable food on food choice in animals or humans though there is much conjecture. This may be due to the fact that CTAs have been reported to have a short duration in human adults tested (Mattes *et al.*, 1991). This is not as contradictory of the animal research as it would appear, however, since these aversions were formed to foods with which the patients had had extensive prior exposure. It is known from animal work that the CTA is generally more robust to novel flavors (Klosterhalfen and Klosterhalfen, 1985). What we do know suggests that aversions may be produced in humans via this simple association and, therefore, may contribute to food choice and intake in humans.

The *conditioned preference procedure* is similar in some respects to conditioned aversion. A taste stimulus is associated with the alleviation of a deficit in energy (Sclafani, 1990) or of a specific nutrient (e.g. an essential amino acid) (Naito-Hoopes *et al.*, 1993) which, after several pairings, produces an increased ingestion of the flavored substance and a preference for it when given a choice to consume it or another flavor. Though this procedure produces a robust preference in rats, little is known of its role in the formation of food preferences in humans though it has been widely assumed that human food preference is largely acquired. Birch and her associates have demonstrated conditioned preferences (and increased food intake) to flavors in children (Birch *et al.*, 1990; Kern *et al.*, 1993) and it has recently been shown that humans will learn to prefer a flavor associated with a protein fortified food when mildly protein deprived (Gibson *et al.*, 1995). These studies, utilizing the methodology of animal research, show that humans do possess the physiological mechanisms that influence food intake known to be present in animals.

A form of taste-to-postingestive consequence learning that has been tested in humans as well as animals and may contribute significantly to control of food intake in humans has been called *conditioned satiety* by Booth (1972). This phenomenon differs from conditioned aversion in that it is a learned reduction in preference or intake based on the pairing of a stimulus with either the bloating or metabolic state that accompanies the transition from repletion to overrepletion of energy flow to the liver towards the end of a meal. Conditioned aversion results in reduction of intake and preference of the stimulus from the start of the meal. It differs from conditioned preference which occurs when the individual is made energetically replete by the food, i.e. it brings the individual out of a state of deprivation (Booth, 1978b). Booth has demonstrated that rats will learn to consume less of a 30% starch liquid meal whose flavor had been previously consumed with 50% starch liquid than of a 30% starch liquid whose flavor had been consumed with a 10% starch liquid (Booth, 1972). Booth has found similar results in humans as have other researchers (Booth *et al.*, 1976, 1982; Chabert *et al.*, 1993). Since the human, like the rat, does appear to be sensitive to caloric manipulations (i.e. the reduction in a meal that follows a high caloric preload when compared to that which follows a low density preload), it is likely that its ability to learn an association between the sensory components of a food and its postingestive consequences will affect meal size and food choice.

3.4.2.2 Social learning. The appropriateness of animal models to study the roles of cognitive and social influence on human food intake is often overlooked. It has been shown that food choices in rats are highly influenced by the behavior of conspecifics and by the choices made by their mothers pre- and post-weaning (see Galef, 1989 and Galef and Beck, 1990 for review) and this behavior could serve as a good model for understanding the social transmission of eating patterns and food choice.

3.4.3 Theory and evidence that food intake is controlled by physiological variables

Animal research has provided both the experimental material for development of physiological theories/mechanisms of food intake control as well as the evidence for them. A brief survey of these mechanisms should convince even the most skeptical reader that physiological control of eating behavior and food intake exists in humans.

3.4.3.1 Neural and neurochemical control. Although observations in humans (see section 3.4.1.1) touched off a flurry of animal research in the early and mid 20th century, the work came full circle with the production of lateral hypothalamic lesions in humans for treatment of obesity at three quarters through the century (Quaade, 1974; Quaade *et al.*, 1974). However, pharmacological manipulation has by far had the greatest impact on human behavior. Because most of the human work has been conducted in conjunction with eating disorders, see section 3.5 for further description of that work and its basis in animal research.

3.4.3.2 Hormonal and metabolic control of eating behavior. Insulin is the most widely studied hormone that affects food intake in both animals and humans. Beginning with the demonstration that insulin injection increases food intake in rats (Mackay *et al.*, 1940), the role of excess and deficient glucose availability has been a lively subject of investigation for more than half a century. The latest thinking (Vanderweele, 1994) is that insulin is actually a satiety hormone in that it reduces food intake by affecting the liver in an unspecified manner.

Provocation of eating by insulin is an indirect effect of either glucoprivation (Ritter *et al.*, 1994) or lack of available substrate for hepatic oxidation (Novin, 1994; Scharrer and Langhans, 1988; Friedman, 1991). A limited amount of work with humans suggests that they at least respond to the glucoprivic stimulus (Thompson and Campbell, 1977) and their lack of reduction in intake after elevation of insulin levels without glucoprivation (Woo *et al.*, 1984) may be attributable to the intravenous route of injection which was also ineffective in reducing food intake in rats (Vanderweele, 1994). Many of the metabolic inhibitors used in animal work (Friedman, 1991) are too toxic to be used in humans. In related studies, Campfield and Smith (1990) have shown that both animals and to a smaller extent humans (Campfield *et al.*, 1992) begin eating when blood glucose levels transiently decline according to a critical pattern. Exercise is another stimulus to changes in food intake. It leads to declines shortly after strenuous exercise in both animals (Routtenberg and Kuznesof, 1967) and humans (Kissileff *et al.*, 1990), but increases when sustained for several days (Mayer *et al.*, 1954; Woo and Pi-Sunyer, 1985). The physiological basis for these effects remains to be discovered.

3.4.3.3 Gastrointestinal control of food intake. As was previously mentioned, both animals and humans reduce food intake when the stomach is sufficiently distended mechanically (see Kissileff and Van Itallie, 1982 for review of animal studies). The ratio of reduction to distention has only been quantified in one human study (Geliebter, 1988) at approximately 0.4 ml/ml when a liquid diet was eaten and a balloon distended in the stomach. The threshold for the effect was approximately 400 ml. Further work on this mechanism is critical because many current treatments for weight loss rely on the notion that any mechanical distention produced, for example, by fiber-containing foods or pharmaceuticals, will automatically reduce intake. Rigorous testing should be done to support such claims and much more work should be done to quantify the effect of mechanical distention in conjunction with other stimuli, such as nutrients and hormones.

3.4.4 The role of gut peptides

Animal research has told us much about the role of gut humoral factors in the control of feeding behavior in animals and humans. A number of peptides of neuroendocrine and paracrine sources in the gut have been implicated in the onset, size, and time course of meals in animals and this research quickly led to speculation about and investigation of their role in human eating behavior. Though a number of these substances are currently under investigation, the peptide whose role in feeding in both animals and humans has been best characterized is cholecystokinin (CCK).

3.4.4.1 The role of CCK. Cholecystokinin was named after its action on gallbladder contraction and has further been shown to simulate enzyme secretion from the pancreas and delay gastric emptying among other gastrointestinal effects (Della-Fera and Baile, 1981; Gibbs and Smith, 1977). It is released by mucosal cells in the duodenum and has been hypothesized to have its effects via a paracrine mechanism (Ritter *et al.*, 1994). CCK was first shown to reduce food intake in rats by Gibbs *et al.* (1973) and has subsequently been shown to be effective in most animal species tested (Silver and Morley, 1991). Likewise, a decrease in food intake by exogenous administration of CCK was demonstrated in humans by a number of investigators (Kissileff *et al.*, 1981; Stacher *et al.*, 1982). An important finding whose basis is still not resolved is that CCK is more effective in reducing intake after larger amounts of food have been consumed than after smaller amounts have been consumed (Antin *et al.*, 1978; Muurahainen *et al.*, 1991) even when the food drains from the stomach (Antin *et al.*, 1978). Other parallels between the action and physiology of CCK in animals and humans were uncovered, usually with the discovery in humans preceded by that in rats. Exogenous administration of CCK was found to slow gastric emptying of nutrients in both rats (Debas *et al.*, 1975) and humans (Muurahainen *et al.*, 1988), and this action was proposed to be the mechanism by which CCK exerted its effect on eating in monkeys (Moran and McHugh, 1982). It has been shown that, in humans, CCK infusion does produce a relaxation of the stomach while increasing feelings of fullness induced by a gastric balloon (Melton *et al.*, 1992) supporting a theory by Muurahainen *et al.* (1988, 1991) that CCK induces satiety, not by increasing stomach distension, but by amplifying afferent signals from the stomach to the central nervous system. Indeed, Raybould *et al.* (1988) and Schwartz and Moran (1994) have shown just that. Afferent vagus neural activity in rats shows a firing frequency, after a combination of exogenous infusion of CCK and stomach distention, at a rate that exceeds the sum of either alone, with a few combinations of distention and CCK level. However, Grundy *et al.* (1995), who find inhibition, rather than excitation, of gastric mechanoreceptive responses when CCK is infused, with larger distentions than those used by Schwartz and Moran (1994), and Raybould *et al.* (1988), have suggested that the results of Schwartz and Moran (1994) and Raybould *et al.* (1988) may have been secondary to local contractile effects of high doses of CCK. Subthreshold levels of exogenous CCK and stomach distention for reduction of food intake reduce food intake in humans when combined (Carretta *et al.*, 1995). These studies show how animal and human studies complement one another and, more importantly, how findings from research with animals can guide the work done with humans, and conversely.

Additionally, CCK action in animals and humans shares other characteristics. The endogenous release of CCK by food was shown in humans (Liddle *et al.*, 1985) and rats (Liddle *et al.*, 1984), however, species differ in which macronutrients are the most potent releasers of CCK with proteins being the most effective in rats (Liddle *et al.*, 1984; Sharara *et al.*, 1993) and fats being the most potent in humans (Hopman *et al.*, 1985; Fried *et al.*, 1991; Mossner *et al.*, 1992). Carbohydrates are rather weak releasers of CCK in all species studied.

Pharmacological manipulations have revealed the role of endogenous CCK on food intake. The CCK-A receptor antagonist lorglumide has been shown to reduce satiety induced by an intestinal fat load in rats (Greenberg *et al.*, 1989; Yox *et al.*, 1989). Loxiglumide, another CCK-A receptor antagonist, has likewise been shown to reduce the satiating quality of intestinal fats in humans (Lieverse *et al.*, 1994). Therefore, endogenous levels of CCK in the periphery do influence satiety level although administration of loxiglumide to humans did not increase food intake significantly (Lieverse *et al.*, 1993; French *et al.*, 1994). Devazepide, a potent CCK-A receptor antagonist, has been shown to increase food intake in rats (Hewson *et al.*, 1988). It is interesting to note, however, that devazepide has been shown to increase hunger ratings in humans (Wolkowitz *et al.*, 1990) while loxiglumide had no effect on hunger or fullness ratings (French *et al.*, 1994). Regardless, there is ample evidence that the intake-reducing effect of exogenous CCK in all species studied is produced by the activation of the CCK-A receptor (Smith and Gibbs, 1992).

Thus, the foundation for studying the role of CCK in both normal and disordered eating behavior in humans is well established. No overt negative side effects have been reported when the peptide dose is infused slowly over a 20 minute period. The primary method by which food intake is reduced is by shortening the meal duration and it appears to have little impact on the initial phase of the meal (Kissileff *et al.*, 1981). Its potential as a therapeutic agent was bolstered by findings that CCK infusion reduced food intake in obese men (Pi-Sunyer *et al.*, 1982). Impaired release of CCK by bulimics has been suggested as a possible mechanism by which they fail to achieve satiety and thereby overeat (Geraciotti and Liddle, 1988). However, administration of CCK to bulimics has failed to reduce binge meals (Mitchell *et al.*, 1986). This failure may have been attributable to the timing and rate of administration of hormone, since it has been shown (Guss *et al.*, 1988) that when low doses of CCK are infused only before, but not during, a meal, food intake is not reduced in humans.

3.4.4.2 Bombesin and gastrin releasing peptide. Bombesin, a peptide first isolated from the skin of a frog, has been shown to be a potent reducer of food intake in rats as well as a stimulator of gastric and pancreatic secretions in a manner similar to CCK though its effects appear to be independent of CCK (Gibbs *et al.*, 1979; for review see Gibbs and Smith, 1988). In addition, blockade of central bombesin receptors in the rat increases food intake (Merali *et al.*, 1993). Bombesin, however, is not present endogenously in mammals: the mammalian form has been identified as gastrin-releasing peptide (GRP) (McDonald *et al.*, 1978). Bombesin/GRP has been implicated in the control of human eating behavior as well. Muurahainen *et al.* (1993) demonstrated a significant reduction in food intake in men as well as a reduction in food palatability ratings, when bombesin was infused intravenously. Similarly, infusion of GRP reduced food intake in men and reduced the feelings of hunger while producing early feelings of fullness (Gutzwiller *et al.*, 1994). No overt side effects were reported in these subjects though there were reports of slightly elevated feelings of sickness in half of the subjects tested (Muurahainen *et al.*, 1993).

3.4.4.3 Other gut peptides. Other peptides that appear to affect food intake in rats include pancreatic polypeptide (PP), neuropeptide Y (NPY), peptide YY (PYY), somatostatin, neurotensin, insulin, glucagon, apolipoprotein A-IV, and the opiates. This is by no means a complete list and we cannot fully describe their action in this chapter. However, we will attempt to describe briefly why they should be further studied. Pancreatic polypeptide and NPY are orexigenic (i.e. they enhance food intake) when administered centrally (Stanley and Leibowitz, 1984; Clark *et al.*, 1984;

Morley *et al.*, 1985; Nakajima *et al.*, 1994). Pancreatic polypeptide has been proposed to act as a peripheral satiety signal as there is a peripheral increase in circulating PP following a meal (Glaser *et al.*, 1983). Of particular clinical interest is a report that PP release in obese rats is retarded, and PP administration has been found to reduce the hyperinsulinemia and hyperphagia of obese rats (Gates and Lazarus, 1977). At the time of this writing, no test of the effect of PP infusion on human eating behavior has been reported, although it is known that circulating PP levels increase in humans in anticipation of meals and as a consequence of ingesting a meal (Meier *et al.*, 1990). There is controversy, however, over whether basal levels of PP and that released by meal consumption in obese humans is abnormal. Some laboratories have found significant differences in PP levels between obese and normal subjects (e.g. Lassman *et al.*, 1980, who measured basal levels) while others have not (e.g. Sirinek *et al.*, 1985, who measured response to oral glucose). In response to a meal, Pieramico *et al.* (1990) have reported no difference in the concentration of PP released by normal and obese humans while Lieverse *et al.* (1994) have reported such a difference.

While most research on NPY has focused on its orexigenic properties and possible neurotransmitter function in the mammalian brain, it is also found throughout the gastrointestinal tract and has known inhibitory actions there (Mannon and Taylor, 1994). Although it has been found in the bloodstream, its level in humans is not affected by meal ingestion (Melchior *et al.*, 1994). Peptide YY, however, is affected by meal ingestion in dogs (Kuvshinoff *et al.*, 1991) and humans (Adrian *et al.*, 1985). Its localization and function in animal gastrointestinal physiology has led some researchers to conclude that it may have a significant satiety function (e.g. its 'anti-CCK' action in the pancreas and its localization in distal small intestine mucosal cells (Mannon and Taylor, 1994).

3.4.4.4 The role of other non-gut peptide hormones. Research with animals continues to reveal new potentially useful satiety agents such as apolipoprotein A-IV which has been isolated in humans and shown to be a potent CNS suppressor of feeding in rats (Fujimoto *et al.*, 1993). Previously known substances are also being shown to have satiety-inducing effects, for example, somatostatin and growth hormone releasing hormone (Feifel and Vaccarino, 1994). Though it is far from certain that these agents and others will have clinical relevance, it is important that their effects in humans be fully investigated. In the case of CCK, animal research gave insights into where and how to look for its possible effects in humans. Though further investigation is necessary, CCK holds significant potential as an agent for the reduction of meal size and, hence, caloric intake. The research into the effects of other gut peptides on human food intake is still in its infancy but it has been demonstrated that agents with known physiological effects on food intake in animals are found to be present endogenously in humans and do possess similar modes of action in humans.

3.5 What animal research tells us about eating disorders in humans

3.5.1 Introduction

The two clinically recognized eating disorders are bulimia nervosa (American Psychiatric Association, 1987), which is characterized by rapid consumption of food in a discrete period of time, and anorexia nervosa (American Psychiatric Association, 1987) whose distinguishing feature is severe emaciation brought on by the patient's restriction of food intake. Although it has been questioned whether there is an eating disorder in the obese (Kissileff, 1989) the answer to this question depends on the level of body weight and norms for eating behavior in humans. Regardless of that answer, it is clear that severely obese individuals eat more than individuals closer to the population median (Yanovski *et al.*, 1992) and that animal research has had a tremendous influence on the understanding of the controls of eating in obese patients. The pathophysiology of severe and morbid obesity, as well as eating disorders, has been studied using anatomical, pharmacological, and behavioral approaches in animals. Animals have been rendered obese via: (1) stimulation and ablation of specific regions of the brain, (2) rearing genetically altered animal strains, and (3) feeding animals specific diets which induce excess or deficient energy intake. Each of these animal models can produce hypo- or hyperphagia, depending on the manipulation.

3.5.2 Historical perspective

Experimental studies of obesity based on anatomical aterations in the brain began in the first half of the 20th century after clinicians described lesions of the hypophysio-hypothalamic region in obese human patients. Hypothalamic obesity was first described by Mohr (1840) in a woman who became remarkably obese within the year preceding her death. At autopsy, a tumor was found in the hypothalamic area, large enough to deform the base of her brain. Fröhlich (1902) subsequently described 9 more similar cases in humans. Originally, Fröhlich proposed that the damaged area responsible for the obesity was the hypophysis, but there was considerable debate about this theory, and through more rigorous experiments in animals (Erdheim, 1904; Aschner, 1912), the actual site in the brain responsible for the subsequent obesity was located within the hypothalamic area of the brain. Brobeck *et al.* (1943) subsequently discovered that the cause of the obesity induced by these hypothalamic lesions was hyperphagia.

These findings led to maps of areas within the hypothalamus in animals involved in the control of food intake. These sites were originally identified in the 1950s using electrical stimulation through electrodes implanted in parts of the brain. Early experiments with rats demonstrated that electrolytic lesions of the ventromedial hypothalamus (VMH) caused overeating and an increase in weight to a new obese level whereas lesions in the lateral hypothalamus (LH) caused hypophagia and a reduction in body weight (Brobeck *et al.*, 1943) (Anand and Brobeck, 1951). The new weight attained after lesioning was defended in both VMH- and LH-lesioned animals. Initially it was suggested that the VMH contained a 'satiety center' that inhibited activity of a 'feeding center' in the LH (Stellar, 1954). Subsequent studies have been more specific in determining what areas of the hypothalamus control food intake (Sclafani and Kirchgessner, 1986). Specific knife cuts have shown no direct neural connection between the LH and VMH (Sclafani, 1984) and that VMH obesity probably results from damage to the ventral noradrenergic bundle (Ricardo, 1983). The paraventricular nucleus (PVN) is apparently the area within the hypothalamus that contains neurons that collectively play the most direct role in the control of feeding in both humans and rats (Leibowitz, 1990).

Investigation of the metabolic status of lesioned animals led some investigators to conclude that hyperphagia was not the primary cause of obesity in these models (Bray and York, 1979; Le Magnen, 1983). They cited studies showing pre-obese post-lesioned animals had lower metabolic rates, excessive insulin production, aggressive behavior, finickiness in eating, and decreased gastric motility. On the other hand, Becker and Kissileff (1974) showed that the overeating associated with hypothalamic lesions in rats occurred immediately after the animals recovered from anesthesia, and they concluded that the hyperphagia was a direct effect of the lesion rather than a secondary effect caused by metabolic alteration.

3.5.3 Neurotransmitters

The electrolytic lesions that induce the LH and VMH syndromes were gross disruptions of neural tissue and were followed by the use of more specific lesions. In the rat, chemical lesions that either destroyed only the cell bodies or axons that pass through each of these regions showed that both the destruction of neurons in these areas and the disruption of communication between other brain areas contributed to the disturbances in eating behavior observed (Stellar, 1990). The fibers of passage destroyed in the early lesions studies provide the hypothalamus and ventral forebrain areas involved in reward with classical neurotransmitters from the hindbrain as well as reciprocal connections. These neurotransmitters include the catecholamines dopamine (DA), epinephrine (EPI), and norepinephrine (NE) and the indolamine serotonin (5-hydroxytryptamine, 5-HT). This led to speculation that these transmitters were involved in the control of food intake. For example, the destruction of fibers of passage through the LH produce nearly the same constellation of effects in rats as total LH destruction (Ungerstedt, 1971) and produces a reduction in the levels of dopamine in the ventral forebrain. The source of these fibers is the ventral tegmental area, an area with a high concentration of dopamine-producing neurons (see Hoebel *et al.*, 1982).

The discovery that the injection of dopamine and epinephrine into the hypothalamus of rats suppresses feeding (Leibowitz and Stanley, 1986) and that norepinephrine (particularly in the paraventricular nucleus, PVN) enhances food intake (Leibowitz and Rossakis, 1978) led to the theory that an opponent neurochemical system was responsible for the onset and end of feeding behavior. There are a large number of NE-containing axon terminals in the PVN and its release has been found to initiate food intake (released by drugs that cause the release of NE from axon terminals) (see Leibowitz and Stanley, 1986). The source of this NE is the hindbrain nuclei that are the primary source of brain NE, in particular the locus coeruleus (see Hoebel *et al.*, 1982). Serotonin and NE are thought to act antagonistically in the medial hypothalamic nucleus to influence food intake (Leibowitz, 1990). In rats, serotonin and its precursors were shown to have an inhibitory effect on food intake when injected into the hypothalamus (Leibowitz and Shor-Posner, 1986) and this is thought to be the mechanism of action for drugs used in the treatment of obesity in humans that are known to act on serotonin receptors (e.g. fenfluramine). Neurochemical treatments that deplete serotonin levels have been found to increase food intake but surgical destruction of serotonin pathways from the midbrain do not (see Sclafani and Kirchgessner, 1986 for review). Microdialysis studies in rats, where the presence of specific transmitters and their metabolites can be directly analyzed in freely behaving rats, have shown a circadian rhythm in the levels of NE in the PVN that appears to follow food intake periodicity: PVN NE levels increase dramatically at the time of dark onset, the time when rats eat most of their food (Stanley *et al.*, 1989). Additionally, it has been shown that there is an increase in DA and serotonin, the feeding-inhibitory neurotransmitters, in the VMH/PVN area during a meal which indicates a possible endogenous role of these substances in satiety (Orosco and Nicolaidis, 1992).

Classical neurotransmitters are not the only chemicals implicated as part of this feeding circuit. Leibowitz (1994), Schwartz *et al.* (1992) and Clark *et al.* (1984) have all suggested that neuropeptide Y, a neuropeptide synthesized and secreted by cells originating in the arcuate nucleus and projecting into the paraventricular nucleus of the hypothalamus, elicits food intake in a variety of situations. When NPY was injected directly into the paraventricular nucleus (PVN) (Stanley and Leibowitz, 1984) or intracerebroventricularly into animals, food intake was immediately and substantially increased (Clark *et al.*, 1984). If indeed NPY is an endogenous agent inducing feeding, one would expect alterations in activity of this neuropeptide during changes in feeding behavior associated with obesity, eating disorders, and disease states that affect feeding. The serotonin agonist fenfluramine, one of the most effective agents used in the treatment of obesity (Leibowitz, 1990), reduces the levels of hypothalamic NPY in the rodent hypothalamus (Dube *et al.*, 1992), suggesting that serotonin could inhibit NPY release in the hypothalamus (Dryden *et al.*, 1994). Williams *et al.* (1988) have demonstrated that NPY levels in the hypothalamus increased in rats when they were induced into a diabetic state. Furthermore, obese strains of rodents show a greater expression of hypothalamic NPY than lean strains (Sahu *et al.*, 1992). NPYs' actions may be mediated by a number of peripheral signals of nutritional status. NPY levels become elevated when animals are food deprived, and diminish when free-feeding is restored (Kalra and Kalra, 1990).

Central insulin levels are correlated with the expression of the NPY gene (Schwartz *et al.*, 1992). Diabetic animals, which do not produce insulin, produce more hypothalamic NPY than do control animals (Schwartz *et al.*, 1991). Food deprived animals also produce more NPY than nondeprived animals. In other words, NPY biosynthesis is stimulated by hypoinsulinemic states (food deprivation or diabetes). This raises the possibility that insulin acts as a central signal of nutritional status in the hypothalamus to modulate the expression of NPY (Schwartz *et al.*, 1992).

Central insulin action is defective in the genetically obese fa/fa Zucker rat (Ikeda *et al.*, 1986). These animals do not reduce food intake or body weight in response to intracerebroventricular (icv) insulin administration. The absence of a satiety effect of insulin in the obese Zucker rat may reflect a failure of insulin to inhibit hypothalamic NPY biosynthesis in the arcuate nucleus neurons. This hypothesis is consistent with the finding that the fa/fa rat over-expresses NPY under free-feeding conditions, and NPY expression is unresponsive to the nutritional status (fed *vs.* fasted) of these animals (Sanacora *et al.*, 1990).

The possibility that the neurotransmitter serotonin may modulate the secretion of NPY has been suggested since serotonin neurons innervate NPY-containing neurons in the arcuate nucleus (Guy *et al.*, 1988) and administration of d-fenfluramine, the serotonin agonist, attenuated NPY induced feeding (Bendotti *et al.*, 1987).

Thus far, little is known about the relevance of NPY to obesity and eating disorders in humans. Human NPY is structurally identical to the rat peptide and is found at high concentrations in the hypothalamic regions analogous to those of the rat (Sanacora *et al.*, 1990). One post-mortem study has suggested that hypothalamic NPY levels were increased in patients who died of respiratory failure but potential changes in obesity have not yet been investigated (Corder *et al.*, 1990). Indirect evidence that NPY in the brain may be altered by changes in nutritional status comes from measurements of cerebrospinal fluid (CSF) levels of NPY in women with anorexia nervosa and bulimia nervosa (Kaye *et al.*, 1990). While severely underweight, their CSF NPY levels were significantly higher than controls', but reverted to normal levels after their body weights reached normal levels. One study of obese subjects did not find any difference in the levels of CSF NPY compared with normal weight subjects (Brunani *et al.*, 1995).

Though there are still some gaps in our knowledge of the neuroanatomical and neurochemical circuitry underlying the control of food intake, what is becoming clear, from this work with rats, is that cell groups of the hypothalamus are part of a circuit including ventral forebrain areas and the sources of monoamine neurotransmitters in the hindbrain and that this system is affected by pharmacological agents which influence food intake in humans. Since many of these anatomical regions in the rat have homologous structures in the human brain (Swaab *et al.*, 1993) it is tempting to speculate that the human may have functional systems similar to those in the rat.

3.5.4 Neurochemical implications for human eating disorders

The involvement of serotonin in the control of feeding has been documented for over 20 years (Angel, 1990; Blundell, 1977; Leibowitz, 1990). Evidence that serotonin plays a role in appetite has been elicited from animals and humans (Leibowitz, 1990). As has already been stated, serotonin is believed to mediate the actions of other neurotransmitters which control food intake. These drugs are part of a class of amine neurotransmitters which reduce appetite and body weight but cause considerable side effects which warrant against their use (Silverstone, 1975). Recently, specific receptor subtypes have been identified for serotonin which may be involved specifically in the control of feeding and not in the other roles that serotonin plays in the brain. However, there is considerable heterogeneity across species concerning the array of receptor subtypes in the brain for serotonin (Leibowitz, 1990). For example, some subtypes found to be directly linked to ingestive behavior in rodents are not present in the human brain.

3.5.5 Eating disturbances in human obesity—application of the animal models

Studies of differences in the feeding behavior of obese and nonobese animals parallel studies of the eating behavior of obese and nonobese humans. Spitzer and Rodin (1981) have reviewed dozens of studies in which the eating behavior of overweight individuals was compared with that of normal weight individuals. In their review, which includes field observations, laboratory eating, dietary record keeping, and manipulations expected to alter eating behavior, they conclude that most of the research on eating behavior of obese humans has been equivocal; i.e. that there are as many studies showing no difference in food intake in overweight subjects as there are studies showing a difference compared with normal weight subjects.

However, in this 'meta-analysis' style of review, a myriad of experimental probes was lumped together to 'tally' which studies show obese subjects eating differently, and which studies show obese subjects eating the same amounts as controls. This over-simplification may have masked some important findings in the earlier research. Across all types of experiments, the palatability of a given meal influenced its consumption more in overweight than in normal weight individuals (Spitzer and Rodin, 1981). Some obese humans tended to show less 'slowing down' toward the end of their meals than did normal weight individuals. This finding was more acutely explored by the work of Kissileff and Thornton (1982) when the microstructure of a meal was studied in humans (see section 3.2.4.3). Recently, the microstructure of food intake was also studied in patients with bulimia nervosa. In that work (LaChaussée *et al.*, 1992; Hadigan *et al.*, 1989) patients with bulimia did not slow down (decelerate) towards the end of their meals, as did control subjects.

3.5.6 Genetic obesity

Obese animal strains such as the Zucker rat and ob/ob mouse (see section 3.3.6) have led to research on the heritability of obesity in humans. Obesity has been shown to cluster in families (Bouchard *et al.*, 1990) and is more closely correlated between mono- than dizygotic twins, although Bouchard (1992) reports that only 25% of the variance in body fat phenotype is heritable. In monozygotic twins reared in separate environments, body weights between twin pairs with identical genetic make-up are more similar than twins with heterogeneous genetic make-up. This observation has advanced research along two lines: (1) exploration of the animal and human genome, in an effort to identify the gene(s) responsible for the obese phenotype, and (2) isolation of substances whose expression may be governed by a gene related to obesity (*cf.* section 3.3.6).

Unlike other inherited diseases such as cystic fibrosis, obesity is a continuous quantitative trait to which a number of genes undoubtedly contribute (*cf.* Herman, this book). These genes probably have effects on ingestive behavior, metabolic energy efficiency, and peripheral insulin metabolism. Since obesity appears to be a polygenic trait, use of an animal model can simplify the isolation of these genes. The Zucker fatty rat mutation is probably a homologue of the mouse db mutation (Leibel *et al.*, 1993). From mapping studies of the mouse ob and db mutations, Leibel and colleagues have found that there is a homologous region of the human chromosome mapping this trait. It may soon be possible to determine susceptibility to some forms of obesity in humans by DNA analysis.

Current research is aimed at determining exactly how genes cause the development of obesity in these animals. One theory is that the brain neuropeptide, NPY, which has been demonstrated to increase food intake in animal models, may be over-expressed by this gene. Obese animal models have now been shown to have increased levels of this peptide in the hypothalamus (Sahu *et al.*, 1992). The effects of NPY have not yet been explored in humans, but this neurotransmitter is likely to play a role in some of the overeating in human, as well as animal, obesity (*cf.* section 3.3.6).

3.5.7 Dietary manipulations inducing obesity

Dietary-induced obesity is another animal model used to investigate obesity. Dietary-induced obesity is usually associated with overeating and overweight. Sclafani (1976) and colleagues have perfected a technique whereby normal weight animals become hyperphagic when served 'supermarket diets' made up of foods typically consumed by humans. His work has stressed the role of a diet's palatability in the generation and maintenance of hyperphagic obesity. Palatability can affect food consumption by affecting food choice. Animals given a choice of foods may select a particular food because of its preferred flavor but overeat that food because of its postingestive effects. High-fat diets, for example, have long been used to produce obesity in laboratory animals (Mickelsen *et al.*, 1955). However, the cafeteria style diet pioneered by Sclafani has been more effective in promoting obesity in animals than single-item diets with a high fat content.

3.5.8 Anorexia nervosa—the use of animal models

Animal models (*cf.* Herman, this book) can mimic the hypophagia and hyperactivity that are both cardinal symptoms of anorexia nervosa (AN), and these models have helped to further our understanding of, and treatment of, this disease. For example, Routtenberg (1967), found that rats starved themselves when given unlimited access to running wheels and were only allowed to feed one hour per day. The rats chose to use the running wheel even when the food was present, and continued to loose weight until the experimenters had to intervene to prevent the animals from starving to death. More recently, Rieg *et al.* (1994) have extended this work by administering various pharmaceutical agents to probe some of the possible mechanisms involved in the pathogenesis of AN.

Marrazzi and Luby (1986) have speculated that the starvation and hyperactivity in AN may stimulate some reward mechanisms in the brain, such as the release of the opiate-like substance, β-endorphin. Self-starving, overactive rodents had higher circulating β-endorphin levels than controls which were fed *ad libitum* (Aravich *et al.*, 1993). However, agents such as naloxone, that block opiate receptors, do not abolish the syndrome of starvation and hyperactivity (Aravich *et al.*, 1993). Therefore, the issue of auto-addiction as a mechanism underlying AN still remains controversial.

On a more promising note, Aravich *et al.* (1993) have also been using a food-restricted exercising rat model to determine whether serotonin abnormalities may be involved in the disorder. Serotonin has been implicated in the pathogenesis of AN because of its intrinsic role in the control of food intake, and because of AN's similarity to obsessive-compulsive disorder which has been successfully treated in humans by administration of serotonin agonists. However, the potential benefit of using a serotonin agonist in AN seems paradoxical, because these agents suppress food intake in overweight humans and animals. Aravich *et al.* (1993) found that the serotonin agonist d-fenfluramine potentiated the weight-loss and hyperactivity in their animal model of AN when it was administered once the syndrome had begun. However, these authors demonstrated that d-fenfluramine had a positive effect on reversing the syndrome, as long as the animals were made tolerant to the intake-suppressing effects of this agent.

3.5.9 Animal models of bulimia nervosa

Bulimia nervosa (BN) is characterized by episodic binge eating, which is defined as eating unusually large quantities of food in a discrete time period (American Psychiatric Association, 1987), followed by some form of purging of the food. However, binge eating also occurs in AN and in obesity. In fact, Binswanger (1944) and Stunkard *et al.* (1955) first reported binge eating as a characteristic of obese patients. Dieting is known to precede BN (Polivy and Herman, 1985; Striegel-Moore *et al.*, 1986). The motivation for dieting, whether it be cultural, such as the desire to mimic ideals of beauty, or symbolic, such as conscientious objectors and hunger strikers, can lead to BN. Keys *et al.* (1950) took a group of normal weight men and restricted their food intake to the extent that they lost 26% of their initial weight. When *ad libitum* food was reinstated, the men frequently took excessively large meals.

Given the known effect of dieting on the development of binge eating, Coscina and Dixon (1983) took normal weight rats and deprived them of food for four days and then observed their response to refeeding. When the rats were given palatable, high-fat foods, they gained weight beyond their pre-dieting level, and importantly, this gain was due solely to hyperphagia, and not a change in their metabolic status. These results suggest that repeated food restriction may in fact contribute to the development of BN.

The sham-fed rat is another animal model which bears some of the features of BN (Van Vort, 1988). Sham-fed animals have a gastric fistula in their stomach which allows ingested foods to drain out of the stomach, thus mimicking the purging of human bulimics. Sham-fed rats also engage in recurrent episodes of binge eating, because they receive little post-ingestive feedback from the foods. Lack of serotonin has been implicated as a possible cause of the abnormal satiety experienced by patients with BN. For example, Robinson *et al.* (1985) demonstrated that increasing central serotonergic activity by the administration of fenfluramine decreased food intake by bulimics. Recently, serotonin-modulating drugs, such as fluoxetine, have successfully treated many patients with BN (Freeman *et al.*, 1988).

The paradigm of studying first the pattern of food intake followed by exploration of the underlying mechanism for that pattern in an eating disturbed animal model such as the ventromedial hypothalamic lesioned, or Zucker fatty rat, was partly behind the strategy adopted by Kissileff and colleagues for a series of studies in patients with bulimia nervosa (Kissileff *et al.*, 1986; Walsh *et al.*, 1989; Hadigan *et al.*, 1989, 1992; LaChaussée *et al.*, 1992) and later obese patients with binge eating disorder (Guss *et al.*, 1994). In these studies, the behavior was first characterized and compared to that of subjects without eating disorders (Kissileff *et al.*, 1986; Walsh *et al.*, 1989). Following that, various manipulations (Hadigan *et al.*, 1992) and measurements were made to determine the locus of the disturbance. The current working hypothesis is that the excessive eating is attributable to a disturbance in the experience of satiety during the middle to three quarters of the meal (Kissileff *et al.*, 1996).

3.6 Conclusions and limitations

Animal research has provided theories, methods, and facts that have driven our knowledge of eating behavior and control of food intake in humans to its present state. It now appears possible to develop an integrated approach to the roles of social, cognitive, environmental and physiological factors in the control of this behavior. Rather than attempting to fractionate them as independent controls, it appears that a learning paradigm in which cues connected with the acts of eating are associated with their physiological after effects may sustain both adaptive and maladaptive habits.

Despite the many parallels between findings in animals and humans, ultimately animal research cannot tell us everything about human eating, because species differ in both their physiological and anatomical constitutions and their ecological niches, all of which affect their food habits. Although animal models cannot completely incorporate all of the complexities of human eating disorders, they are still important for learning more about eating disorder syndromes and their possible treatments.

Within the past two decades, we have already seen major advances in our understanding and treatment of obesity, AN, and BN. Continued use of animal probes will advance our knowledge even further until these disorders can hopefully be targeted and treated early and effectively.

The questions for the next generation of investigators of human eating behavior are: 1) How are these physiological controls integrated with cognitive, social and environmental controls? 2) Does the physiological imperative to maintain body weight, for example, override the individual's ability to choose where, when, what, and how much to eat? 3) Alternatively, does the individual's choice to be obese or starve override physiological signals to reduce or increase food intake? 4) And overarching both of these prior questions, what will be the role of genetic factors in controlling both the physiological autonomic and the cognitive volitional responses to food?

Acknowledgements

We thank Tim Wentzlaff and Julie Carretta for help in preparing the bibliography and Joseph R. Vasselli for constructive comments on the manuscript. Work on this project was partially supported by the New York Obesity Research Center grant DK-26687, and NIMH grant MH42206 and NIH grant DK-36507.

References

Adolph, E.F. (1947) Urges to eat and drink in rats. *American Journal of Physiology*, **151**, 110–125.

Adrian, T.E., Ferri, G.-L., Bacarese-Hamilton, A.J., Fuessl, H.S., Polak, J.M. and Bloom, S.R. (1985) Human distribution and release of a putative new gut hormone, peptide YY. *Gastroenterology*, **89**, 1070–1077.

Ainslie, G.W. (1974) Impulse control in pigeons. *Journal of the Experimental Analaysis of Behavior*, **21**, 485–489.

Akabayashi, A., Koenig, J.I., Watanabe, Y., Alexander, J.T. and Leibowitz, S.F. (1994) Galanin-containing neurons in the paraventricular nucleus: a neurochemical marker for fat ingestion and body weight gain. *Neurobiology*, **91**, 10375–10379.

American Psychiatric Association (1987) *Diagnostic and Statistical Manual of Mental Disorders, 3rd ed, Revised (DSM-IIIR)*, American Psychiatric Association, Washington, DC.

Anand, B.K. and Brobeck, J.R. (1951) Localization of a 'feeding center' in the hypothalamus of the rat. *Proceedings of the Society for Experimental Biology and Medicine*, **77**, 323–324.

Angel, I. (1990) Central receptors and recognition sites mediating the effects of monoamines and anorectic drugs on feeding behavior. *Clinical Neuropharmacology*, **13**, 361–391.

Antin, J., Gibbs, J. and Smith, G.P. (1978) Cholecystokinin interacts with pregastric food stimulation to elicit satiety in the rat. *Physiology and Behavior*, **20**, 67–70.

Aravich, P.F., Rieg, T.S., Ahmed, I. and Lauterio, T.J. (1993) Fluoxetine induces vasopressin and oxytocin abnormalities in food-restricted rats given voluntary exercise: relationship to anorexia nervosa. *Brain Research*, **612** (1–2), 180–189.

Aschner, B. (1912) Über die Funktion der Hypophyse. *Pflügers Archiv für die Gesamte Physiologie des Menschen und der Tiere*, **146**, 1–146.

Baile, C.A. (1974) Data acquisition system (DAS) for studying feeding and related behavioral and physiological responses of sheep and cattle. *Federation Proceedings*, **33**, 364 (abstract).
Bard, P. (1928) A diencephalic mechanism for the expression of rage with special reference to the sympathetic nervous system. *American Journal of Physiology*, **84**, 490–515.
Barry, W.S. and Bray, G.A. (1969) Plasma triglycerides in genetically obese rats. *Metabolism*, **18**, 833–839.
Basiotis, P.P., Welsh, S.O., Cronin, F.J., Kelsay, J.L. and Mertz, W. (1987) Number of days of food intake records to estimate individual and group nutrient intakes with defined confidence. *Journal of Nutrition*, **117**, 1638–1641.
Beaudoin, R. and Mayer, J. (1953) Food intakes of obese and non-obese women. *Journal of the American Dietetic Association*, **29**, 29–33.
Becker, E.E. and Kissileff, H.R. (1974) Inhibitory controls of feeding by the ventromedial hypothalamus. *American Journal of Physiology*, **226** (2), 383–396.
Bellisle, F. (1989) Quantifying palatability in humans. *Annals of the New York Academy of Sciences*, **575**, 363–375.
Bellisle, F. and Le Magnen, J. (1980) The analysis of human feeding patterns: the Edogram. *Appetite*, **1**, 141–150.
Bellisle, F. and Le Magnen, J. (1981) The structure of meals in humans: Eating and drinking patterns in lean and obese subjects. *Physiology and Behavior*, **27**, 649–658.
Bellisle, F., Lucas, F., Amrani, R. and Le Magnen, J. (1984) Deprivation, palatability and the micro-structure of meals in human subjects. *Appetite*, **5**, 85–94.
Bellows, R.T. and Van Wagenen, W.P. (1938) The relationship of polydipsia and polyuria in diabetes insipidus. *Journal of Nervous Mental Disease*, **88**, 417–473.
Bendotti, C., Garattini, S. and Samanin, R. (1987) Eating caused by neuropeptide Y injection in the paraventricular hypothalamus: response to (+)-amphetamine in rats. *Journal of Pharmacy and Pharmacology*, **39**(11), 900–903.
Berkun, M.M., Kessen, M.L. and Miller, M.E. (1952) Hunger-reducing effects of food by stomach fistula versus food by mouth measured by a consummatory response. *Journal of Comparative Physiological Psychology*, **45**, 550–554.
Bernstein, I.L., Zimmerman, J.C., Czeisler, C.A. and Weitzman, E.D. (1981) Meal patterns in 'free-running' humans. *Physiology and Behavior*, **27**, 621–623.
Binswanger, L. (1944) Der Fall Ellen West. *Schweizer Archiv Fuer Neurologie und Psychiatrie*, **54**, 69–71.
Birch, L.L. and Deysher, M. (1985) Conditioned and unconditioned caloric compensation: evidence for self-regulation of food intake in young children. *Learning and Motivation*, **16**, 341–355.
Birch, L.L., McPhee, L., Steinberg, L. and Sullivan, S. (1990) Conditioned flavor preferences in young children. *Physiology and Behavior*, **47**, 501–505.
Bjorntorp, P. and Yang, M.U. (1982) Refeeding after fasting in the rat: effects on body composition and food efficiency. *American Journal of Clinical Nutrition*, **36**, 444–449.
Bjorntorp, P., Destrom, S., Kral, J.G., Lundholm, K., Presta, E., Walks, D. and Yang, M.U. (1982) Refeeding after fasting in the rat: energy substrate fluxes and replenishment of energy stores. *American Journal of Clinical Nutrition*, **36**, 450–456.
Block, G., Dresser, C.M., Hartman, A.M. and Carroll, M. D. (1985) Nutrient sources in the American diet: Quantitative data from the NHANES II survey. *American Journal of Epidemiology*, **122**, 27–40.
Blundell, J.E. (1977) Is there a role for serotonin (5-hydroxytryptamine) in feeding? *International Journal of Obesity*, **1**, 15–42.
Bobroff, E.P. and Kissileff, H.R. (1986) Effects of change in palatability on food intake and the cumulative food intake curve in man. *Appetite*, **6**, 85–96.
Bolles, R.C. (1975) *Theory of Motivation*, 2nd edn, Harper & Row, New York.
Booth, D.A. (1972) Conditioned satiety in the rat. *Journal of Comparative and Physiological Psychology*, **81**, 457–471.
Booth, D.A. (1976) Approaches to feeding control, in *Appetite and Food Intake Report of the Dahlem Workshop on Appetite and Food Intake Berlin 1975 December 8 to 12* (ed. Silverstone, T.), Abakon Verlagsgesellschaft, Berlin, pp. 414–478.
Booth, D.A. (1978a) Prediction of feeding behavior from energy flows in the rat, in *Hunger*

Models: Computable Theory of Feeding Control (ed. D.A. Booth), Academic Press, New York, pp. 227–278.
Booth, D.A. (1978b) Acquired behavior controlling energy intake and output, in *The Psychiatric Clinics of North America*, **1**(3), 545–579.
Booth, D.A. (1989a) The effect of dietary starches and sugars on satiety and on mental state and performance, in *Dietary Starches and Sugars in Man: A Comparison* (ed. J. Dobbing), Springer-Verlag, New York, pp. 225–249.
Booth, D.A. (1989b) Mood- and nutrient-conditioned appetites—Cultural and physiological bases for eating disorders. *Annals of the New York Academy of Sciences*, **575**, 122–135.
Booth, D.A. (1994) *Psychology of Nutrition*, Taylor and Francis, London.
Booth, D.A. and Weststrate, J.A. (1994) Concepts and methods in the psychobiology of ingestion, in *Food Intake and Energy Expenditure* (eds M.S. Westerterp-Plantenga, E.W.H.M. Fredrix and A.B. Steffens), CRC Press, Boca Raton, pp. 31–46.
Booth, D.A., Lovett, D. and McSherry, G.M. (1972) Postingestive modulation of the sweetness gradient in the rat. *Journal of Comparative and Physiological Psychology*, **78**, 485–512.
Booth, D.A., Lee, M. and McAleavy, C. (1976) Acquired sensory control of satiation in man. *British Journal of Psychology*, **2**, 137–147.
Booth, D.A., Mather, P. and Fuller, J. (1982) Starch content of ordinary foods associatively conditions human appetite and satiation, indexed by intake and eating pleasantness of starch-paired flavors. *Appetite*, **3**, 163–184.
Booth, D.A., Gibson, E.L. and Baker, B.J. (1986) Behavioral dissection of the intake and dietary selection effects of injection of fenfluramine, amphetamine or PVN norepinephrine. *Society For Neuroscience Abstracts*, **15**, 593 (Abstract).
Bouchard, C. (1992) Genetics and adaptation to overfeeding and negative energy balance, in *The Biology of Feast and Famine: Relevance to Eating Disorders* (eds G.H. Anderson and S.H. Kennedy), Academic Press, Inc, New York, pp. 183–193.
Bouchard, C., Tremblay, A., Depres, J., Nadeau, A., Lupien, P.J., Theriault, G., Dussault, J., Moorjani, S., Pinault, S. and Fournier, G. (1990) The response to long-term overfeeding in identical twins. *New England Journal of Medicine*, **322**, 1477–1482.
Bray, G.A. (1994) Appetite Control in Adults, in *Appetite and Body Weight Regulation: Sugar, Fat, and Macronutrient Substitutes* (eds J.D. Fernstrom and G.D. Miller), CRC Press, Boca Raton, pp. 17–34.
Bray, G.A. and York, D.A. (1979) Hypothalamic and genetic obesity in experimental animals; an autonomic and endocrine hypothesis. *Physiological Reviews*, **59**, 719–809.
Brobeck, J.R. (1955) Neural regulation of food intake. *Annals of the New York Academy of Sciences*, **63**, art. 1, 44–55.
Brobeck, J.R. (1965) Exchange, control and regulation, in *Physiological Controls and Regulations* (eds W.S. Yamamoto and J.R. Brobeck), W.B. Saunders, Philadelphia, pp. 1–13.
Brobeck, J.R., Tepperman, J. and Long, C.N.H. (1943) Experimental hypothalamic hyperphagia in the albino rat. *Yale Journal of Biology and Medicine*, **15**, 831–853.
Brunani, A., Invitti, C., Dubini, A., Piccoletti, R., Bendinelli, P., Maroni, P., Pezzoli, G., Ramella, G., Calogero, A. and Cavagnini, F. (1995) Cerebrospinal fluid and plasma concentrations of SRIH, beta-endorphin, CRH, NPY and CHRH in obese and normal weight subjects. *International Journal of Obesity*, **19**, 17–21.
Burke, B.S. (1947) The dietary history as a tool in research. *Journal of the American Dietetic Association*, **23**, 1041–1046.
Burley, V.J., Cotton, J.R., Weststrate, J.A. and Blundell, J.E. (1994) Effect on appetite of replacing natural fat with sucrose polyester in meals or snacks across one whole day, in *Obesity in Europe 1993* (eds H. Ditschuneit, F.A. Gries, H. Hauner, V. Schusdziarra and J.G. Wechsler), John Libbey & Company, Ltd., London, pp. 227–233.
Cabanac, M. (1971) Physiological role of pleasure. *Science*, **173**, 1103–1107.
Cabanac, M. (1989) Palatability of food and the ponderostat. *Annals of the New York Academy of Sciences*, **575**, 340–352.
Campbell, R., Hashim, S. and Van Itallie, T.B. (1971) Studies of food-intake regulation in man. Responses to variations in nutritive density in lean and obese subjects. *New England Journal of Medicine*, **285**, 1402–1407.
Campfield, L.A. and Smith, F.J. (1990) Systemic factors in the control of food intake: Evidence for patterns as signals, in *Handbook of Behavioral Neurobiology, Volume 10: Neurobiology*

of food and Fluid Intake (ed. E.M. Stricker), Plenum, New York, pp. 183–206.
Campfield, L.A., Smith, F.J. and Rosenbaum, M. (1992) Human hunger: Is there a role for blood glucose dynamics? *Appetite*, **18**, 244.
Campfield, L.A., Smith, F.J., Guisez, Y., Devos, R. and Burn, P. (1995) Recombinant mouse OB protein: Evidence of a peripheral signal linking adiposity and central neural networks. *Science*, **269**, 546–549.
Cannon, W.B. and Washburn, A.L. (1912) An explanation of hunger. *American Journal of Physiology*, **29**, 444–454.
Carretta, J.C., Kissileff, H.R., Pi-Sunyer, F.X., Oberman, L. and Geliebter, A. (1995) Gastric distention and CCK combine to reduce food intake in humans. *Society for Neuroscience Abstracts*, **21** (1), 7 (abstract).
Castonguay, T.W. (1981) Dietary dilution and intake in the cat. *Physiology and Behavior*, **27**, 547–549.
Chabert, M., Bonnefond, F.E. and Louis-Sylvestre, J. (1993) Conditioned satiation occurrence depends on the taste and nutrient. *International Journal of Obesity and Related Metabolic Disorders*, **17**, Suppl. 2, 62 (abstract).
Clark, J.T., Kalra, P.S., Crowley, W.R. and Kalra, S.P. (1984) Neuropeptide Y and human pancreatic polypeptide stimulate feeding behavior in rats. *Endocrinology*, **115**, 427–429.
Clifton, P. (1987) Analysis of feeding and drinking patterns, in *Techniques in the Behavioral Sciences, Volume 1: Feeding and Drinking* (eds F.M. Toates and N.E. Rowland), Elsevier, New York, pp. 19–35.
Cohn, C. and Joseph, D. (1962) Influence of body weight and body fat on appetite of 'normal' lean and obese rats. *Yale Journal of Biology and Medicine*, **34**, 598–607.
Collier, G. (1987) Operant methodologies for studying feeding and drinking, in *Techniques in the Behavioral and Neural Sciences, Volume 1: Feeding and Drinking* (eds F.M. Toates and N.E. Rowland), Elsevier, New York, pp. 37–75.
Collier, G. (1989) The economics of hunger, thirst, satiety, and regulation. *Annals of the New York Academy of Sciences*, **575**, 136–154.
Cooper, S.J. (1980) Benzodiazepines as appetite-enhancing compounds. *Appetite*, **1**, 7–19.
Cooper, S.J. (1987) Drugs and hormones: Their effects on ingestion, in *Techniques in the Behavioral and Neural Sciences, Volume 1: Feeding and Drinking* (eds F.M. Toates and N.E. Rowland), Elsevier, New York, pp. 231–269.
Corder, R., Pralong, P.F., Muller, A.F. and Gaillard, R.C. (1990) Regional distribution of neuropeptide Y-like immunoreactivity in human hypothalamus measured by immunoradiometric assay: possible influence of chronic respiratory failure on tissue levels. *Neuroendocrinology*, **51**, 23–30.
Coscina, D.V. and Dixon, L.M. (1983) Body weight regulation in anorexia nervosa: insights from an animal model, in *Anorexia Nervosa: Recent Developments in Research* (eds P.L. Darby, P.E. Garfinkel, D.M. Garner and D.V. Coscina), Alan R. Liss, Inc., New York, pp. 207–220.
Coscina, D.V. and Garfinkel, P.E. (1991) Animal models of eating disorders: a clinical perspective, in *Behavioural Models in Psychopharmacology: Theoretical, Industrial and Clinical Perspectives* (ed. P. Willner), Cambridge University Press, New York, pp. 237–250.
Cruce, J.A.F., Greenwood, M.R.C., Johnson, P.R. and Quartermain, D. (1974) Genetic versus hypothalamic obesity: studies of intake and dietary manipulations in rats. *Journal of Comparative and Physiological Psychology*, **87** (2), 295–301.
Davis, C.M. (1928) Self selection of diet by newly weaned infants. *American Journal of Diseases of Children*, **36**, 651–679.
Davis, C.M. (1939) Results of the self-selection of diets by young children. *Canadian Medical Association Journal*, **41**, 257–261.
Davis, J.D. (1989) The microstructure of ingestive behavior. *Annals of the New York Academy of Sciences*, **575**, 106–121.
Davis, J.D. and Levine, M.W. (1977) A model for the control of ingestion. *Psychological Review*, **84**, 379–412.
Davis, J.D. and Smith, G.P. (1990) Learning to sham feed: behavioral adjustments to loss of physiological postingestional stimuli. *American Journal of Physiology*, **259**, R1228–R1235.
de Castro, J.M. (1994) Methodology, correlational analysis and interpretation of diet diary records of the food and fluid intake of free-living humans. *Appetite*, **23** (2), 179–192.

de Castro, J.M. (1995) Social facilitation and inhibition of eating, in *Not Eating Enough*, National Academy Press, Washington, DC, pp. 373–392.
de Castro, J.M. and de Castro, E.S. (1989) Spontaneous meal patterns of humans: Influence of the presence of other people. *American Journal of Clinical Nutrition*, **50**, 237–247.
de Silva, P. and Rachman, S. (1987) Human food aversions: Nature and acquisition. *Behavior Research and Therapy*, **25**, 457–468.
Debas, H.T., Farooq, O. and Grossman, M.I. (1975) Inhibition of gastric emptying is a physiological action of cholecystokinin. *Gastroenterology*, **68**, 1211–1217.
Della-Fera, M.A. and Baile, C.A. (1981) Peptides with CCK-like activity administration intracranially elicit satiety in sheep. *Physiology and Betavior*, **26**, 979–983.
Deutsch, J.A., Moore, B.O. and Heinrichs, S.C. (1989) Unlearned specific appetite for proteins. *Physiology and Behavior*, **46**, 619–624.
Drewnowski, A. (1992) Nutritional perspectives on biobehavioral models of dietary change, in *Proceedings Promoting Dietary Change in Communities: Applying Existing Models of Dietary Change to Population-Based Interventions* (eds M.M. Henderson, D.J. Bowen and K.K. DeRoos), Fred Hutchinson Cancer Research Center, Seattle, WA, pp 96–112.
Drewnowski, A. (1994) Human preferences for sugar and fat, in *Appetite and Body Weight Regulation: Sugar, Fat, and Macronutrient Substitutes* (eds J.D. Fernstrom, and G.D. Miller), CRC Press, Inc., Boca Raton, pp. 137–147.
Drewnowski, A., Krahn, D.D., Demitrach, M.A., Nairn, K. and Gosnell, B.A. (1995) Naloxone, an opiate blocker, reduces the consumption of sweet high-fat foods in obese and lean female binge eaters. *American Journal of Clinical Nutrition*, **61** (6), 1206–1212.
Dryden, S., Frankish, H., Wang, Q. and Williams, G. (1994) Neuropeptide Y and energy balance: one way ahead for the treatment of obesity? *European Journal of Clinical Investigation*, **24** (5), 293–308.
Dube, M.G., Sahu, A., Phelps, C.P., Kalra, P.S. and Kalra, S.P. (1992) Effect of d-fenfluramine on neuropeptide Y concentration and realease in the paraventricular nucleus of food-deprived rats. *Brain Research Bulletin*, **29** (6), 865–869.
Eibl-Eibesfeldt, I. (1970) *Ethology the Biology of Behavior*, Holt, Rinehart and Winston, New York.
Elkins, R.L. (1980) A reconsideration of the relevance of recent animal studies for development of treatment procedures for alcoholics. *Drug & Alcohol Dependence*, **5**, 101–113.
Epstein, A.N. and Teitelbaum, P. (1962) Regulation of food intake in the absence of taste, smell, and other oropharyngeal sensations. *Journal of Comparative and Physiological Psychology*, **55** (5), 753–759.
Erdheim, J. (1904) Uber Hypophyseganggeschwulste und Hirncholesteatome. *S.-B. Akad. Wissencchaft Wien*, **113** Abt. III, 537–726.
Etscorn, F. and Stephens, R. (1973) Establishment of conditioned taste aversions with a 24-hour CS-US interval. *Physiological Psychology*, **1**, 251–253.
Evvard, J.M. (1915) Is the appetite of swine a reliable indication of physiological needs? *Proceedings of the Iowa Academy of Sciences*, **22**, 375–403.
Faust, I.M., Johnson, P.R. and Hirsch, J. (1977) Surgical removal of adipose tissue alters feeding behavior and development of obesity in rats. *Science*, **197**, 393–396.
Feifel, D. and Vaccarino, F.J. (1994) Growth hormone-regulatory peptides (GHRH and Somatostatin) and feeding: A model for the integration of central and peripheral function. *Neuroscience and Biobehavioral Reviews*, **18** (3), 421–433.
Fitzsimons, J.T. and Le Magnen, J. (1969) Eating as a regulatory control of drinking. *Journal of Comparative and Physiological Psychology*, **67**, 273–283.
Foltin, R.W. and Fischman, M.W. (1990) Effects of caloric manipulations on food intake in baboons. *Appetite*, **15** (2), 135–149.
Foltin, R.W. and Moran, T.H. (1989) Food intake in baboons: Effects of a long-acting cholecystokinin analog. *Appetite*, **12** (2), 145–152.
Foltin, R.W., Fischman, M.W., Emurian, C.S. and Rachlinski, J.J. (1988) Compensation for caloric dilution in humans given unrestricted access to food in a residential laboratory. *Appetite*, **10**, 13–24.
Fomon, S.J., Owen, G.M. and Thomas, L.N. (1964) Milk or formula volume ingested by infants fed ad libitum. *American Journal of Diseases of Children*, **108**, 601.
Fomon, S.J., Filer, L.J., Jr., Thomas, L.M., Rogers, R.R. and Proksch, A.M. (1969) Relation-

ship between formula concentration and rate of growth of normal infants. *Journal of Nutrition*, **98**, 241–254.
Freeman, C.P.L., Morris, J.E. and Cheshire, K.E. (1988) A double-blind controlled trial of fluoxetine versus placebo for bulimia nervosa. *Abstracts of the Third International Conference of Eating Disorders*, Abstract no. 129 (Abstract).
French, S.J., Bergin, A., Sepple, C.P., Read, N.W. and Rovati, L. (1994) The effects of loxiglumide on food intake in normal weight volunteers. *International Journal of Obesity and Related Metabolic Disorders*, **18** (11), 738–41.
Fried, M., Erlacher, U., Schwizer, W., Lochner, C., Koerfer, J., Beglinger, C., Jansen, J.B., Lamers, C.B., Harder, F., Bischof-Delaloye, A., Stalder, G.A. and Rovati, L. (1991) Role of cholecystokinin in the regulation of gastric emptying and pancreatic enzyme secretion in humans. *Gastroenterology*, **101**, 503–511.
Friedman, M.I. (1991) Metabolic control of calorie intake, in *Chemical Senses, Volume 4: Apppetite and Nutrition* (eds M.I. Friedman, M.G. Tordoff and M.R. Kare), Marcel Dekker, New York, pp. 19–38.
Frier, H.I. (1993) Design of new foods, in *Mechanisms of Taste Transduction* (eds S.A. Simon and S.D. Roper), CRC Press, Boca Raton, FL, pp 479–490.
Fröhlich, A. (1902) Dr. Alfred Fröhlich stellt einen Fall von Tumor der Hypophyse ohne Akromegalie vor. *Wiener Klinische Rundschau*, **15**, 883.
Fujimoto, K., Machidori, H., Iwakiri, R., Yamamoto, K., Fujisaki, J., Sakata, T. and Tso, P. (1993) Effect of intravenous administration of apolipoprotein A-IV on patterns of feeding, drinking and ambulatory activity of rats. *Brain Research*, **608**, 233–237.
Fuller, J.L. (1972) Genetic aspects of regulation of food intake, in *Advances in Psychosomatic Medicine, Volume: 7 Hunger and Satiety in Health and Disease* (ed. F. Reichsman), S. Karger, Basel, pp. 2–23.
Fuller, J.L. (1981) Genetics of eating behavior in animals, in *The Body Weight Regulatory System: Normal and Disturbed Mechanisms* (eds L.A. Cioffi, W.P.T. James and T.B. Van Itallie), Raven Press, New York, NY, pp. 197–204.
Galef, B.G. (1989) Enduring social enhancement of rats' preferences for the palatable and the piquant. *Appetite*, **13**, 81–92.
Galef, B.G. and Beck, M. (1990) Diet selection and poison avoidance by mammals individually and in social groups, in *Handbook of Behavioral Neurobiology, Volume 10: Neurobiology of Food and Fluid Intake* (ed. E.M. Stricker), Plenum, New York, pp. 329–349.
Galef, B.G. (1991) Social factors in diet selection and poison avoidance by Norway rats, in *Chemical Senses, Volume 4: Appetite and Nutrition* (eds M.I. Friedman, M.G. Tordoff and M.R. Kare), Marcel Dekker, Inc., New York, pp. 177–194.
Garcia, J., Kimeldorf, D.J. and Koelling, R.A. (1955) Conditioned aversion to saccharin resulting from exposure to gamma radiation. *Science*, **122**, 157–158.
Gates, R.J. and Lazarus, N.R. (1977) The ability of pancreatic polypeptides (APP and BPP) to return to normal the hyperglycemia, hyperinsulinemia, and weight gain of New Zealand obese mice. *Hormone Research*, **8**, 189–202.
Geliebter, A. (1988) Gastric distension and gastric capacity in relation to food intake in humans. *Physiology and Behavior*, **44**, 665–668.
Geraciotti, T.D. and Liddle, R.A. (1988) Impaired cholecystokinin secretion in bulimia nervosa. *New England Journal of Medicine*, **319**, 683–688.
Gibbs, J. and Smith, G.P. (1977) Cholecystokinin and satiety in rats and rhesus monkeys. *American Journal of Clinical Nutrition*, **30**, 758–761.
Gibbs, J. and Smith, G.P. (1988) The actions of bombesin-like peptides on food intake. *Annals of the New York Academy of Sciences*, **547**, 210–216.
Gibbs, J., Young, R.C. and Smith, G.P. (1973) Cholecystokinin elicits satiety in rats with open gastric fistulas. *Nature*, **245**, 323–325.
Gibbs, J., Fauser, D.J., Rowe, E.A., Rolls, B.J., Rolls, E.T. and Maddison, S.P. (1979) Bombesin suppresses feeding in rats. *Nature*, **282**, 208–210.
Gibson, E.L., Wainwright, C.J. and Booth, D.A. (1995) Disguised protein in lunch after low-protein breakfast conditions food-flavor preferences dependent on recent lack of protein intake. *Physiology and Behavior*, **58**, 363–371.
Gil, K.M., Sikei, B., Kvetan, V., Friedman, M.I. and Askenazi, J. (1990) Parenteral nutrition and oral intake: Effect of branched chain amino acids. *Nutrition*, **6**, 291–295.

Glaser, B., Floyd, J.C. and Vinik, A.I. (1983) Secretion of pancreatic polypeptide in man in response to beef ingestion is mediated in part by an extravagal cholinergic mechanism. *Metabolism*, **32**, 57–61.

Goodall, J. (1986) *The Chimpanzees of Gombe: Patterns of Behavior*, Harvard University Press, Cambridge, MA.

Gosnell, B.A., Levine, A.S. and Morley, J.E. (1983) The effects of aging on opioid modulation of feeding in rats. *Life Science*, **32**, 2793–2799.

Green, J. and Tapp, W.N. (1986) Feeding cycles in smokers, exsmokers and nonsmokers. *Physiology and Behavior*, **36**, 1059–1063.

Green, J., Pollak, C.P. and Smith, G.P. (1987) Meal size and intermeal interval in human subjects in time isolation. *Physiology and Behavior*, **41** (2), 141–147.

Greenberg, D., Torres, N.I., Smith, G.P. and Gibbs, J. (1989) The satiating effect of fats is attenuated by the cholecystokinin antoagonist lorglumide. *Annals of the New York Academy of Sciences*, **575**, 517–520.

Greenwood, M.R.C. and Vasselli, J.R. (1981) The effects of nitrogen and caloric restriction on adipose tissue, lean body mass, and food intake of genetically obese rats: The LPL hypothesis, in *Nutritional Factors: Modulating Effects on Metabolic Processes* (eds R.F. Beers, Jr. and E.G. Basset), Raven Press, New York, pp. 323–335.

Grill, H.J. and Berridge, K.C. (1985) Taste reactivity as a measure of the neural control of palatability. *Progress in Psychobiology and Physiological Psychology*, **11**, 1–61.

Grill, H.J., Spector, A.C., Schwartz, G.J., Kaplan, J.M. and Flynn, F.W. (1987) Evaluating taste effects on ingestive behavior, in *Feeding and Drinking* (eds F.M. Toates and N.E. Rowland), Elsevier, New York, pp. 151–188.

Grill, H.J., Roitman, M.F. and Kaplan, J.M. (1995) Food deprivation does not affect taste reactivity to concentrations of glucose equivalently. *Looking at Ingestive Consumption & Kinetics Symposium*, **33** (Abstract).

Grinker, J.A. (1981) Behavioral and metabolic factors in chidhood obesity, in *The Uncommon Child* (eds M. Lewis and L.A. Rosenblum), Plenum Press, New York, pp. 115–150.

Grossman, M.I. (1955) Integration of current views on the regulation of hunger and appetite. *Annals of the New York Academy of Sciences*, **63** Art 1, 76–91.

Grossman, S.P. (1960) Eating or drinking elicited by direct adrenergic or cholinergic stimulation of hypothalamus. *Science*, **132**, 301–302.

Grossman, S.P. (1962) Direct adrenergic and cholinergic stimulation of hypothalamic mechanisms. *American Journal of Physiology*, **202**, 872–882.

Gruen, R.K. and Greenwood, M.R.C. (1981) Adipose tissue lipoprotein lipase and glycerol release in fasted Zucker (fa/fa) rats. *American Journal of Physiology*, **241**, E76–E83.

Grundy, D., Bagaev, V. and Hillsley, K. (1995) Inhibition of gastric mechanoreceptor discharge by cholecystokinin in the rat. *American Journal of Physiology*, **268**, G355–G360.

Guss, J., Kissileff, H.R., Pierson, R. and Pi-Sunyer, F.X. (1988) The effects of CCK-8 infusion on intake and gastric emptying in nonobese women. *Society For Neuroscience Abstracts*, **14** (2), 1196 (abstract).

Guss, J.L., Kissileff, H.R., Walsh, B.T. and Devlin, M.J. (1994) Binge eating behavior in patients with eating disorders. *Obesity Research*, **2**, 335–363.

Gutzwiller, J.-P., Drewe, J., Hildebrand, P., Rossi, L., Lauper, J.Z. and Beglinger, C. (1994) Effect of intravenous human gastrin-releasing peptide on food intake in humans. *Gastroenterology*, **106**, 1168–1173.

Guy, J., Pelletier, G. and Bosler, O. (1988) Serotonin innervation of neuropeptide Y-containing neurons in the rat arcuate nucleus. *Neuroscience Letters*, **85**, 9–13.

Hadigan, C.M., Kissileff, H.R. and Walsh, B.T. (1989) Patterns of food selection during meals in women with bulimia. *American Journal of Clinical Nutrition*, **50**, 759–766.

Hadigan, C.M., Walsh, B.T., Kissileff, H.R., LaChaussée, J.L. and Devlin, M.J. (1992) Behavioral assessment of satiety in bulimia nervosa. *Appetite*, **18**, 233–242.

Hafez, E.S.E. and Schein, M.W. (1962) The behavior of cattle, in *The Behaviour of Domestic Animals* (ed. E.S.E. Hafez), Balliere, Tindall, and Callel, Limited, London, pp. 247–296.

Halaas, J.L., Gajiwala, K.S., Maffei, M., Cohen, S.L., Chait, B.T., Rabinowitz, D., Lallone, R.L., Burley, S.K. and Friedman, J.M. (1995) Weight-reducing effects of the plasma protein encoded by the obese gene. *Science*, **269**, 543–546.

Hamilton, C.L. (1967) Food and temperature, in *Handbook of Physiology, Section 6, Alimen-*

tary Canal, Vol. 1 (ed. C.F. Code), American Physiological Society, Washington, DC, pp. 303–317.
Hamilton, C.L. (1972) Long term control of food intake in the monkey. *Physiology and Behavior*, **9**, 1–6.
Hansen, B.C., Jen, K.-L.C. and Kripps, P. (1981) Regulation of food intake in monkeys: Response to caloric dilution. *Physiology and Behavior*, **26**, 479–486.
Harris, R.B.S. (1993) Factors influencing body weight regulation. *Digestive Diseases*, **11**, 133–145.
Hashim, S.A. and Van Itallie, T.B. (1964) An automatically monitored food dispensing apparatus for the study of food intake in man. *Federation Proceedings*, **23** (1), 82–84.
Heitmann, B.L., Lissner, L., Sorensen, T.I.A. and Bengtsson, C. (1995) Dietary fat intake and weight gain in women genetically predisposed for obesity. *American Journal of Clinical Nutrition*, **61**, 1213–1217.
Herman, C.P., Polivy, J. and Silver, R. (1979) Effects of an observer on eating behavior: The induction of 'sensible' eating. *J. Personality*, **47**, 85–99.
Hervey, G.R. (1959) The effects of lesions in the hypothalamus in parabiotic rats. *Journal of Physiology*, **145**, 336–352.
Hetherington, M. and Rolls, B.J. (1987) Methods of investigating human eating behavior, in *Techniques in the Behavioral and Neural Sciences, Volume 1: Feeding and Drinking* (eds N.E. Rowland and F.M. Toates), Elsevier, New York, pp. 77–109.
Hewson, G., Leighton, G.E., Hill, R.G. and Hughes, J. (1988) The cholecystokinin receptor antagonist L364,718 increases food intake in the rat by attenuation of the action of endogenous cholecystokinin. *British Journal of Pharmacology*, **93**, 79–84.
Hill, A.J., Leathwood, P.D. and Blundell, J.E. (1987) Short-term caloric compensation in man: The effects of high and low energy meals on ratings of motivation to eat, food preferences and food intake. *Human Nutrition*, **41** (A), 244–257.
Hill, J.H. and Stellar, E. (1951) An electronic drinkometer. *Science*, **114**, 43–44.
Hirsch, E. (1995) The effects of ration modifications on energy intake, body weight change, and food acceptance, in *Not Eating Enough*, National Academy Press, Washington, DC, pp. 151–174.
Hoebel, B.G., Hernandez, L., McLean, S., Stanley, B.G., Aulissi, E.F., Glimcher, P. and Margolin, D. (1982) Catecholamines, Enkephalin, and Neurotensin in feeding and reward, in *The Neural Basis of Feeding and Reward* (eds B.G. Hoebel and D. Novin), Haer Institute for Electrophysiological Research, Brunswick, ME, pp. 465–478.
Hoebel, B.G., Hernandez, L., Schwartz, D., Mark, G.P. and Hunter, G.A. (1989) Microdialysis studies of brain norepinephrine, serotonin, and dopamine release during ingestive behavior: Theoretical and clinical implictions. *Annals of the New York Academy of Sciences*, **575**, 171–193.
Hopman, W.P.M., Jansen, J.B.M.J. and Lamers, C.B.H.W. (1985) Comparative study of the effects of equal amounts of fat, protein, and starch on plasma cholecystokinin in man. *Scandinavian Journal of Gastroenterology*, **20**, 843–847.
Houpt, K.A. and Houpt, T.R. (1975) Effects of gastric loads and food deprivation on subsequent food intake in suckling rats. *Journal of Comparative and Physiological Psychology*, **88** (2), 764–772.
Hulshoff, T., De Graaf, C. and Westrate, J.A. (1993) The effects of preloads varying in physical state and fat content on satiety and energy intake. *Appetite*, **21**, 273–286.
Ikeda, H., West, D.B., Pustek, J.J., Figlwicz, D.P., Greenwood, M.R.C., Porte, D.J. and Woods, S.C. (1986) Intraventricular insulin reduces food intake and body weight of lean but not obese Zucker rats. *Appetite*, **7**, 381–386.
Janowitz, H.D. and Grossman, M.I. (1949) Some factors affecting food intake of normal dogs and dogs with esophagostomy and gastric fistula. *American Journal of Physiology*, **159**, 143–148.
Jordan, H.A. (1969) Voluntary intragastric feeding: Oral and gastric contributions to food intake and hunger in man. *Journal of Comparative and Physiological Psychology*, **68**, 498–506.
Jordan, H.A., Wieland, W.F., Zebley, S.P., Stellar, E. and Stunkard, A.J. (1966) Direct measurement of food intake in man: A method for the objective study of eating behavior. *Psychosomatic Medicine*, **28**, 836–842.

Kalra, S.P. and Kalra, P.S. (1990) Nuropeptide Y: A novel peptidergic signal for the control of feeding behavior. *Current Topics in Neuroendocrinology*, **10**, 192–217.
Kanarek, R.B. (1976) Energetics of meal patterns in rats. *Physiology and Behavior*, **17**, 395–399.
Kaye, W.H., Berrettini, W., Gwirtsman, H. and George, D.T. (1990) Altered cerebro-spinal fluid neuropeptide Y and peptide YY immuno-reactivity in anorexia and bulimia nervosa. *Archives of General Psychiatry*, **47**, 548–556.
Keesey, R.E. (1986) A set-point theory of obesity, in *Handbook of Eating Disorders* (eds K.D. Brownell and J.P. Foreyt), Basic Books, New York, pp. 63–87.
Kennedy, G.C. (1953) The role of depot fat in the hypothalamic control of food intake. *Proceedings of the Royal Society of London Series (B)*, **140**, 578–592.
Kern, D.L., McPhee, L., Fisher, J., Johnson, S. and Birch, L.L. (1993) The postingestive consequences of fat condition preferences for flavors associated with high dietary fat. *Physiology and Behavior*, **54**, 71–76.
Keys, A., Brozek, J., Henschel, A., Mickelsen, O. and Taylor, H.L. (1950) *The Biology of Human Starvation*, University of Minnesota Press, Minneapolis, MN.
Kim, J.Y. and Kissileff, H.R. (1996) The effect of social setting on response to a preloading manipulation in nonobese women and men. *Appetite* (in press).
Kissileff, H.R. (1969) Oropharyngeal control of prandial drinking. *Journal of Comparative and Physiological Psychology*, **67**, 284–300.
Kissileff, H.R. (1970) Free feeding in normal and 'recovered lateral' rats monitored by a pellet-detecting eatometer. *Physiology and Behavior*, **5**, 163–173.
Kissileff, H.R. (1984) Satiating efficiency and a strategy for conducting food loading experiments. *Neuroscience and Biobehavioral Reviews*, **8**, 129–135.
Kissileff, H.R. (1989) Is there an eating disorder in the obese? *Annals of the New York Academy of Sciences*, **575**, 410–419.
Kissileff, H.R. (1990) Some suggestions on dealing with palatability: Response to Ramirez. *Appetite*, **14**, 162–166.
Kissileff, H.R. (1991) Chance and necessity in ingestive behavior. *Appetite*, **17**, 1–22.
Kissileff, H.R. (1992) Where should human eating be studied and what should be measured? *Appetite*, **19**, 61–68.
Kissileff, H.R. and Thornton, J. (1982) Facilitation and inhibition in the cumulative food intake curve in man, in *Changing Concepts of the Nervous System* (eds A.J. Morrison and P. Strick), Academic Press, New York, pp. 585–607.
Kissileff, H.R. and Van Itallie, T.B. (1982) Physiology of the control of food intake. *Annual Review of Nutrition*, **2**, 371–418.
Kissileff, H.R., Klingsberg, G. and Van Itallie, T.B. (1980) Universal eating monitor for continuous recording of solid or liquid consumption in man. *American Journal of Physiology*, **238**, R14–R22.
Kissileff, H.R., Pi-Sunyer, F., Thornton, J. and Smith, G. (1981) C-terminal octapeptide of cholecystokinin decreases food intake in man. *American Journnmal of Clinical Nutrition*, **34**, 154–160.
Kissileff, H.R., Thornton, J. and Becker, E. (1982) A quadratic equation adequately describes the cumulative food intake curve in man. *Appetite*, **3**, 255–272.
Kissileff, H., Walsh, T., Kral, J. and Cassidy, S. (1986) Laboratory studies of eating behavior in women with bulimia. *Physiology and Behavior*, **38**, 563–570.
Kissileff, H.R., Pi-Sunyer, F.X., Segal, K., Meltzer, S. and Foelsch, P.A. (1990) Acute effects of exercise on food intake in obese and nonobese women. *American Journal of Clinical Nutrition*, **52**, 240–245.
Kissileff, H.R., Wentzlaff, T.H., Guss, J.L., Walsh, B.T., Devlin, M.J. and Thornton, J.C. (1996) A direct measure of satiety disturbance in patients with bulimia nervosa. *Physiology and Behavior* (in press).
Klosterhalfen, S. and Klosterhalfen, W. (1985) Conditioned taste aversion and traditional learning. *Psychological Research*, **47**, 71–94.
Krauchi, K. and Wirz-Justice, A. (1992) Seasonal patterns of nutrient intake in relation to mood, in *The Biology of Feast and Famine: Relevance to Eating Disorders* (eds G.H. Anderson and S.H. Kennedy), Academic Press, Inc., New York, pp. 157–182.
Kron, R.E., Stein, M. and Goddard, K.A. (1963) A method of measuring sucking behavior of newborn infants. *Psychosomatic Medicine*, **25**, 181–191.

Kuvshinoff, B.W., Rudnicki, M. and McFadden, D. (1991) The effect of SMS 201-995 on meal and CCK-stimulated peptide YY release. *Journal of Surgical Research*, **50**, 425–429.

LaChaussée, J.L., Kissileff, H.R., Walsh, B.T. and Hadigan, C.M. (1992) The single item meal as a measure of binge-eating behavior in patients with bulimia nervosa. *Physiology and Behavior*, **51**, 593–600.

Lashley, K.S. (1938) An experimental analysis of instinctive behavior. *Psychological Review*, **45**, 445–771.

Lassman, V., Vague, P., Vialettes, B. and Simon, M.C. (1980) Low plasma levels of pancreatic polypeptide in obesity. *Diabetes*, **29**, 428–430.

Le Magnen, J. (1966) La Péridodicité spontanée de la prise d'aliments ad libitum du rat blanc. *Journal of Physiology (Paris)*, **58**, 323–349.

Le Magnen, J. (1981) The metabolic basis of the dual periodicity of feeding in rats. *Behavioral and Brain Sciences*, **4**, 561–607.

Le Magnen, J. (1983) Body energy balance and food intake: a neuro-endocrine regulatory mechanism. *Physiological Reviews*, **63**, 314–386.

Le Magnen, J. (1992) *Neurobiology of Feeding and Nutrition*, Academic Press, Orlando, FL.

Le Magnen, J. and Tallon, S. (1963) Enregistrement et analyse préliminaire de la 'périodicité alimentaire spontanée' chez le rat blanc. *Journal of Physiologie* (*Paris*), **55**, 286–287.

Leibel, R.L., Bahary, N. and Friedman, J.M. (1993) Strategies for the molecular genetic analysis of obesity in humans. *Critical Reviews in Food Science and Nutrition*, **33** (4/5), 351–358.

Leibowitz, S.F. (1990) The role of serotonin in eating disorders. *Drugs*, **39** (3), 33–48.

Leibowitz, S.F. (1994) Specificity of hypothalamic peptides in the control of behavioral and physiological processes. *Annals of the New York Academy of Sciences*, **739**, 12–35.

Leibowitz, S.F. and Rossakis, C. (1978) Pharmacological characterization of perifornical hypthalamic beta-adrenergic receptors mediating feeding inhibition in the rat. *Neuropharmacology*, **172**, 692–702.

Leibowitz, S.F. and Shor-Posner, G. (1986) Brain serotonin and eating behavior. *Appetite*, **7** (Suppl), 1–14.

Leibowitz, S.F. and Stanley, B.G. (1986) Neurochemical control of appetite, in *Feeding Behavior: Neural and Humoral Controls* (eds R.C. Ritter, S. Ritter and C.D. Barnes), Academic Press, Orlando, FL, pp. 191–234.

Levin, B.E. (1986) Neurological regulation of body weight. *Clinical Neurobiology*, **2**, 1–60.

Levine, A.S., Tallman, J.A., Grace, M.K., Parker, S.A., Billington, C.J. and Levitt, M.D. (1989) Effect of breakfast cereals on short-term food intake. *American Journal of Clinical Nutrition*, **50**, 1303–1307.

Liddle, R.A., Goldfine, I.D. and Williams, J.A. (1984) Bioassay of plasma cholecystokinin in rats: Effects of food, trypsin inhibitor, and alcohol. *Gastroenterology*, **87**, 542–549.

Liddle, R.A., Goldfine, I.D., Rosen, M.S., Taplitz, R.A. and Williams, J.A. (1985) Cholecystokinin bioactivity in human plasma: Molecular forms, responses to feeding, and relationship to gallbladder contraction. *Journal of Clinical Investigation*, **75**, 1144–1152.

Lieverse, R.J., Jansen, J.B.M.J., Masclee, A.A.M. and Lamers, C.B.H.W. (1993) Satiety effects of a physiologic dose of cholecystokinin in man. *Gastroenterology*, **104**, A632.

Lieverse, R.J., Jansen, J.B.M.J., Masclee, A.A.M. and Lamers, C.B.H.W. (1994) Role of cholecystokinin in the regulation of satiation and satiety in humans. *Annals of the New York Academy of Sciences*, **713**, 268–272.

Linscheer, W.G. and Vergroesen, A.J. (1988) Lipids, in *Modern Nutrition in Health and Disease*, 7th edn (eds M. Shils and V. Young), Lea & Febiger, Philadelphia, pp. 72–107.

Logue, A.W. and King, G.R. (1991) Self-control and impulsiveness in adult humans when food is the reinforcer. *Appetite*, **17**, 105–120.

Louis-Sylvestre, J., Tournier, A., Verger, P., Chabert, M., Delorme, B. and Hossenlopp, J. (1989) Learned caloric adjustment of human intake. *Appetite*, **12** (2), 95–104.

Mackay, E.M., Calloway, J.W. and Barnes, R.H. (1940) Hyperalimentation in normal animals produced by protamine zinc insulin. *Journal of Nutrition*, **20**, 59–66.

Mackinnon, J. (1974) The behavior and ecology of wild orang-utans (Pongo pygmaeus). *Animal Behavior*, **22**, 3–74.

Mannon, P. and Taylor, I.L. (1994) The pancreatic polypeptide family, in *Gut Peptides: Biochemistry and Physiology* (eds J.H. Walsh and G.J. Dockray), Raven Press, Ltd., New York, pp. 341–371.

Marrazzi, M.A. and Luby, E.D. (1986) An auto-addiction opioid model of chronic anorexia nervosa. *International Journal of Eating Disorders*, **5**, 191–208.
Mattes, R.D., Curran, W.J., Powlis, W. and Whittington, R. (1991) A descriptive study of learned food aversions in radiotherapy patients. *Physiology and Behavior*, **50**, 1103–1109.
Matthews, J.W., Gibson, E.L. and Booth, D.A. (1985) Norepinephrine-facilitated eating: Reduction in saccharin preference and conditioned flavor preferences with increase in quinine aversion. *Pharmacology, Biochemistry and Behavior*, **22**, 1045–1052.
Mayer, J. (1955) Regulation of energy intake and body weight. *Annals of the New York Academy of Sciences*, **63** Art. 1, 15–43.
Mayer, J., Marshall, N.B., Vitale, J., Christensen, J.H., Mashayekhi, M.B. and Stare, F.J. (1954) Exercise, food intake and body weight in normal rats and genetically obese adult mice. *American Journal of Physiology*, **177**, 544–548.
McCleery, R.H. (1977) On satiation curves. *Animal Behavior*, **25**, 1005–1015.
McDonald, T.J., Nilsson, G., Vagne, M., Ghatei, M., Bloom, S.R. and Mutt, V. (1978) A gastrin releasing peptide from the porcine non-antral gastric tissue. *Gut*, **19**, 769–774.
McHugh, P.R., Moran, T.H. and Killiea, M. (1989) The approaches to the study of human disorders in food ingestion and body weight maintenance. *Annals of the New York Academy of Sciences*, **575**, 1–12.
Meier, R., Hildebrand, P., Thumshirn, M., Albrecht, C., Studer, B., Gyr, Klaus. and Beglinger, C. (1990) Effect of loxiglumide, a cholecystokinin antagonist, on pancreatic polypeptide release in humans. *Gastroenterology*, **99**, 1757–1762.
Meiselman, H.L. (1992) Methodology and theory in human eating research. *Appetite*, **19**, 49–55.
Meiselman, H.L., Staddon, S.L., Hedderley, D., Pierson, B.J. and Symonds, C.R. (1994) Effect of effort on meal selection and meal acceptability in a student cafeteria. *Appetite*, **23** (1), 43–56.
Melchior, J.-C., Rigaud, D., Chayvialle, J.-A., Colas-Linhare, N., Laforest, M.-D., Petiet, A., Comoy, E. and Apfelbaum, M. (1994) Palatability of a meal influences release of beta-endorphin, and of potential regulators of food intake in healthy human subjects. *Appetite*, **22**, 233–244.
Melton, P.A., Kissileff, H.R. and Pi-Sunyer, F.X. (1992) Cholecystokinin (CCK-8) affects gastric pressure and ratings of hunger and fullness in women. *American Journal of Physiology*, **263**, R452–R456.
Merali, Z., Moody, T.W. and Coy, D. (1993) Blockade of brain bombesin/GRP receptors increases food intake in satiated rats. *American Journal of Physiology*, **264**, R1031–R1034.
Mickelsen, O., Takahashi, S. and Craig, C. (1955) Experimental obesity I: Production of obesity in rats by feeding high-fat diets. *Journal of Nutrition*, **57**, 541–554.
Mitchell, J.E., Laine, D.C., Morley, J.E. and Levine, A.S. (1986) Naloxone but not CCK-8 may attenuate binge-eating behavior in patients with bulimia syndrome. *Biological Psychiatry*, **21**, 199–206.
Mohr, B. (1840) Hypertropohie der Hypophyse cerebri und dadurch bedingter Druch auf die Heohengrundflaeche insbesondere auf die Sehnerven, dass Chiasma derselben, und dem laengseitigen Hoehenschenkel. *Wochenschrift fuer die Gesamte Heilkunde*, **6**, 565–574.
Mook, D.G. (1989) Oral factors in appetite and satiety. *Annals of the New York Academy of Sciences*, **575**, 265–280.
Moran, T.H. and McHugh, P.R. (1982) Cholecystokinin suppresses food intake by inhibiting gastric emptying. *American Journal of Physiology*, **242**, R491–R497.
Morley, J.E., Levine, A.S., Grace, M. and Kneip, J. (1985) Peptide YY (PYY) a potent orexigenic agent. *Brain Research*, **341**, 200–203.
Mossner, J., Grumann, M., Zeeh, J. and Fischbach, W. (1992) Influence of various nutrients and their mode of application on plasma cholecystokinin (CCK) bioactivity. *Clinical Investigator*, **70**, 125–129.
Mrosovsky, N. and Faust, I.M. (1985) Cycles of body fat in hibernators. *International Journal of Obesity*, **9**, 93–98.
Muurahainen, N.E., Kissileff, H.R., De Rogatis, A.J. and Pi-Sunyer, F.X. (1988) Effects of cholecystokinin-octapeptide (CCK-8) on food intake and gastric emptying in man. *Physiology and Behavior*, **44**, 645–649.

Muurahainen, N.E., Kissileff, H.R., LaChaussée, J. and Pi-Sunyer, F.X. (1991) Effect of a soup preload on reduction of food intake by cholecystokinin in humans. *American Journal of Physiology*, **260**, R272–R280.
Muurahainen, N.E., Kissileff, H.R. and Pi-Sunyer, F.X. (1993) Intravenous bombesin reduces food intake in man. *American Journal of Physiology*, **264**, R350–R354.
Myer, J.S. and Kowell, A.P. (1971) Eating patterns and body weight change of snakes when eating and when food deprived. *Physiology and Behavior*, **6** (1), 71–74.
Naito-Hoopes, M., McArthur, L.H., Gietzen, D.W. and Rogers, Q.R. (1993) Learned preference and aversion for complete and isoleucine-devoid diets in rats. *Physiology and Behavior*, **53**, 485–494.
Nakajima, M., Inui, A., Teranishi, A., Miura, M., Hirosue, Y., Okita, M., Himori, N., Baba, S. and Kasuga, M. (1994) Effects of pancreatic polypeptide family peptides on feeding and learning behavior in mice. *Journal of Pharmacology and Experimental Therapeutics*, **268**, 1010–1014.
NCHS (National Center for Health Statistics) (1979) *Dietary Intake Source Data*, DHEW Publ. No. 79-1221, NCHS, Hyattsville, MD.
Nicolaidis, S., Danguir, J. and Mather, P. (1979) A new approach of sleep and feeding behaviors in the laboratory rat. *Physiology and Behavior*, **23**, 717–722.
Novin, D. (1994) Regulatory control of food and water intake and metabolism by the liver, in *Neurophysiology of Ingestion* (ed. D.A. Booth), Pergamon Press, New York, pp. 19–32.
Obarzanek, E. and Levitsky, D.A. (1985) Eating in the laboratory: is it representative? *American Journal of Clinical Nutrition*, **42**, 323–328.
Odum, E.P. (1960) Premigratory hyperphagia in birds. *American Journal of Clinical Nutrition*, **8**, 621–629.
Orosco, M. and Nicolaidis, S. (1992) Spontaneous feeding-related monoaminergic changes in the rostromedial hypothalamus revealed by microdialysis. *Physiology and Behavior*, **52**, 1015–1019.
Osborne, T.B. and Mendel, L.B. (1918) The choice between adequate and inadequate diets, as made by rats. *Journal of Biological Chemistry*, **35**, 19–27.
Pelchat, M.L. and Rozin, P. (1982) The special role of nausea in the acquisition of food dislikes by humans. *Appetite*, **3**, 341–351.
Pelleymounter, M.A., Cullen, M.J., Baker, M.B., Hecht, R., Winters, D., Boone, T. and Collins, F. (1995) Effects of the obese gene product on body weight regulation in ob/ob mice. *Science*, **269**, 540–543.
Phifer, C.B. and Hall, W.G. (1987) Developement of ingestive behavior, in *Techniques in the Behavioral and Neural Sciences, Volume 1: Feeding and Drinking* (eds F.M. Toates and N.E. Rowland), Elsevier, Amsterdam, pp. 189–230.
Pi-Sunyer, F.X., Kissileff, H.R., Thornton, J. and Smith, G.P. (1982) C-terminal octapeptide of cholecystokinin decreases food intake in obese men. *Physiology and Behavior*, **29**, 627–630.
Pieramico, O., Malfertheiner, P., Nelson, D.K., Glasbrenner, B. and Ditschuneit, H. (1990) Interdigestive cycling and post-prandial release of pancreatic polypeptide in severe obesity. *International Journal of Obesity*, **14**, 1005–1011.
Polivy, J. and Herman, C.P. (1985) Dieting and binging. *American Psychologist*, **40**, 193–201.
Polivy, J., Herman, C.P., Younger, J.C. and Erskine, B. (1979) Effects of a model on eating behavior: The induction of a restrained eating style. *Journal of Personality*, **47**, 100–117.
Porikos, K.P., Hesser, M.F. and Van Itallie, T.B. (1982) Caloric regulation in normal weight men maintained on a palatable diet of conventional foods. *Physiology and Behavior*, **29**, 293–300.
Prentice, A.M., Whitehead, R.G., Roberts, S.B. and Paul, A.A. (1981) Long-term energy balance in child-bearing Gambian women. *American Journal of Clinical Nutrition*, **34**, 2790–2799.
Pruvost, M., Duquesnel, J. and Cabanac, M. (1973) Injection de glucose dans le territoire porte chez l'homme, absence d'alliesthesie negative en response à des stimulus sucres. *Physiology and Behavior*, **11**, 355–358.
Pudel, V. (1971) Food-Dispenser: eine methode zur untersuchung des spontanen appetitverhaltens. *Zeitschrift für Ernahrungswissenschaft*, **10**, 382–393.

Quaade, F. (1974) Untraditional treatment of obesity, in *Obesity* (eds W.C. Burland, P.D. Samuel and J. Yudkin), Churchill Livingstone, Edinburgh, pp. 338–352.

Quaade, F., Vaernet, K. and Larsson, S. (1974) Stereotaxic stimulation and electrocoagulation of the lateral hypothalamus in obese humans. *Acta Neurochirurgica*, **30**, 111–117.

Raybould, HE, Gayton, R.J. and Dockray, G.J. (1988) Mechanisms of action of peripherally administered cholecystokinin octapeptide on brain stem neurons in the rat. *The Journal of Neuroscience*, **8**, 3018–3024.

Ricardo, J.A. (1983) Hypothalamic pathways involved in metabolic regulatory functions, as identified by track-tracing methods. *Advances in Metabolic Disorders*, **10**, 1–30.

Richter, C.P. (1927) Animal behavior and internal drives. *Quarterly Review of Biology*, **2**, 307–343.

Richter, C.P. (1942) Total self-regulatory functions in animals and human beings. *Harvey Lecture Series*, **38**, 63–103.

Richter, C.P. (1965) *Biological Clocks in Medicine and Psychiatry*, C.C. Thomas, Publisher, Springfield, IL.

Richter, C.P., Holt, L.E. and Barelare, B. (1938) Nutritional requirements for normal growth and reproduction in rats studied by the self-selection method. *American Journal of Physiology*, **122**, 734–744.

Rieg, T.S., Maestrello, A.M. and Aravich, P.F. (1994) Weight cycling alters the effects of D-fenfluramine on susceptibility to activity-based anorexia. *American Journal of Clinical Nutrition*, **60** (4), 494–500.

Ritter, R.C., Brenner, L.A. and Tamura, C.S. (1994) Endogenous CCK and the peripheral neural substrates of intestinal satiety. *Annals of the New York Academy of Sciences*, **713**, 225–267.

Roberts, S.B., Fuss, P., Heyman, M.B., Evans, W.J., Tsay, R., Rasmussen, H.R., Fiatarone, M., Cortiella, J., Dallal, G. and Young, V.R. (1994) Control of food intake in older men. *Journal of the American Medical Association*, **272**, 1601–1606.

Robinson, P.H., Checkley, S.A. and Russell, G.F.M. (1985) Suppression of eating by fenfluramine in patients with bulimia nervosa. *British Journal of Psychiatry*, **146**, 169–176.

Rolls, B.J., Rowe, A., Rolls, E.T., Kingston, B., Megson, A. and Gunary, R. (1981) Variety in a meal enhances food intake in man. *Physiology and Behavior*, **26**, 215–221.

Rolls, B.J., Pirraglia, P.A., Jones, M.B. and Peters, J.C. (1992) Effects of olestra, a noncaloric fat substitute, on daily energy and fat intakes in lean men. *American Journal of Clinical Nutrition*, **56**, 84–92.

Routtenberg, A. and Kuznesof, A.W. (1967) 'Self-starvation' of rats living in activity wheels on a restricted feeding schedule. *Journal of Comparative and Physiological Psychology*, **64**, 414–421.

Rozin, P. (1968) Are carbohydrate and protein intakes separately regulated? *Journal of Comparative and Physiological Psychology*, **65**, 23–29.

Rozin, P. (1976) Psycho-biological and cultural determinants of food choice, in *Dahleon Workshop on Appetite and Food Intake* (ed. T. Silverstone), Dahlem Konferenzen, Berlin, pp. 285–312.

Rozin, P. (1977) The significance of learning mechanisms in food selection: Some biology, psychology, and sociology of science, in *Learning Mechanisms in Food Selection* (eds L.M. Barker, M.R. Best and M. Domjan), Baylor University Press, Austin, TX, pp. 557–592.

Rusak, B. and Zucker, I. (1979) Neural regulation of circadian rhythms. *Physiological Reviews*, **59**, 449–526.

Sahu, A., Sninsky, C.A., Phelps, C.P., Dube, M.G., Kalra, P.S. and Kalra, S.P. (1992) Neuropeptide Y release from the paraventricular nucleus in association with hyperphagia in streptozotocin-induced diabetic rats. *Endocrinology*, **131**, 2979–2985.

Sanacora, G., Kershaw, M., Finkelstein, J.A. and White, J.D. (1990) Increased hypothalamic content of preproneuropeptide Y mRNA in genetically obese Zucker rats and its regulation by food deprivation. *Endocrinology*, **127**, 730–737.

Sargent, F. (1954) Season and the metabolism of fat and carbohydrate: a study of vestigial physiology. *Meteorological Monographs*, **2** (8), 68–80.

Schachter, S. (1968) Obesity and Eating. *Science*, **161**, 751–756.

Schachter, S. (1971) Some extraordinary facts about obese humans and rats. *American Psychologist*, **26**, 129–144.

Schachter, S. and Rodin, J. (1974) *Obese Humans and Rats*, Erlbaum-Halstead, Washington, D.C.

Scharrer, E. and Langhans, W. (1988) Metabolic and hormonal factors controlling food intake. *International Journal of Vitamin and Nutrition Research*, **58**, 249–261.

Schemmel, R., Mickelsen, O. and Gill, J.L. (1970) Dietary obesity in rats: body weight and body fat accretion in seven strains of rat. *Journal of Nutrition*, **100**, 1941–1948.

Schneider, L., Cooper, S.J. and Halmi, K. (1989) Orosensory self-stimulation by sucrose involves brain dopaminergic mechanisms. *Annals of the New York Academy of Sciences*, **575**, 307–320.

Schwartz, G.J. and Moran, T.H. (1994) CCK elicits and modulates vagal afferent activity arising from gastric and duodenal sites. *Annals of the New York Academy of Sciences*, **713**, 121–128.

Schwartz, M.W., Marks, J., Sipols, A.J., Baskin, D.G., Woods, S.C., Kahn, S.E. and Porte, D.J. (1991) Central insulin administration reduces neuropeptide Y mRNA expression in the arcuate nucleus of food-deprived lean (Fa/Fa) but not obese (fa/fa) Zucker rats. *Endocrinology*, **128**, 2645–2647.

Schwartz, M.W., Figlewicz, D.P., Baskin, D.G., Woods, S.C. and Porte Porte, D.P., Jr. (1992) Insulin in the brain: A hormonal regulator of energy balance. *Endocrinology Review*, **13**, 387–414.

Sclafani, A. (1976) Dietary obesity in adult rats: Similarities to hypothalamic and human obesity syndromes. *Physiology and Behavior*, **17**, 461–471.

Sclafani, A. (1984) Animal models of obesity: classification and characterization. *International Journal of Obesity*, **8**, 491–508.

Sclafani, A. (1989) Dietary-induced overeating. *Annals of the New York Academy of Sciences*, **575**, 281–291.

Sclafani, A. (1990) Nutritionally based learned flavor preferences in rats, in *Taste, Experience, and Feeding* (eds E.D. Capaldi and T.L. Powley), American Psychological Association, Washington, D.C., pp. 139–156.

Sclafani, A. (1991) The hedonics of sugar and starch, in *The Hedonics of Taste* (ed. R.C. Bolles), Lawrence Erlbaum Associates, Inc., Hillsdale, NJ, pp. 59–87.

Sclafani, A. and Kirchgessner, A. (1986) The role of the medial hypothalamus in the control of food intake: An update, in *Feeding Behavior: Neural and Humoral Controls* (eds R.C. Ritter, S. Ritter and C.D. Barnes), Academic Press, New York, pp. 27–63.

Sharara, A.I., Bouras, E.P., Misukonis, M.A. and Liddle, R.A. (1993) Evidence for indirect dietary regulation of cholecystokinin release in rats. *American Journal of Physiology*, **265**, G107–G112.

Shide, D.J., Caballero, B., Reidelberger, R. and Rolls, B.J. (1995) Accurate energy compensation for intragastric and oral nutrients in lean males. *American Journal of Clinical Nutrition*, **61**, 754–764.

Shor-Posner, G., Ian, C., Brennan, G., Cohn, T., Moy, H., Ning, A. and Leibowitz, S.F. (1991) Self-selecting albino rats exhibit specific macronutrient preferences: Characterization of three subpopulations. *Physiology and Behavior*, **50**, 775–777.

Shor-Posner, G., Brennan, G., Ian, C., Jasaitis, R., Madhu, K. and Leibowitz, S.F. (1994) Meal patterns of macronutrient intake in rats with particular dietary preferences. *American Journal of Physiology*, **35**, R1395–R1402.

Siegel, P.S. and Stuckey, H.L. (1947) The diurnal course of water and food intake in the normal mature rat. *Journal of Comparative and Physiological Psychology*, **40**, 365–370.

Silver, A.J. and Morley, J.E. (1991) Role of CCK in regulation of food intake. *Progress in Neurobiology*, **36**, 23–34.

Silver, M. and Flood, J.F. (1988) Effect of gastrointestinal peptides on ingestion in old and young mice. *Peptides*, **9**, 221–225.

Silverstone, T. (1975) Anorectic drugs, in *Obesity, Its Pathogenesis and Management* (ed. T. Silverstone), Publishing Sciences Group, Inc., Acton, MA, pp. 193–227.

Silverstone, T., Fincham, J. and Brydon, J. (1980) A new technique for the continuous measurement of food intake in man. *American Journal of Clinical Nutrition*, **33**, 1852–1855.

Simson, P.C. and Booth, D.A. (1973) Olfactory conditioning by association with histidine-free or balanced amino-acid loads in rats. *Journal of Experimental Psychology*, **25**, 354–359.

Simson, P.C. and Booth, D.A. (1974) The rejection of a diet which has been associated with a single administration of an histidine-free amino acid mixture. *British Journal of Nutrition*, **31**, 285–296.

Sirinek, K.R., O'Dorisio, T.M., Howe, B. and McFee, A.S. (1985) Pancreatic islet hormone response to oral glucose in morbidly obese patients. *Annals of Surgery*, **201**, 690–694.

Skinner, B.F. (1932) Drive and reflex strength. *Journal of General Psychology*, **6**, 22–47.

Skinner, B.F. (1953) *Science and Human Behavior*, Macmillan, New York.

Smith, G.P. (1982) The physiology of the meal, in *Drugs and Appetite* (ed. T. Silverstone), Academic Press, London, pp. 1–21.

Smith, G.P. (1989) Animal models of human eating disorders. *Annals of the New York Academy of Sciences*, **575**, 63–74.

Smith, G.P. and Gibbs, J. (1992) The development and proof of the CCK hypothesis of satiety, in *Multiple Cholecystokinin Receptors in the CNS* (eds C.T. Dourish, S.J. Cooper, S.D. Iversen and L.L. Iversen), Oxford University Press.

Spiegel, T.A. (1973) Caloric regulation of food intake in man. *Journal of Comparative and Physiological Psychology*, **84**, 24–37.

Spiegel, T.A. and Jordan, H.A. (1978) Effects of simultaneous oral-intragastric ingestion on meal patterns and satiety in humans. *Journal of Comparative and Physiological Psychology*, **92**, 133–141.

Spiegel, T.A., Shrager, E.E. and Stellar, E. (1989) Responses of lean and obese subjects to preloads, deprivation, and palatability. *Appetite*, **13**, 45–69.

Spitzer, L. and Rodin, J. (1981) Studies of human eating behavior: A critical review of normal weight and overweight individuals. *Appetite*, **2**, 293–330.

Stacher, G., Steinringer, H., Schmierer, G., Schneider, C. and Winklehner, S. (1982) Cholecystokinin octapeptide decreases intake of solid food in man. *Peptides*, **3**, 133–136.

Stanley, B.G. and Leibowitz, S.F. (1984) Neuropeptide Y: Stimulation of feeding and drinking by injection into the paraventricular nucleus. *Life Sciences*, **35**, 2635–2642.

Stanley, B.G., Schwartz, D.H., Hernandez, L., Hoebel, B.G. and Leibowitz, S.F. (1989) Patterns of extracellular norepinephrine in the paraventricular hypothalamus: Relationship to circadian rhythm and deprivation-induced eating behavior. *Life Sciences*, **45**, 275–282.

Stellar, E. (1954) The physiology of motivation. *Psychological Review*, **61**, 5–22.

Stellar, E. (1967) Hunger in man: Comparative and physiological studies. *American Psychologist*, **22** (2), 105–117.

Stellar, E. (1976) The CNS and appetite: Historical introduction, in *Dahlem Workshop on Appetite and Food Intake* (ed. T. Silverstone), Abakon Verlagsgesellschaft, Berlin, pp. 15–20.

Stellar, E. (1989) Long-term perspectives on the study of eating behavior. *Annals of the New York Academy of Sciences*, **575**, 478–486.

Stellar, E. (1990) Brain and behavior, in *Handbook of Behavioral Neurobiology, Volume 10: Neurobiology of Food and Fluid Intake* (ed. E.M. Stricker), Plenum Press, New York, pp. 3–22.

Stellar, E. (1992) Real eating and the measurement of real physiological and behavioral variables. *Appetite*, **19** (1), 78–79.

Stellar, E. and Schrager, E.E. (1985) Chews and swallows and the microstructure of eating. *American Journal of Clinical Nutrition*, **42**, 973–982.

Stern, J.S., Castonguay, T.W. and Rogers, Q.R. (1988) Physiological regulation of intakes of carbohydrate, fat, and protein, in *Diet and Obesity* (eds G.A. Bray, J. LeBlanc, S. Inoue and M. Suzuki), Japan Sci. Soc. Press/S. Karger, Tokyo/Basel, pp. 51–60.

Story, M. and Brown, J.E. (1987) Do young children instinctively know what to eat? *New England Journal of Medicine*, **316**, 103–106.

Striegel-Moore, R.H., Silberstein, L.R. and Rodin, J. (1986) Toward an understanding of risk factors for bulimia. *American Psychologist*, **41**, 246–263.

Strubbe, J.H. (1994) Circadian rhythms of food intake, in *Food Intake and Energy Expenditure* (eds M.S. Westerterp-Plantenga, E.W.H.M., Fredrix and A.B. Steffens), CRC Press, Boca Raton, FL, pp. 155–174.

Strubbe, J.H., Keyser, J., Dijkstra, T. and Prins, A.J.A. (1986) Interaction between circadian and caloric control of feeding behavior in the rat. *Physiology and Behavior*, **36**, 489–493.

Stubbs, R.J., Prentice, A.M. and Harbron, C.G. (1996) Covert manipulation of the dietary fat to carbohydrate ratio of isoenergetically dense diets: effect on food intake in ad libitum feeding men. *International Journal of Obesity, in press.*

Stunkard, A.J. and Kaplan, D.L. (1977) Eating in public places: A review of the reports of the direct observation of eating behavior. *International Journal of Obesity*, **1**, 89–101.

Stunkard, A.J., Grace, W.J. and Wolff, H.G. (1955) The night-eating syndrome. *American Journal of Medicine*, 78–86.

Swaab, D.F., Hofman, M.A., Lucasson, P.J., Purba, J.S., Raadsheer, F.C. and Van de Nes, J.A.P. (1993) Functional neuroanatomy and neuropathology of the human hypothalamus. *Anatomy and Embryology*, **187**, 317–330.

Taggart, N. (1962) Diet, activity and body-weight. A study of variations in women. *British Journal of Nutrition*, **16**, 223–235.

Thomas, P.R. (1991) *Improving America's Diet and Health From Recommendations to Action*, National Academy Press, Washington, DC.

Thompson, D.A. and Campbell, R.G. (1977) Hunger in humans induced by 2-Deoxy-D-glucose: Glucoprivic control of taste preference and food intake. *Science*, **198**, 1065–1067.

Tinbergen, N. (1969) *The Study of Instinct*, 2nd edn, Oxford University Press, New York.

US Dept. of Agriculture (1984) *Food intakes: Individuals in 48 states, year 1977–78. Nationwide food consumption survey report No. 1-2*, Nutrition Monitoring Division, Human Nutrition Information Service, USDA, Hyattsville, MD.

Ungerstedt, U. (1971) Stereotaxic mapping of monoamine pathways in the rat brain. *Acta Physiologiae Scandinavia Supplementum*, **367**, 1–48.

Van Itallie, T.B. and Kissileff, H.R. (1985) Physiology of energy intake: an inventory control model. *American Journal of Clinical Nutrition*, **42**, 914–923.

Van Itallie, T.B. and Kissileff, H.R. (1990) Human obesity: A problem in body energy economics, in *Handbook of Behavioral Neurolobiology, Volume 10: Neurobiology of Food and Fluid Intake* (ed. E.M. Stricker), Plenum Publishing Corp., New York, pp. 207–240.

Van Itallie, T.B., Gale, S. and Kissileff, H.R. (1978) The control of food intake in the regulation of depot fat: An overview, in *Advances in Modern Nutrition, Volume 2* (eds H.M. Katzen and R.J. Mahler), Hemisphere Publishing Corporation, Washington, DC.

Van Staveren, W.A., Deurenberg, P., Burema, J., DeGroot, L.P.B.M. and Hautvast, J.G.A.J. (1986) Seasonal variation in food intake, pattern of physical activity, and change in body weight in a group of young adult Dutch women consuming self-selected diets. *International Journal of Obesity*, **10**, 133–145.

Van Vort, W.B. (1988) Is sham feeding an animal model of bulimia? *International Journal of Eating Disorders*, **7**, 797–806.

Vanderweele, D.A. (1994) Insulin is a prandial satiety hormone. *Physiology and Behavior*, **56** (3), 619–622.

Walsh, B.T., Kissileff, H.R., Cassidy, S.M. and Dantzic, S. (1989) Eating behavior of women with bulimia. *Arch. Gen. Psychiatry*, **46**, 54–58.

Warner, K.E. and Balagura, S. (1975) Intrameal eating patterns of obese and nonobese humans. *Journal of Comparative and Physiological Psychology*, **89**, 778–783.

Warwick, Z.S. (1996) Probing the causes of high-fat diet hyperphagia: A mechanistic and behavioral dissection. *Neuroscience and Biobehavioral Reviews* (In Press).

Welch, I., Saunders, K. and Read, N.W. (1985) Effect of ileal and intravenous infusions of fat emulsions on feeding and satiety in human volunteers. *Gastroenterology*, **89**, 1293–1297.

Welch, I.M., Davison, P.A., Worldling, J. and Read, N.W. (1988) Effect of ileal infusion of lipid on jejunal motor patterns after a nutrient and nonnutrient meal. *American Journal of Physiology*, **255**, G490–G497.

Wentzlaff, T., Guss, J.L. and Kissileff, H.R. (1995) Subjective ratings as a function of amount consumed: A preliminary report. *Physiology and Behavior*, **57** (6), 1209–1214.

West, D.B., York, B., Goudey-Lefevre, J. and Truett, G.E. (1996) Genetics and physiology of dietary obesity in the mouse, in *Molecular and Genetic Aspects of Obesity, Penningtion Nutrition Series, Volume 5* (eds G.A. Bray and D. Ryan), Louisana State University Press, Baton Rouge, LA (In press).

Westerterp, K.R. (1994) Body weight regulation, in *Food Intake and Energy Expenditure* (eds M.S. Westerterp-Plantenga, E.W.H.M. Fredrix and A.B. Steffens), CRC Press, Inc, Boca Raton, FL, pp. 321–338.

Westerterp-Plantenga, M.S., Van den Heuvel, E., Wouters, L. and Ten Hoor, F. (1992) Diet

induced thermogenesis and cumulative food intake curves, as a function of familiarity with food and dietary restraint in humans. *Physiology and Behavior*, **51** (3), 457–465.
Williams, G., Steel, J.H., Cardoso, H.M., Lee, Y.C., Gill, J.S. and Burrin, J.M. (1988) Increased hypothalamic neuropeptide Y concentrations in the diabetic rat. *Diabetes*, **37**, 763–772.
Wilson, G.T. (1993) Assessment of binge eating, in *Binge Eating: Nature, Assessment, and Treatment* (eds C.G. Fairburn and G.T. Wilson), Guilford Press, New York, pp. 227–249.
Wolkowitz, O.M., Gertz, B., Weingartner, H., Beccaria, L., Thompson, K. and Liddle, R.A. (1990) Hunger in humans induced by MK-329, a specific peripheral-type cholecystokinin receptor antagonist. *Biological Psychiatry*, **28**, 169–173.
Woo, R. and Pi-Sunyer, F. (1985) Effect of increased physical activity on voluntary intake in lean women. *Metabolism*, **34** (9), 836–841.
Woo, R., Kissileff, H.R. and Pi-Sunyer, F.X. (1984) Elevated postprandial insulin levels do not induce satiety in normal weight humans. *American Journal of Physiology*, **247**, R745–R749.
Wrangham, R.W. (1974) Artificial feeding of chimpanzees and baboons in their natural habitat. *Animal Behavior*, **22**, 83–93.
Yanovski, S.Z., Leet, M., Flood, M., Yanovski, J.A., Gold, P.W., Kissileff, H.R. and Walsh, B.T. (1992) Food intake and selection of obese women with and without binge-eating disorder. *American Journal of Clinical Nutrition*, **56**, 975–980.
Yox, D.P., Stokesberry, H. and Ritter, R.C. (1989) Suppression of sham feeding by intraintestinal oleate: blockade by a CCK antagonist and reversal of blockade by exogenous CCK-8, in *The Neuropeptide Cholecystokinin (CCK), Anatomy and Biochemistry, Receptors, Pharmacology and Physiology* (eds V. Huges, G. Dockray and G.N. Woodruff), Ellis Horwood, Chichester.
Zhang, Y., Proenca, R., Maffei, M., Barone, M., Leopold, L. and Friedman, J.M. (1994) Positional cloning of the mouse obese gene and its human homologue. *Nature*, **372**, 425–431.

4 The development of children's eating habits

LEANN L. BIRCH, JENNIFER ORLET FISHER and KAREN GRIMM-THOMAS

Energy expenditure and the demand for nutrients is continuous, but eating is periodic. Eating is a discrete behavioral event, and the quantity and quality of the child's diet depends upon (i) meal frequency, (ii) the amount consumed at each meal, and (iii) which foods the child prefers and selects for consumption. Variations in one or a combination of these three behavioral parameters can dramatically alter food intake and dietary quality. The review that follows includes three major subsections, addressing the developing controls of (i) meal frequency, (ii) meal size, and (iii) food preferences and food selection. The largest body of evidence concerns factors influencing children's food preferences, and the space allocated in this review reflects this. While a review of children's food habits could be restricted to children's food preferences and food selection, the emerging evidence suggests that early experience, prior to the onset of weaning influences the acquisition of food habits. Our review begins with discussions of the developing control of meal frequency and meal size during infancy. Even prior to weaning, when only one food is being consumed, and food selection is not an issue, the infant can control intake via modulating meal frequency and meal size. Evidence indicates that factors influencing the developing controls of meal frequency and meal size also shape children's food preferences and food selection.

During the first years of life, dramatic changes in eating behavior occur. As mammals, human infants begin life consuming a single food: milk. However, humans are also omnivores, and this means that by early childhood, dietary variety is needed to obtain adequate nutrition. The infant comes into the world with a set of predispositions (including the acceptance of sweet, rejection of sour and bitter tastes, 'preparedness' to learn to associate food cues with the consequences and contexts of eating, and neophobic reactions to new foods) (Cardello, 1996) and reflexive behaviors (sucking, rooting) that allow the infant to obtain needed nutrients, first during the suckling period and later during the transition to the omnivorous diet. Because children are prepared to learn to associate foods' sensory cues with the social contexts and physiological consequences of eating, the infant's predispositions and reflexive behaviors are soon transformed.

Although the evidence is very incomplete, results indicate that child feeding practices play an influential role in the shaping of individual differences in (i) the extent to which the child controls meal frequency, (ii) the child's control of meal size, and (iii) the child's developing food preferences. By the second half of the first year of life, weaning typically begins and the child must begin to eat solid foods, and culture and individual experience play powerful roles in shaping food preferences and the timing and number of meals that are taken, and in shaping food preferences and food selection. By the time the child is four of five years old, the reflexive, depletion driven, eating of early life has been transformed: children are taking meals at socially determined times, and consuming at least a subset of the foods of the adult diet of their culture. While we all begin life consuming the same milk diet, by early childhood, children of different cultural groups are consuming diets that are composed of completely different foods, sharing no foods in common. This observation points to the essential role of early experience and the social and cultural context of eating in shaping food habits.

The dramatic change that occurs in food habits during early life is highlighted in each of the three major sections of the review. Many of these changes (the transition from suckling to consuming solid foods) are obvious to any casual observer of children's development. Our review of the literature reveals that we have descriptive information on the development of food habits during the early years, but that we understand relatively little regarding how nature and nurture interact as development proceeds. Such understanding could make both theoretical and practical contributions; it would inform our understanding of the control of food intake in humans and could provide a framework for designing strategies to prevent the development of obesity and eating disorders that have become increasingly pervasive. For an excellent review on eating behavior and its relationship to growth and obesity, see Hammer (1992).

The review is divided into three major subsections, dealing in turn with the development of meal frequency, meal size and finally with food preferences and food selection. Each section begins with a discussion of what we know about the infant's innate predispositions, followed by a description of what is known regarding how these predispositions interact with early experience to shape food habits, including the emerging controls of meal frequency, meal size and the development of food preferences and food selection.

4.1 The frequency and timing of meals

4.1.1 Innate predispositions: biological constraints and meal frequency

When unlimited food is available, free feeding animals take more or fewer meals as the primary means of adjusting food intake (Le Magnen, 1985).

The adult's opportunities for adjusting meal timing may be limited due to social and cultural constraints on the timing of meals; however, for the infant, taking more or fewer meals is the major mechanism for controlling total daily intake. Meal frequency is a particularly important control of food intake during early infancy because other mechanisms are not yet fully functioning. During this period, the infant is consuming an exclusive milk diet, food preferences are not an issue; and meal size is limited by the infant's gastric capacity. With respect to the infant's control of the timing and frequency of meals, across cultures and throughout history, the prevailing approach to infant feeding has been to allow the infant to determine the timing of meals, and to feed 'on demand' when distress cues seem to indicate hunger. An implicit assumption underlying demand feeding is that the young infant 'knows' when she is hungry and when she is full, and given the opportunity to control meal timing and meal size, the infant will consume the quantities of milk needed to maintain growth and health. The infant enters the world with a set of reflexes (sucking, rooting, swallowing) that allow the neonate to suckle. However, there are no human data that speak to the issue of whether the newborn infant has any innate ability to respond to internal signals of hunger and satiety to regulate meal frequency.

Prevailing advice to feed 'on demand' and allow the infant control over meal frequency can be difficult for formula feeding parents, who can see how much the infant has consumed from the bottle, and can attempt to control how much the infant consumes at a feeding. In contrast, breast feeding mothers cannot readily determine how much the infant has consumed, and in the absence of such information, the mother may allow the infant relatively greater control over the size and timing of feedings. Evidence presented below suggests that differences in patterns of intake do emerge between breast fed and formula fed infants, and that these early differences could be precursors of later individual differences in intake.

Fomon (1993) points out that allowing the infant control of the size and timing of feedings creates a circumstance that is 'conducive to establishing habits of eating in moderation' (p. 114). Put in other terms, demand feeding allows the infant to learn to associate meal initiation with feelings of hunger, and meal termination with normal satiety. It should be emphasized that little attention has been given to whether parents who profess to demand feed actually allow the child to control the timing and size of feedings. Fomon (1993) argues persuasively that few infants are truly fed on demand. Failure to feed on demand may occur because the parent is not really motivated to do so, due to scheduling conflicts, or because the parent has difficulty accurately differentiating whether the infant's distress cues signal hunger or other discomfort. Despite the centrality of demand feeding, we have been unable to locate any systematic studies that have investigated the effects of demand *vs.* schedule feeding on infants' physical growth, health, or later food habits.

The timing of meals is not independent of meal size: large meals are associated with long intervals between meals, and small meals with shorter intermeal intervals. However, the sequential nature of these relationships changes with development (Le Magnen, 1985). In mature, free-feeding animals and in adult humans living in environments without the usual social constraints (Bernstein, 1981), the size of a meal is positively related to the time that elapses until the next meal. In this case, there is a *postprandial* relationship between meal size and intermeal interval: the size of the meal determines the period of satiation that follows the meal. However, in young animals, there are strong *preprandial* relationships between meal size and meal intervals. In this case, the size of a meal is strongly related to the duration of the previous interval; the time since the last feeding seems to dictate the volume consumed at a feeding.

In examining the meal patterns of exclusively breast fed infants, we found that intermeal intervals were related to meal size, and these findings were consistent with developmental data from animal models: we found positive relationships between meal interval and meal size. Strong preprandial relationships were seen, at least in the first three months of life (Matheny *et al.*, 1990; Pinillia and Birch, 1993). In a recent study, we explored behavioral procedures to teach breast fed infants to self-soothe and to sleep through the night (Pinilla and Birch, 1993). In response to our procedures, as intermeal increased with increasing periods of sleep between midnight and 5 am, the size of first morning feeding increased dramatically. As infants began to sleep through the night, 8-week-old infants took the largest meal of the day at their early morning feeding, following the overnight fast, consistent with a preprandial relationship between meal interval and meal size.

Peter Wright and colleagues (Wright *et al.*, 1980) also reported similar positive relationships between meal interval and meal size in another sample of exclusively breast fed infants. At about 8 weeks, Wright's breast fed infants' diurnal meal patterns revealed that their largest feeding was taken early in the morning, following the relatively long overnight fast. By the time the infants were 6 months old, they showed the more adult-like postprandial pattern: the size of a meal predicts the size of the following meal interval. These older infants were taking a large feeding late at night, just prior to the long overnight fast. Wright and colleagues (1980) have proposed that via early learning, the relationships between feeding interval and feeding size change during the first year of life, from a predominantly preprandial pattern, to the more mature postprandial, adult-like pattern. In this view, at about 6 months, infants are learning that certain environmental cues predict a long overnight fast, and they begin taking a large meal prior to a long period without food; the size of a meal anticipates the size of the subsequent intermeal interval. In summary, at least among breast fed infants who are fed on demand, meal interval is positively related to meal size during infancy, but the sequential patterning of the relationship may change during early development. These findings suggest that while meal size is initially a reaction to prior period of deprivation, later in development, the child learns to adjust intake at a meal in anticipation of how long it will be to the next meal.

The positive relationships between meal interval and meal size observed in breast fed infants were not obtained for formula fed infants (Wright *et al.*, 1980). Formula fed infants' meal sizes were not related to intermeal intervals, and they did not show the diurnal patterns in meal size similar to those of breast fed infants. Wright and colleagues point out that the formula fed infants were not fed on demand; formula fed infants' scheduled feedings meant that meal intervals and feeding volumes were relatively fixed, not free to vary, as they were determined by the mother. Therefore, although there is some evidence that infants adjust meal intervals and meal size to regulate intake, in the presence of parental controls, evidence for such regulation disappears.

As infants become young children, they must begin to forgo demand feeding as they are socialized into the temporal patterns of meals for their culture, and their control over the timing of meals becomes more limited. With mealtimes fixed, the child's strategies are limited to adjustments in meal size and food selection. However, in our culture and others, children are allowed some flexibility in this regard: while children partake of scheduled meals, snacking on demand is relatively common. Unfortunately, there is almost no information on how the timing of meals and snacks might impact on well nourished children's food intake and growth patterns. Even Clara Davis' (Davis, 1928, 1939) classic research on self selection of diets by young children (see section 4.2) provides no evidence on this point, because children were fed three or four meals at regularly scheduled times.

Recent research by Dewey and colleagues has explored relationships between the timing and number of meals and snacks taken by children and their total daily energy intake (Garcia *et al.*, 1990a,b). Their findings revealed that young children's caloric intake was directly related to feeding frequency. In addition to the two regularly scheduled meals each day, the 3- to 5-year-old Mexican children in their sample requested food frequently and were usually given snacks when they requested them, so that the children had a mean of 13.5 eating occasions each day. They reported that children who ate more than the average number of meals had significantly higher energy intakes than those who ate less frequent meals and snacks. They also reported that far more food was available to the children than they ate, so that intake was not limited by availability. Because many of the foods offered to the children were of relatively low energy density, the bulkiness of the food may have been a factor limiting intake within an eating occasion. Their results indicate that when given the opportunity to have frequent snacks, the low energy density of the foods did not limit children's total daily energy intake. However, their data did not speak to the question of whether variations in meal frequency might produce differences in growth even when diets are of equivalent total energy content. Because their sample was drawn from chronically malnourished children, extrapolation to well nourished populations must be made with caution.

By early childhood, the timing of meals taken 'on demand' is not exclusively controlled by hunger cues, as environmental and social cues begin to play a role in initiating eating (Birch *et al.*, 1989b). In research investigating how social and environmental cues play a role in initiating meals during childhood, preschool children in day care had repeated opportunities to play in two different playrooms. In one playroom, snack foods were always available, while in the other playroom food was never present. Following these opportunities to associate food with one environment but not the other, we tested whether children would be more likely to initiate a meal in the playroom previously associated with food, *even when they weren't hungry*. We made sure the children were not hungry by feeding them just prior to the test session. Right after eating, the children went into the playrooms, and during this session food was readily available in both rooms. We found that when the children were in the room previously associated with food and eating, they initiated a meal more quickly, and took a larger meal than in the room that had never been associated with food and eating. These findings indicate that the 'depletion driven' (Weingarten, 1985) eating of infancy has been altered via learning and experience at least by the time children are 3 to 5 years old, and that even in the absence of internal cues signalling depletion, meals can be initiated by environmental cues.

4.2 Learning, experience and meal size

In the following section, contributions of learning and experience to the regulation of meal size are discussed. More specifically, we address how meal size is influenced by four factors: innate predispositions, dietary variety, the energy content of foods, and child feeding practices.

4.2.1 Evidence of innate predispositions in the control of meal size

Satter (1986, 1987, 1990) advocates a division of responsibility in feeding, and argues persuasively that while parents need to take responsibility for providing children with an array of healthful foods, the child must assume responsibility for how much she eats (or whether she eats at all). This advice is based on an assumption that, left to their own devices, children will adjust intake to maintain positive energy balance and sustain growth and health. What is the evidence that children can regulate meal size to control energy intake? Clara Davis, based on her classic studies of the self selection of diets by newly weaned infants, concluded that her results, showing children's success at self-selecting healthful diets, were evidence for '. . . the existence of some *innate*, automatic mechanism for its accomplishment, of which appetite is a part' (Davis, 1939). In fact, her data are consistent with the idea that the children, who had many opportunities to self select meals, were learning food preferences and the control of food intake. There are no data to indicate that infants are responsive to the satiety signals that can control meal size in adults, although Fomon suggests that by about 6 weeks of age, infants begin to adjust intake in response to the energy density of formula (see Fomon, 1993). The animal literature does not support the view that there are innate, unlearned controls of meal size; in fact, Hall and colleagues report that in the first days after birth, rat pups will 'virtually drown themselves at the nipple', and are not sensitive to physiological feedback which, later in development, signal satiety (Hall and Rosenblatt, 1977).

The infants and toddlers who participated in Davis' studies had no control over meal frequency; meals were served at fixed times, but children were free to control meal size and to select whatever foods they liked. Children's participation began at weaning, so that all the solid foods served were initially novel to them. Most children continued in the study for a minimum of several months; some participants were followed for several years. In addition to looking at the infants' food selections, meal patterns, and dietary intake, data on growth and morbidity were obtained. Davis' reports of infants' and children's energy and macronutrient intakes are consistent with current guidelines (Fomon, 1993). The children in Davis' studies obtained about 35% of energy from fat, and 17% from protein, and they grew well and had few childhood illnesses. Meal size and composition of individual meals varied dramatically, with children rarely consuming more than two or three of the 10 to 12 foods presented at any given meal. The children also went on 'food jags', eating only one or a few foods for several meals and then abandoning that food in favor of others. There were large individual differences in what was consumed, and in the patterning of meal size, but all children developed well during the course of the study.

4.2.2 *Dietary variety and meal size*

The 'food jags' of Davis' children are quite different from typical mixed meal patterns of American adults. Work by Barbara Rolls and colleagues (Rolls, 1986) has revealed that for adults, the variety of foods available influences meal size, with greater variety stimulating greater intake. When a variety of palatable foods are present, meals are larger than when only one or a few palatable foods are present; increased variety leads to increased intake (*cf.* Rolls, 1986). This increased meal size is a result of the specificity of satiety. That is, preference for a food eaten declines during eating, but preference for uneaten foods remains relatively high, so that if a variety of foods is available, and satiety increases for the food eaten, the individual can simply switch to eating another food. Despite their propensity to consume meals consisting of one or only a few foods, we know that 3- to 5-year-olds also show sensory specific satiety (Birch and Deysher, 1986), although we have not investigated whether sensory specific satiety increases meal size in children as it does in adults.

4.2.3 The energy content of foods and meal size

The energy density of the diet is an important determinant of meal size in later infancy and early childhood. Perhaps because maintaining positive energy balance is so critical during the rapid growth of the early years of life, infants and young children adjust meal size based on the energy density of food available, eating relatively large amounts of energy dilute diets, and much smaller amounts of energy dense ones. Research conducted by Fomon and colleagues (1993) with human infants revealed evidence for such adjustments in the volume of formula by infants as young as six weeks. Infants compensated for energy differences among formulas and adjusted the volume of intake, consuming more energy dilute (54 kcal/dl) than energy dense formula (100 kcal/dl), so that total energy intake was very similar to that of infants fed standard formula of 67 kcal/dl energy density. Adjustments in intake were not immediate but took days or weeks to emerge, suggesting that the infants may have been learning about the physiological consequences of ingesting foods of different energy density and adjusting intake accordingly.

We have conducted research to determine the extent to which the energy density of foods influences young children's food intake, particularly their energy intake. We have monitored children's energy intake within two-course meals and across 24 h periods. In the experiments investigating children's ability to regulate energy intake within meals, children consume a fixed amount of a first course 'preload', that was either high or low in energy density. We have varied energy density by varying carbohydrate or fat content, keeping protein constant. These preloads are consumed on different days, followed in each case by the same array of palatable foods as a second course. The child self selects an *ad libitum* meal from among these foods and the data of interest are derived from intake in the second course, and include the energy and macronutrient content of the foods selected, as well as which foods are selected. In all cases, the food consumption data are derived from pre- and post-weighing of all foods, which are analyzed using nutrient composition data. Children are responsive to these changes in the energy content of the first course; they adjust their food intake in the *ad libitum* second course, eating more following the energy dilute than following an energy dense first course (Birch and Deysher, 1985, 1986; Birch *et al.*, 1987a, 1989a, 1990).

Additional evidence for energy content in the control of meal size comes from research investigating whether children can learn to associate the sensory cues of food with the physiological consequences of eating. In this research, children have repeated opportunities to consume fixed volumes of high and low energy density versions of a food, such as yogurts or puddings. The high and low energy versions of the food typically contain about 170 kcal and 50 kcal/4 oz serving, respectively, are distinctively flavored, with a flavor cue consistently paired with each energy density. For example, a child would have repeated opportunities to consume low energy vanilla yogurt on some days, and high energy almond yogurt on others. Following several opportunities to consume each one, we tested whether they had learned to associate the flavor cue with the postingestive consequences of the differences in energy content. In the test situation, children again consumed fixed volumes of the two flavors of yogurt, this time prepared so that both flavors were of the same energy density. The yogurts were consumed as first course preloads on two different days, followed in each case by the same self selected, *ad libitum* lunch. Results confirmed that children were learning associations between foods' flavor cues and the postingestive consequences of eating those foods. Children ate less *ad libitum* following the flavor previously paired with high energy density, indicating that they had learned to associate the high energy paired flavor with the postingestive consequences of high energy density, and adjusted subsequent food intake accordingly (Johnson *et al.*, 1991; Kern *et al.*, 1993). These findings, consistent with those obtained with animal models (Kissileff, 1996; Sclafani, 1990), indicate that satiety is influenced by learning and can be conditioned (Booth, 1985). Learning about the satiety value of familiar foods can provide an explanation of how meals can be terminated before postingestive cues signalling satiety have a chance to develop. As a result of repeatedly consuming familiar foods, we have learned about the 'fillingness and fatteningness' of familiar foods, and make anticipatory adjustments in intake (Stunkard, 1975).

The research just described has provided evidence that meal size is strongly influenced by energy density, and revealed that infants and children can be responsive to cues resulting from the energy density of the diet in learning to adjust meal size. An obvious next question is whether energy intake is regulated over 24 h periods. Is meal size influenced by energy intake at previous meals as well as by the energy content of the current meal? To investigate these questions, we measured the 24-hr food intake of fifteen 2- to 5-year-old children for 6 days (Birch *et al.*, 1991). The same menus were offered to the children on all 6 days, and food intake was not limited by availability. Energy intake was derived from pre- and post-weighing all foods consumed by each child.

One of the clearest findings emerging from this research concerned the variability of individual children's intake (i) at individual meals, and

(ii) over 24-h periods. In Davis' research, she reported an apparently paradoxical finding: children self selecting their diets grew well, were healthy, and had normal energy intake (Davis, 1928, 1939), but she also indicated that their intake at individual meals was highly unconventional, erratic, and unpredictable. We compared the variability of meal size and variability in intake for total 24 h periods, and used coefficients of variation as an index of variability. These coefficients of variation were calculated for each child for meal size, and for 24 h total energy intake. Results revealed that while the size of individual meals was highly variable, total 24 h energy intake was relatively constant for each child (see Figure 4.1). The mean coefficient of

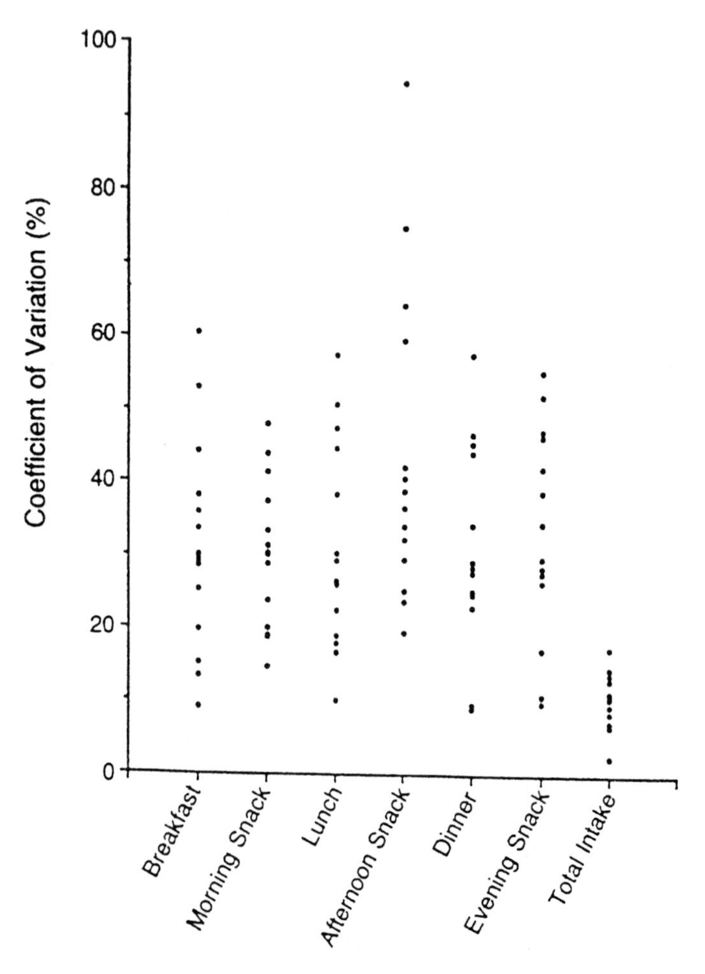

Figure 4.1 Coefficients of variation for total energy intake at the six meals and snacks for individual children (within subject variation).

variation for meal size was 33.6%, but in contrast, the mean coefficient of variation for children's 24 h energy intake was 10.4%. In most cases, there was evidence for adjustments in energy intake *across* successive meals; intake at successive meals was negatively correlated so that high energy intake at one meal was followed by low energy intake at the next, or vice versa. Therefore, while the size of meals was highly variable and seemingly erratic, the adjustments in meal size across successive meals produced relatively tight regulation of energy intake for 24 h periods.

Subsequently, we covertly altered the energy content of several foods in the diet by substituting olestra for about 14 g of dietary fat in several foods served during the first three meals of the day (Birch *et al.*, 1993a). Olestra (Procter & Gamble, Co., Cincinnati, OH) is a nonenergy fat substitute that has the sensory characteristics of fat but is not absorbed, and hence has no energy value. Participants were twenty-nine 2- to 5-year olds, whose food intake at all eating occasions was monitored for 8 days by pre- and post-weighing of all foods. Again, intake was not limited by availability, and the same menus were used throughout the study. Results revealed adjustments in meal size in response to the manipulation of energy density, so that total energy intake across 2-day blocks was nearly identical across conditions (13 673 *vs.* 13 573 kJ in the full fat *vs.* fat substitute conditions) (see Figure 4.2). As an index of variability, coefficients of variation showed the same pattern as in the previous work: for meal size, coefficients of variation were large (around 40%), but the coefficient of variation for 24 h energy intake was much smaller (around 10%). Consistent with our previous findings and confirming parents' informal observations, meal size was highly variable; however, we again noted evidence for adjustments in intake across successive meals, with large meals followed by small ones and vice versa, so that the total 24 h energy intake was relatively consistent. This pattern, in conjunction with the evidence for adjustments in meal size across successive meals, suggests that children regulate energy intake, and that regulation of meal size in response to the energy density of foods constitutes one mechanism controlling children's food intake.

4.2.4 Child feeding practices and meal size

The findings described in the preceding section indicate that children can regulate energy intake within and across meals; however, we also know that early experience with feeding can shape the extent to which children are responsive to the energy content of the diet. Specifically, child feeding practices can profoundly influence children's responsiveness to energy density and meal size. To investigate the impact of contrasting child feeding practices on children's responsiveness to energy density as a control of meal size, children participated in single meal protocols very similar to those previously described, in which the energy content of a first course was either

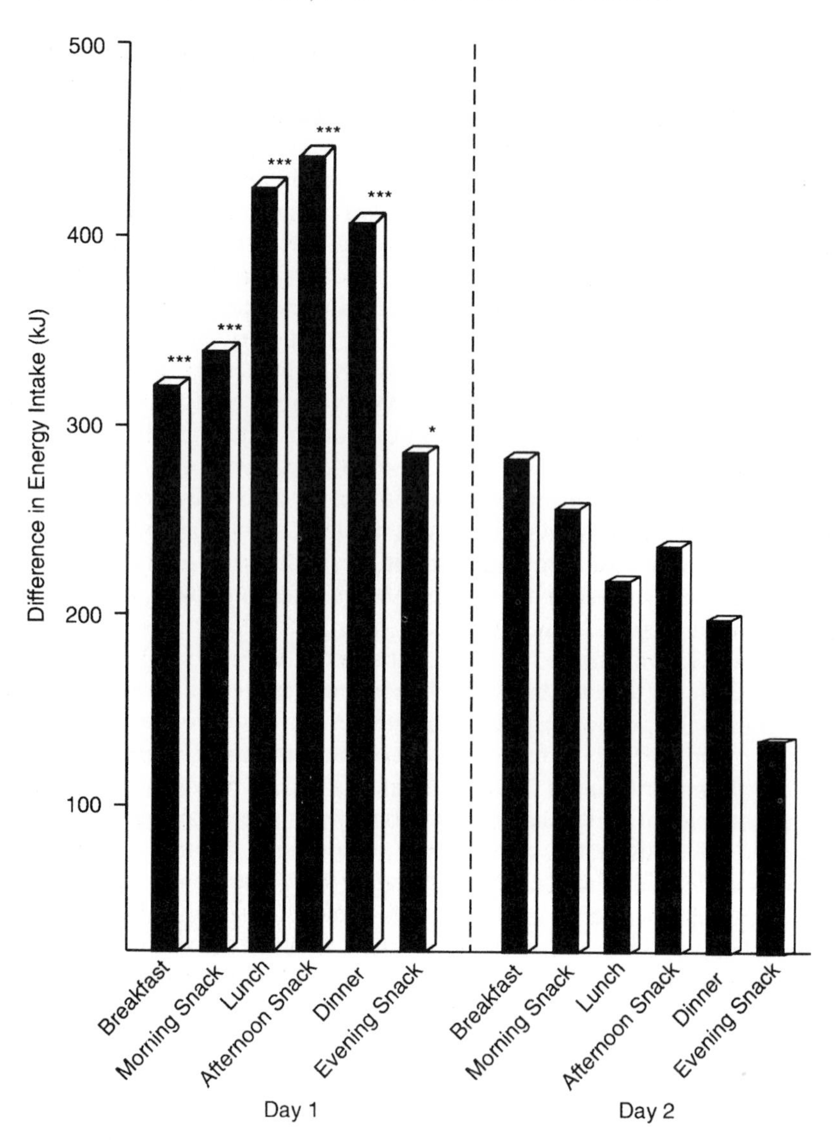

Figure 4.2 Cumulative difference in children's energy intake between placebo and fat-substitute conditions as a percentage of cumulative placebo intake, across eating occasions, over 2-d blocks. ■, Test meals. ***, $p < 0.001$; **, $p < 0.01$; *, $p < 0.05$.

high or low, and children then consumed a second course *ad libitum*. We varied the child feeding strategies used by the adults who were present at mealtimes (Birch *et al.*, 1987a). In one condition, adults focused the children on their internal hunger and satiety cues and discussed how those feelings help us know how much to eat. In the contrasting condition, children were focused on cues other than hunger and satiety that can control eating:

children ate at specific times, they were focused on how much food remained on the plate, and they were rewarded for finishing their portions. The children who were focused on internal cues of hunger and satiety showed clear evidence of adjusting energy intake in response to the energy content of foods, consistent with our previous findings. In contrast, when children were focused on external factors, *ad libitum* meal size did not differ with preload size and no evidence of responsiveness to energy density was noted. These findings emphasize that child feeding practices influence children's responsiveness to the energy content of foods and meal size.

We have begun to investigate how young children's responsiveness to energy density is influenced by parental control of child feeding (Johnson and Birch, 1994). We developed a laboratory measure of children's responsiveness to energy density, based on the single meal protocol described previously. On two different occasions, children consumed two-course meals, consisting of a first course 'preload' and an *ad libitum* self selected meal. From the child's energy intake on these two eating occasions, we calculated a compensation index for each child. This compensation index, COMPX, is the difference in *ad libitum* energy intake on the two occasions divided by the difference in the preloads' energy content, transformed to a percentage. A value of 100% would indicate 'calorie for calorie' adjustments in intake; smaller values reflect partial, incomplete compensation and larger values overcompensation.

Results indicated that the best predictor of the child's responsiveness to energy density was parenting style in the feeding context, in particular the child's ability to regulate energy intake was negatively related to the imposition of authoritarian parental controls on children's eating. Children's ability to regulate energy intake was also negatively related to their adiposity, with heavier children showing the least precise regulation of energy intake. Several investigators have noted relationships between parenting behavior and overweight in children (Agras *et al.*, 1988; Birch *et al.*, 1980a; Klesges *et al.*, 1983, 1986; Waxman and Stunkard, 1980). Results of research by Sjoden and colleagues (Koivisto *et al.*, 1994) investigating relationships between parental mealtime practices and children's food intake showed that children's intake was positively related to parental encouragement to eat, but intake was negatively related to parents' negative behaviors.

These recent findings are consistent with Hilda Bruch's extensive case history data on the etiology of childhood obesity (*cf.* Bruch, 1961) and with Ainsworth and Bell's (1969) pioneering study relating infant feeding styles to child outcomes are consistent with this view. Based on her work with families with an obese child, Bruch hypothesized that the quality of parent–child interaction was central in the development of the child's ability to regulate food intake and in determining the child's effective reactions to food. She concluded that hunger awareness was not innate, but resulted from learning experiences early in life, and that the failure of parents to give appropriate responses to child-initiated cues indicating need was of primary importance in the development of deficits in the self regulation of food intake. When a mother learns to offer food in response to signals indicating nutritional need, the infant will develop 'the engram of hunger' as a sensation distinct from other emotions, tensions, or needs. If on the other hand, the mother's reaction is consistently inappropriate 'be it neglectful, oversolicitous, inhibiting, or indiscriminatingly permissive, the outcome for the child will be perplexing confusion' and will result in a child who cannot determine whether hunger or some other discomfort is the cause of distress. Such a child needs 'outer criteria' to know when to eat and how much to eat.

To investigate links between parenting style and infants' feeding patterns, Ainsworth and Bell (1969) classified the feeding styles of 26 mother–child pairs during the first three months into one of nine patterns and related them to (i) feeding and digestive problems in the child, to (ii) other maternal care variables, and to (iii) the child's response to the Strange Situation procedure (a measure of attachment to the mother). Feeding patterns varied in the extent to which the mother intended to gratify the baby and respond to the baby's rhythms. Feeding patterns (described in such terms as 'thoroughgoing demand', 'flexible schedule', 'overfeeding to satisfy the baby') were associated with low incidence of feeding and digestive problems in the infants, and with securely attached infants at one year. In contrast, patterns indicative of little maternal responsiveness to the infant's needs, labeled as 'underfeed', 'overfeed', 'schedule', 'rigid', 'arbitrary' were associated with feeding problems and with infants showing attachment problems at one year. Feeding problems included spitting up, unhappy feedings, gastric distress, and underweight or overweight. Mothers were rated on their perceptions of the baby, acceptance, appropriateness and physical contact. Higher ratings on these were associated with the first set of feeding patterns. Ainsworth and Bell argued that the intent of the mother is most important; feeding patterns that have as objectives the gratification of the baby and the regulation of rhythms, and that recognize the baby as an autonomous participant were associated with the more positive infant outcomes. These patterns would optimize the chances that the child could learn that internal cues of hunger and satiety are relevant to the control of food intake.

The evidence is scant, but it appears that infants are predisposed to be responsive to the energy content of the diet, and to learn to associate food flavor cues with the consequences of ingestion, and both predispositions play an important role in the child's emerging ability to control meal size. Child feeding practices are a critical factor in shaping the extent to which meal size is regulated in response to internal cues of hunger and satiety, or in response to external environmental cues.

4.3 Children's food preferences and food selection

Humans are omnivores and, as such, have few innate food preferences. This maximizes adaptive potential and assigns a central role to learning and experience in establishing food preferences and consumption patterns. Several questions regarding the role of learning and experience in the development of food preferences will be addressed. First, what innate propensities constrain the effects of experience on the development of food preferences? The infant's early taste and olfactory abilities are discussed and evidence regarding innate preferences is presented. Second, the effects of experience on the formation of food preferences are addressed, and children's food neophobia and the effects of repeated exposure on the acceptance of new foods are described. Finally, evidence for the role of learning in the acquisition of preferences and aversions is described, followed by a discussion of research indicating how children's food preferences influence food intake regulation. We will see that even when children have little control over what foods are offered to them, their preferences have a profound impact on the quality of their diets. Relatively few studies address children's food preferences and the processes involved in the acquisition and modification of these preferences. Many more descriptive studies of children's consumption patterns appear in the literature (*cf.* Dierks and Morse, 1965; Dudley *et al.*, 1960; Dunshee, 1931; Lamb and Ling, 1946). These descriptive studies of children's intake patterns will not be reviewed here because they contribute little to our understanding of where children's food preferences and food habits come from. In addition, many of the studies of children's consumption patterns are compromised by the use of maternal reports of children's consumption patterns. Maternal report is a notoriously unreliable source of information regarding children's behavior in general (Wenar, 1963) and their food preferences in particular (Birch, 1980a; Glaser, 1964).

One reason for the reliance on maternal reports was the belief held by many nutrition researchers that children could not provide reliable and valid information regarding their food preferences (Bryan and Lowenberg, 1958; McCarthy, 1935; Sanjur and Scoma, 1971). A statement by Bryan and Lowenberg is representative of this attitude:

> 'Because children of this age were not considered reliable enough to furnish the desired information on their food preferences, the mother was queried on the child's specific reactions to . . . foods' (p. 31).

Subsequent research, however, has demonstrated that children can indeed provide reliable and valid preference data (*cf.* Birch, 1979b, 1980a; Phillips and Kolasa, 1980). Our procedure for obtaining preference data from young children is reliable (Birch, 1979a) and predicts consumption in a self selection setting (Birch, 1979b). In one study designed to investigate the relationship between preference and consumption (Birch, 1979b), the obtained correlation was 0.80, higher than the correlations typically obtained for adult subjects (*cf.* Peryam, 1963).

4.3.1 Biological constraints on the effects of experience

A variety of innate predispositions constrain the impact of experience on the development of food preferences. Included here are the constraints on (i) what is perceived, (ii) what is preferred prior to experience, and (iii) the learning meachanisms that are functional in the acquisition of food preferences, together with the kinds of associations and information that are acquired.

4.3.2 What is perceived?

The infant's ability to taste and smell places limits on food preferences, especially as they limit the flavors the infant can experience. Infant taste and olfactory perception is discussed only briefly here, and the interested reader is referred to two excellent reviews on this topic (*cf.* Bartoshuk and Beauchamp, 1994; Cardello, 1996; Cowart, 1981). In humans, both the taste and olfactory systems are functional at birth and even during the late fetal period. For example, DeSnoo (1937, cited in Cowart, 1981) demonstrated that the fetal gulp rate varied systematically with the presence or absence of a sweet tastant in the amniotic fluid. After birth, the infant shows differential responsiveness to several of the basic taste qualities and to variations in the concentration of some of these tastants. A variety of measures have been used to investigate taste perception in the newborn, including sucking rate and pattern (Crook, 1978), total intake (Desor *et al.*, 1975), heart rate (Lipsitt, 1977), and facial expressions (Steiner, 1977), all of which have their advantages and disadvantages. Based on these measures of ingestion, newborns appear to be able to distinguish between sweet solutions and water (Desor *et al.*, 1973), among different sweet tastants such as sucrose, fructose, lactose and glucose, and among different concentrations of these tastants. In addition, infants respond differently to sweet, sour and bitter tastants (Steiner, 1977), avidly accepting sweet and rejecting sour and bitter substances. Differential responsiveness to salt appears to develop at about four months postnatally; prior to that there is no clear evidence that salt is perceived or preferred (Beauchamp *et al.*, 1994).

By the time they are several months old, infants can distinguish among various tastants in a manner similar to that of adults. However, we do not have any evidence regarding possible developmental differences in the suprathreshold functions relating stimulus concentration to perceived intensity (see Bartoshuk, 1979, for a discussion of this point). With respect to evidence regarding any physiological basis for developmental differences in perception and preference, recent research refutes the contention that infants are more sensitive to tastants because they have more and more widely distributed taste buds than older individuals. There appears to be little change in the number of taste receptors from birth to middle adulthood (Cowart, 1981). With respect to individual differences in perception that can also influence preferences, there is some evidence for genetic differences in responsiveness to bitter (6-n-propylthiouracil; PROP) as early as the preschool period (Anliker *et al.*, 1991). Further, these genetic differences in sensitivity to bitter are also related to the children's preferences for common foods, including cheese and milk: PROP tasters appeared to like cheese less than nontasters, while liking milk more than nontasters.

Several methodological difficulties hinder the investigation of the development of taste and olfactory perception in infancy. First, given their limited response repertoires and cognitive abilities, standard psychophysical methods cannot be used with infants and very young children. When differential responses to stimuli are noted, it is difficult or impossible to determine whether infants' differential responsiveness is attributable to preference or perception. A second methodological difficulty concerns the use of intake and indices of sucking as dependent measures; it appears that the infant's ability to inhibit the sucking reflex is limited even in the presence of stimuli that adults find noxious. This limitation was tragically illustrated by the fatal salt poisonings that occurred among newborns when salt was accidently substituted for sugar in their formula preparations (Finberg *et al.*, 1963). For a review on factors influencing the formation of food preferences, a third difficulty is that most of the research on taste perception during infancy and early childhood has investigated only responses to basic tastants, not to complex, multidimensional food stimuli. It is not clear how relevant work with simple tastants is to our understanding of preferences for complex food stimuli. For example, recent research (Beauchamp, 1980) suggests that as early as two years, preference for salt concentration differs dramatically as a function of whether the diluent is water or soup, a finding that calls into question any extrapolations from data on simple tastants to preferences for foods. Finally, although much of flavor is a result of the smell of food, less is known about olfaction than about taste during infancy. For example, the most salient quality of odor is its pleasantness or unpleasantness, and the unpleasantness and pleasantness of odors is learned, probably early in life (see Bartoshuk and Beauchamp, 1994, for a discussion of this issue). We know very little about this early learning about odors; Engen (1982) has reviewed what is known about the development of olfaction during the first years of life. The recent work of Mennella and colleagues (see below) on the effects of infants' exposure to flavors in breast milk has contributed to our understanding of early learning about odors and flavors.

4.3.3 What is preferred? Innate preferences and neophobia

In omnivores such as man, food preferences result from an interaction of genetic and experiential factors, with learning playing a critical role (see Rozin, 1977). However, a central question concerns what, if any, preferences exist prior to experience with food. As in the case of research on taste perception during early life, a variety of measures have been employed in studies of taste preference, including amount ingested (Desor *et al.*, 1975; Nisbett and Gurwitz, 1970), tongue movements (Nowlis, 1973), facial expressions (Steiner, 1977) and heart rate (Lipsitt, 1977).

The only well documented unlearned preference is for the sweet taste; results of all the studies cited above, regardless of the measure used, consistently reveal a preference for the sweet taste. Lipsitt (1977) reports that heart rate increases and the sucking rate within bursts slows in response to sweet tastants. His interpretation is that the sucking pattern slows to allow the infant to savor the taste, while the heart rate change is seen as an indicator of 'the joy that the infant . . . derives' from the sweet tastant. Nowlis (1973) used patterns of tongue movements to investigate newborns' preferences and reported that proportionate increases in the amplitude of anterior tongue movements occurred in response to sweet tastants, which he took as evidence for a preference, and that a proportionate increase in posterior tongue movement amplitude occurred in response to bitter tastants, indicating an aversion. When Desor and colleagues (Desor *et al.*, 1975) used differential ingestion as an indicator of preference, they noted that newborns ingested more of sweet solutions than of plain water, or sour or bitter solutions; they also found that the infants ingested more solution as the sweetness concentration increased (Desor *et al.*, 1973), which they interpreted as a preference for sweet and for increased preference with increased concentration. The widespread and apparently innate preference for sweet has led to the suggestion that sweetness is an unconditioned stimulus for ingestion (Le Magnen, 1977). The evidence cited above also suggests innate aversions to bitter and sour tastants. Responses to salt are more ambiguous.

One of the most significant features of taste and olfaction is that they elicit very strong affective reactions that appear to be innate. From birth, the gustofacial reflex (GFR) is elicited by basic tastants. The GFR consists of well differentiated motor reactions of the facial muscles to taste and smell stimuli (Steiner, 1974). Steiner (1974) and Peiper (1963) have argued that these reactions are rigidly fixed features, differing in appearance in response to sweet, sour, and bitter stimuli. They are innate and are controlled by the brainstem, and do not necessarily involve the cortex, as evidenced by the presence of the GFR in anencephalic infants (Steiner, 1977). The GFR can be viewed as rudimentary nonverbal communication that indicates the organism's willingness to accept or to reject a stimulus. These reflexive expressions are consistently interpreted as indicative of the infant's hedonic reactions by adults.

Initially, Steiner noted the GFR in studies of adults using high concentrations of tastants. Subsequently, he investigated the GFR in a study of 100 term newborns, four of whom had CNS malformations. The procedure involved rinsing the tongue of the neonate with the four basic tastants (sweet, salty, sour, and bitter). Subjects were filmed during this procedure. A set of adult judges, blind to the tastant used on a particular trial, categorized the infants' GFRs in terms of the facial expressions and frequently provided labels of the effect conveyed by the expressions. The response to the sweet stimulus (sucrose) was labelled 'liking', 'enjoyment' and involved a slight smile, sucking movements, and the licking of the upper lip. To the sour stimulus (citric acid) the infants showed 'sour' lip pursing, often wrinkling of the nose and blinking of the eyes. Stimulation with bitter (quinine sulfate) elicited descriptors such as 'dislike', 'disgust', 'rejection' and involved the opening of the mouth with the upper lip elevated, and protrusion of the tongue in a flat position. It was often followed by spitting or vomiting. These responses were reliably discriminated by the adult judges. However, there did not seem to be a characteristic and consistent response to salt.

Chiva (1982) has also investigated the GFR in infants and has traced the course of the GFR during infancy. He argues persuasively that the GFR forms the basis of preferential responses to a wide variety of social as well as taste stimuli, and that it develops during infancy. Based on a longitudinal study of 40 infants from birth to two years, he describes three phases in the evolution of the GFR from reflex to social response. This evolution occurs because, although the GFR is initially reflexive, it is interpreted by parents as communicating effect, and it is the parent–child interactions triggered by the GFR that are responsible for this evolution. First, from birth to six months, taste stimulation triggers the standard GFR, which is not directed at anyone in particular. Second, from 9 to 14 months, the response tends to be more modulated, and can occur to more dilute stimulation. Finally, in the third stage, from around 16 months or so, the child begins to initiate the response in the absence of anything that resembles the unconditioned stimulus, and begins to use these expressions deliberately and to orient them to specific individuals. Chiva does not discuss in detail the process that produces this shift in the stimuli that elicits these characteristic expressions, but it seems likely that associative conditioning is implicated. Associative conditioning appears to be central in the acquisition of evaluative responses (Martin and Levey, 1978), and the evidence for the contributions of associative processes to the acquisition of food preferences will be discussed in the subsequent section on the contributions of learning and associative processes.

4.3.4 Neophobia and effects of repeated exposure on food preference

In addition to an innate preference for sweet and an unlearned rejection of sour and bitter tastants, there seems to be a built-in bias to respond negatively to novel tastes; this fear and dislike of novel tastants has been termed neophobia. Although there is no research in the literature demonstrating the existence of neophobia in the neonate, the pervasiveness of the neophobic response across human cultural groups and across species of omnivores suggests that the neophobic response may be innate. There is evidence for neophobic responding by humans as early as 6 months (Sullivan and Birch, 1994), and neophobia is quite strong among children by 2 years (Birch and Marlin, 1982). Adult humans are neophobic (Pliner, 1982), as are rats (Rozin, 1976), Japanese macaques (Itani, 1958) and chimps (Weiskrantz and Cowey, 1963).

Children don't readily accept new foods, with the notable exception of foods high in sugar, which children don't have to learn to like. This neophobia, or fear of the new, is normal in children, and neophobia is found among other omnivorous species. In his discussion of the 'omnivore's dilemma' Rozin (1976) has pointed out that this neophobia is normal and adaptive: as omnivores, we need a variety of foods, and must compose the diet from available edible substances. However, ingesting new substances is a risky business; the substance may prove to be toxic. This view suggests an evolutionary basis for neophobia, which can protect against the ingestion of potentially dangerous items. Typically, neophobia is reduced by repeated consumption of a new food, which is not followed by any negative gastrointestinal consequences. For the child who is just being introduced to the adult diet, all foods are initially novel, and the young child's neophobic response can have a particularly powerful effect on their food selection and eating behavior.

In her research on dietary self selection, Davis (1939) described the dramatic changes that occurred in food acceptance as children were introduced to new foods and had repeated opportunities to sample them. When the weanling infants first began to self select their diets, they initially tasted many of the foods, sampling widely from among the new foods. She described how with repeated opportunities to eat the foods, she saw food preferences develop. They began to avidly seek out some foods and reject others; food preferences emerged before her eyes. Davis was careful to emphasize that the 'trick' of her experiment was the array of foods that were offered to the children: a set of healthy foods, simply prepared without additional salt or sugar. These foods bear little resemblance to many of the foods available today, especially to many of those foods that are marketed for children. As Davis stated:

> '. . . leave the selection of the foods to be made available to young children in the hands of their elders where everyone has always known it belongs'.

When the children self selected their diets from a limited array of healthful foods, the children did very well. Davis' work is often misquoted and used to support the 'wisdom of the body': children 'know' what nutrients they need, and will seek out the foods containing those needed nutrients. In fact, there are no human data to support this view. Except for a few substances (e.g. sodium), the evidence for such wisdom of the body and for 'specific hungers' is weak, even in experimental animals who are severely deprived of the essential nutrient (Galef, 1991).

We have investigated the effects of repeated exposure to new foods on children's preferences for those foods, and found that with repeated exposure, many new foods that children initially rejected are accepted (Birch and Marlin, 1982). However, acceptance does not come immediately, but may take 8 to 10 exposures, and must involve tasting the food; looking at and smelling it is not sufficient to induce increased acceptance (Birch *et al.*, 1987b). Unfortunately, parents do not often appreciate that the child's initial rejection of a new food (i) is normal, (ii) reflects an adaptive process, and (iii) can be transformed to acceptance following repeated opportunities to eat the new food. The commonly held view is that the child's initial rejection of a food reflects a fixed, immutable dislike for the food. As a result, the child may be viewed as finicky, and the new food may not be offered to the child again, eliminating any opportunity for the child to learn to like the food. The child's neophobia plays a central role in early food acceptance. The fact that early and repeated opportunities to eat new foods can change initial rejection to acceptance underscores the critical role of parents in selecting the array of foods offered to their children.

We recently investigated infants' responses to their first solid foods and whether their acceptance of new foods was enhanced with repeated exposure (Sullivan and Birch, 1994). Infants 4 to 6 months old were fed a novel vegetable on 10 occasions, several times each week by their mothers, and intake of the vegetable was measured before, during, and after their opportunities to eat the food. Infants were videotaped while eating, and adults rated the videotapes for infants' acceptance of the foods. Based on facial expressions and intake data, infants showed some reluctance to ingest the foods when they were first presented. Over the exposure series, infants showed dramatic increases in intake of the vegetables, doubling their intake from about 30 g to about 60 g. About half our sample had been formula fed, while the other infants had been breast fed. The results differed for formula fed and breast fed infants; increases in intake were most dramatic for the breast fed infants. We hypothesize that this greater acceptance of a novel food by the breast fed infants is due to their greater experience with a variety of flavors. It has now been confirmed that flavors from the maternal diet pass into breast milk (Mennella and Beauchamp, 1991a). Recent research reveals that volatile flavor compounds ingested by mothers are present in breast milk, and that infants respond systematically to such flavors, by altering sucking rate, time spent at the breast, and in some cases amount ingested. Flavors employed in this research include alcohol (Mennella and Beauchamp, 1991b), garlic (Mennella and Beauchamp, 1993), and vanilla (Mennella and Beauchamp, in press). Because research with animal models has shown that young animals who experience dietary variety show much more ready acceptance of novel diets than do animals whose dietary experience is limited to a single diet, additional research is needed to determine ways in which human infants' early experience during the suckling period influences subsequent preferences and food acceptance patterns. Specifically, we know that rat pups learn to prefer their mother's diet via exposure to cues present in her milk. Do human infants learn to prefer flavors that are routinely present in their mother's milk? Do these learned preferences have an impact on the acceptance of solid foods? At this point, we don't know.

Our findings indicating that the reduction of neophobia with exposure differs as a function of the infant's prior feeding experience suggests avenues for the development of individual differences in neophobia and food preferences, and we will return to the question of individual differences in neophobia in older children and adults in the next paragraph. Numerous questions regarding developmental trends in neophobia remain and although systematic studies remain to be done, there is some evidence suggesting that neophobia may either increase or decrease with age. For example, in some of the work on the modification of food preferences done in our laboratory, multidimensional scaling analyses of the preference data indicated that one very salient dimension of preference was familiarity, with more familiar foods being more preferred. When the data for older (4-yr-olds) and younger (3-yr-olds) children were analyzed separately, familiarity was found to account for a greater proportion of the variance in the data of the younger group (Birch, 1979a). Itani (1958) also reported that new foods were more readily accepted by the younger members of the troop of Japanese macaques. In contrast, Otis (1984) reported that among adults from 17 to 50, the older ones were more willing to try unfamiliar foods than the younger ones. We have argued elsewhere (Birch and Marlin, 1982) that minimal neophobia would be adaptive during the weaning period when children are being exposed to foods of the adult diet of their culture and must accept these new foods in order to maintain growth and health. Fischler and Chiva (1985) have also pointed out that increasing neophobia with age during early childhood would be adaptive in that such neophobia would minimize the likelihood that unknown, untried substances would be consumed at a time of increased independence and reduced adult supervision. Pliner (1994) and her colleagues reported a relatively high degree of neophobia among their school age subjects (5- to 11-year-olds), and no age differences across this period of development. Cashdan (1994) has recently made a similar argument for a 'sensitive period' during the first 2 or 3 years of life, when children are learning which foods are safe to eat. However, the data to support a 'sensitive period' are very incomplete. Pliner and colleagues (Pliner and Hobden, 1992) have developed a scale for measuring neophobia and have shown it to be relatively stable during adulthood, and related to other individual characteristics such as trait anxiety and sensation seeking, and prior experience with novel foods. However, situational factors, such as the novelty of the eating situation, and exposure to models who are more or less neophobic, can also influence the neophobic response (Pliner *et al.*, 1993).

It is clear that neophobia exerts an effect on food preference and food acceptance patterns, and that neophobia is typically reduced with exposure. However, children's experience with food does not occur in isolation, but in a social context. Adults, siblings, and peers can serve as models, and can also attempt to exert direct control over children's eating. Evidence for the effects of such social context on the acquisition of food preferences appears in the next section.

4.3.5 Which learning processes? What is learned?

Rozin and Zellner (1985) have argued convincingly that Pavlovian or associative conditioning plays a central role in the acquisition of food preferences. This view is consistent with others in the psychological literature in assigning conditioning a central role in the acquisition of affective reactions to objects (*cf.* Martin and Levey, 1978). Rozin and Zellner's (1985) broad definition of Pavlovian or associative conditioning will be adopted here:

> 'changes in response or attitude to stimuli resulting from their contingent occurrence (temporally and/or spatially) with other, more potent stimuli. The potency of these other stimuli may be genetically programmed (the traditional US) or result from an acquisition process, as in higher order conditioning. . . .'

Research will be presented that suggests that flavor cues in foods are readily associated with the effect generated by the physiological consequences and social contexts of eating. Since this effect can be either positive or negative, such conditioning can result in positive or negative shifts in preference.

There is large body of evidence indicating that in adult organisms, food flavor cues are very readily associated with the physiological consequences of eating, particularly when those physiological consequences are nausea and vomiting. In fact, flavor cues are more readily associated with physiological consequences of eating than are other potential CSs that are present, such as the food dish, or the location of eating (see Barker *et al.*, 1977, for discussions of this general point). These conditioned aversions have been studied extensively in the rat and other organisms, and this research provides examples of selective associations (see, for example Garcia *et al.*,

1974). These associations are also unusual in that they can form after only a single pairing of the food flavor with physiological consequences that follow ingestion. Furthermore, they are long lasting, and resistant to extinction.

There are a few reports of conditioned aversions in humans in the literature (Bernstein and Webster, 1980; Garb and Stunkard, 1974; Logue *et al.*, 1981), and there is some evidence that such aversions are most commonly established during childhood. In contrast to the voluminous literature on conditioned aversions in other animals, there are relatively few studies on conditioned aversions in humans. Based on the retrospective reports of approximately 700 individuals, ranging in age from preschoolers to old age, Garb and Stunkard (1974) found that more than half of their subjects reported having at least one food aversion and that the onset of such aversions was greatest during childhood. More recently, Logue *et al.* (1981) questioned 517 undergraduates regarding their acquisition of illness-induced aversions to foods and drinks. Sixty-five percent of the subjects reported having at least one aversion to a food's taste. Such aversions were notoriously long lasting, and subjects reported that extinction rather than forgetting was responsible for the disappearance of the aversion. In about 20% of their cases, the subjects reported that they were certain that something else other than the aversive food (flu, consumption of alcohol) had caused the nausea and vomiting, but they formed an aversion to the food anyway. This evidence is consistent with the view that what is learned in this associative conditioning is not a contingency, 'If I eat this food, then I'll get sick'. Instead, the food takes on characteristics of the unconditioned stimulus, evoking feelings of nausea and revulsion when the food is encountered on subsequent occasions.

Research on experimentally induced taste aversions in humans is sparse. However, Bernstein and her colleagues have investigated the development of aversions in both adults and children receiving chemotherapy treatment for cancer (Bernstein, 1978; Bernstein and Webster, 1980). In the initial study, they asked whether children receiving drugs that produced nausea and vomiting would acquire aversions to foods consumed in temporal proximity of the effects of the drugs. Children between 2 and 16 years were given either a novel ice cream flavor or no ice cream prior to their treatment. A control group that received either no drug treatment or treatment that did not produce nausea was also included. Two to four weeks later, subjects were offered a choice between eating the ice cream or playing a game. The differences among the groups in their choices were striking; while only 21% of the children in the experimental group chose to eat the ice cream; 67% and 71% of the children in the two control groups did so. The subjects were also offered a choice between the ice cream associated with treatment and another novel ice cream. They were asked to taste each ice cream and then choose the one they preferred, and eat as much of it as they wanted. Both the choice data and the consumption data indicated that preference for the presented ice cream was significantly lower in the experimental group of children. Her subsequent research with adults has shown similar findings, suggesting that such aversions may be a contributing factor to the anorexia frequently seen in patients undergoing chemotherapy. For our purposes, this work provides evidence that children do form aversions that could influence their patterns of preference. However, despite the fact that aversions are very powerful deterrents to consumption, most individuals report the existence of only one or a few such aversions (35% of the subjects in the research by Logue *et al.* (1981) reported no aversions). Thus, aversions may account for why a few foods are not consumed, but they probably have a limited effect on the range of foods consumed by the individual.

There is also some evidence that conditioned *preferences* can be acquired in a manner analogous to conditioned aversions. These result from the association of food cues with positive consequences of ingestion, including recovery from vitamin deficiency produced illness (Zahorik *et al.*, 1974), and cessation of drug withdrawal (Parker *et al.*, 1973). While these conditioned preferences seem to be less potent than conditioned aversions, they may be more widespread and play a much more pervasive role in determining patterns of food preference and consumption. Given repeated opportunities to consume foods, rats learn to associate foods' sensory cues with positive postingestive consequences of ingested nutrients, and they subsequently learned to prefer energy dense foods over low energy versions of the same foods. These learned preferences based on high energy density result whether the high energy content is due to carbohydrate or fat (Sclafani, 1990). Booth has reported research with human adults indicating that satiety can serve as a reinforcer to produce conditioned preferences (Booth *et al.*, 1982), and that satiety can be conditioned (Booth *et al.*, 1976).

Children consume foods they like and many of these well liked, preferred foods are energy dense foods, high in fat or carbohydrate content. Why do children and adults like these energy dense foods? There are a variety of reasons, ranging from cultural to sensory: 'rich' foods are high fat foods, often foods reserved for feast, holidays and special occasions. From a sensory perspective, many of the volatile substances that impart flavor to food are fat soluble, so that high fat foods are often flavorful ones (Birch, 1992; Mela, 1992). In addition, to the extent that the energy content of a food comes from simple sugars, the foods are also sweet, and the preference for sweetness is strong, unlearned, and well established at birth. Other high energy foods are high in salt content. As previously indicated, an apparently unlearned preference for salt emerges at about 4 months after birth (Beauchamp *et al.*, 1986), and infants and children tend to prefer higher levels of salt in food than do adults. High energy, high fat and high carbohydrate foods also have specific physiological effects (Smith and Greenberg, 1992) that can produce pleasurable feelings of satiety.

Recently, we have shown that preferences for energy dense foods also have a learned component; the gastrointestinal consequences of ingestion can apparently serve as powerful unconditioned stimuli in associative conditioning of food preferences. Inspired by the animal research on these learned food preferences, we have conducted a number of experiments that have confirmed that such learned associations between foods sensory cues and the postingestive consequences of high energy content contribute to children's preferences for high energy foods (Birch *et al.*, 1990; Johnson *et al.*, 1991; Kern *et al.*, 1993). These protocols are very similar to those described earlier to explore children's responsiveness to energy density. Children are given repeated opportunities to consume high and low energy versions of a food, such as yogurts, drinks, or puddings, that vary in energy density. The high and low energy versions of the foods are distinctively flavored, while other sensory characteristics of the two versions, such as texture and mouthfeel, are very similar. After assessing the child's initial preferences for the foods, on different days, a child would consume equivalent fixed amounts of a high energy, almond flavored yogurt, and on other days, a low energy peppermint yogurt (other children have the reverse energy density/flavor pairing). Following several opportunities to eat the high and low energy versions, the child's preferences for both flavors are assessed. Children's preferences are assessed by having the child tell us how well they like a food based on tasting small food samples, using a procedure that has been shown to yield reliable, valid preferences that are good predictors of children's consumption in a self selection setting (Birch, 1979a,b, 1980a). Using this preference measure, results confirm that via association of flavor cues with the postingestive consequences of high energy density, children learn to prefer flavors repeatedly associated with high energy density. Learned preferences could serve an adaptive function, allowing the child to learn that particular flavor cues predict pleasurable feelings of satiety that result from high energy dense foods.

Children and other omnivores seem predisposed to learn associations between foods' sensory cues and the consequences of ingesting those foods. In addition, the social context of eating provides a rich complex of cues that are apparently also very central in the development of the young omnivore's food acceptance patterns. For the most well documented evidence on this point, the reader is referred to Galef and colleagues' very elegant studies on the social transmission of diet preferences in rats (Galef, 1976). Research on how social learning and associative conditioning to the social contexts of eating shape children's food preferences appear in the next section.

4.3.6 The social and cultural context of eating and learned food preferences

Rozin (1984) has pointed out that if you want to know about an individual's food acceptance patterns and can ask that person only one question, it should be 'what is your cultural or ethnic group?' From among the wide array of potentially edible substances available in the environment, the members of a particular cultural group eat only a relative few. For example, Americans don't eat sea slugs and insects, while for orthodox Jews and Muslims, pork is not an acceptable food. In addition, within a culture, there are frequently food taboos for particular age and sex groups that restrict food experience for those subgroups. The child must also acquire rules of cuisine regarding what foods are eaten at which meals during the day, and what food combinations are appropriate and inappropriate. It's often easier to see these cultural rules in cultures outside our own, but we also have such taboos and rules of cuisine. For example, Americans don't view coffee, lobster, and spicy foods to be appropriate for young children, and the combination of pickles and ice cream is bizarre and inappropriate except for pregnant women. With respect to time of day rules, most of us eat pizza for dinner, not for breakfast. The few studies that have investigated the impact of cuisine rules and food taboos on children's food acceptance and rejection and rules of cuisine are described below.

Various potential edibles may be rejected as food for different reasons. Rozin and his colleagues have proposed a taxonomy of food rejections and have conducted research on the development of children's understanding of this classification system (Fallon and Rozin, 1984; Rozin and Fallon, 1980, 1981). They have explored people's motivation for food acceptance and rejection in the USA and have concluded that by young adulthood, there are three basic types of reasons for acceptance or rejection of edibles: (i) sensory affective factors; (ii) anticipated consequences; and (iii) ideational factors. Each of the three categories of reasons has a positive pole that motivates acceptance and a negative pole that motivates rejection In the case of the individual's reaction to a particular food, it may be based on one or more than one of these reasons. Their work on the development of children's understanding and use of these categories has revealed a developmental sequence in the acquisition of these three categories (Fallon *et al.*, 1984). Before reviewing this developmental sequence, the three categories will be described in more detail.

First, many foods are accepted or rejected primarily in response to the foods' sensory characteristics, including the taste, odor, mouthfeel and appearance of food. Positive affect is produced by good tastes, negative affect by distastes. Affect may be generated simply as unconditioned responses, as in the case of the sweet taste, or may also be modified by associative conditioning, so that the response to a food is a result of both unconditioned and conditioned components. Second, some edibles are accepted or rejected because of the consequences they are thought to produce. These consequences may be immediate, as when you are hungry and eat something to obtain the pleasant feelings of satiation, or more delayed, as when you eat something because it's 'good for you', or even because of an anticipated change in social status that might result from eating a particular food. Finally, there are ideational factors. This category concerns knowledge of where foods come from and what they are. This knowledge about the sources and nature of foods operates more commonly in rejection than acceptance. Some substances are inappropriate as food; grass and sand are examples. Inappropriate substances are rejected because they are simply not food. The second subcategory is that of disgust, which is strongly affectively loaded. These disgusting items, although usually never tasted, are thought to taste bad and may produce nausea by just thinking about eating them. Disgust items are usually of animal origin. They contaminate and pollute; contact with even the tiniest amounts makes otherwise palatable food inedible. Many of these disgusts appear to be culturally specific; a Hindu may find the thought of eating beef disgusting, while we find the thought of eating horsemeat disgusting. Although the disgust reaction to feces appears to be universal (Angyal, 1941), it is not present until around three years of age leading to speculation that it is attributable to experiences surrounding toilet training (Rozin and Fallon, 1981).

What is the developmental progression in children's use and understanding of these various motivations for accepting or rejecting a food? In an initial study of the child's conceptions of foods Fallon and Rozin (1984) used the psychological taxonomy they developed from adults' classification of foods and nonfoods based on motivation for ingestion (or rejection). Children ranging in age from 3.5 to 12 years responded in structured interviews regarding their acceptance and rejection of a variety of edible and nonedible substances, and supplied reasons for their responses. The authors were particularly interested in the development of disgust reactions based on contamination sensitivity. Recall that disgust objects are typically contaminants; even small amounts of disgusting objects can render an otherwise preferred substance inedible. They noted a developmental sequence in the appearance of these categories of rejection. The first category to appear, present in the 3- to 5-year-olds, was distaste. A substance is not eaten based on sensory affective considerations; it tastes bad. This applies to objects that will later be categorized as dangerous (poison), inappropriate (grass), or disgusting (grasshopper, 'doggie doo'); the youngest subjects did not differentiate among distaste, danger and inappropriate as reasons for rejection. The youngest subjects seemed to believe that all things that are bad for them will taste bad. Such a simple state of affairs might be desirable from an adaptive point of view, but obviously does not reflect reality.

In their investigation of the development of contamination sensitivity, the children were told several stories of the same general structure. A preferred beverage is presented, and several different target substances (something either disgusting [grasshopper], inappropriate [leaf], or dangerous [poison], a preferred food [M&Ms], or a distasteful food) are first placed in proximity to the juice, fall into the juice and are removed in a series of stages. At each of these points in the story the child is asked to give a hedonic response to the beverage. For many of the youngest subjects, the simple removal of the potentially contaminating substance (grasshopper) was sufficient to return the beverage to its initial hedonic value. In contrast, at the other end of the age range, pouring out that sample and replacing it was necessary to produce increases in the hedonic rating of the beverage. Even washing the glass and pouring a new sample of the beverage into the glass was not sufficient to return the ratings of some of the older subjects to their initial level; contamination persisted in an ideational form. The mere idea of the association of the contaminant with the beverage was sufficient to reduce liking.

The idea that contamination persists when the contaminating item is removed appeared in the data of the older children (older than 6 years), but not the younger ones. This idea seems to require knowledge about the physical properties of objects; the concept of diffusion in particular would seem necessary. When questioned, the younger children did not seem to understand this concept and believed that removing the offending object would also remove any of the substance that might have dissolved in the beverage. Fallon and Rozin (1984) speculate that this developmental shift is related to the transition from preoperational to operational thinking proposed by Piaget, which involves the ability to distinguish between reality and how things appear to be. Their data indicate that by the age of about 4 years, children reject foods using the same categories that adults use, but that they do not differentiate distaste from disgust and inappropriateness. This work is an excellent first step in our understanding of the development of motivation for food acceptance and rejection. But as Fallon and Rozin point out, it raises the question of how the cognitive–affective linkage develops that is necessary for the child to come to see disgust as a separate motivation for rejection.

Every culture also specifies rules of cuisine regarding foods' appropriateness for particular occasions and mealtimes. We were interested to see whether young children had acquired sociocultural cuisine rules regarding the time of day when particular foods are eaten, and whether these rules regarding food appropriateness influenced children's food preference (Birch *et al.*, 1984a). We hypothesized that rules of cuisine could be strengthened and perpetuated by conditioned preferences that resulted from repeated consumption in a particular meal context; certain foods are eaten at particular times of day and occasions, not just because it is appropriate to do so, but also because that's when they taste best. To investigate this point, adults and young children (3- and 4-year-olds) categorized a set of foods according to when it was most appropriate to eat them: 'for breakfast' and 'for dinner'. They did this categorization twice; once at breakfast time and once at dinner time. On both occasions they also gave us preferences for the foods. Included in the set were foods appropriate in our culture for breakfast (e.g. cereal) or dinner (e.g. pizza).

Both the adults' and the children's categorizations of the foods were consistent across the times of day, but their preferences were not. Preferences showed systematic shifts with time of day, with foods being more preferred at the time when it was most appropriate to eat them. Cereal was more preferred in the morning than in the afternoon; pizza was more preferred in the afternoon than in the morning. We were surprised to note that there were no age differences in this effect; both children and adults showed this pattern of preference shifts with time of day. Children and adults showed no clear differences in their categorization of the foods. This pattern of results indicates that by 3 or 4 years old, children have already acquired knowledge about rules of food appropriateness in their culture; and this knowledge is related to patterns of preference.

TV is a major source of information about popular culture in the USA. However, of the large number of studies investigating the impact of television on children, only a few deal with effects on children's food preferences and eating behavior. This research indicates that television provides a vehicle for transmitting sociocultural constraints regarding food, including beliefs, attitudes and preferences. Galst and White (1976) found that the greater the amount of commercial television to which children were exposed, the greater the number of purchase requests they made during trips to the supermarket. Most of these requests were for cereals, the most frequently advertised foods in commercials directed at children. In a subsequent study, Goldberg *et al.* (1978) found that when children were exposed to commercials for highly sugared foods, they selected more of these foods in comparison to children who viewed public service announcements on nutrition. Galst (1980) showed 3 to 6-year-olds programs with commercials that were for foods with added sugar, without added sugar, or pro-nutrition public service announcements. The conditions also varied in whether these messages were accompanied by adult comments regarding the messages. Each day following the viewing, the children then selected their snacks. The condition that was most effective in reducing the children's selection of highly sugared foods for snacks was the combination of commercials for foods without added sugar and pro-nutritional messages accompanied by adult comments.

Unwritten rules of cuisine regarding when certain foods are appropriate and should be eaten shape our dietary patterns and place additional constraints on children's food preferences. Television has been used as a vehicle for advertising messages that also suggest that particular foods taste good, are fun to eat, and are good for you. However, social factors also influence food preferences. As discussed in the next section, social interaction at mealtimes with adults and with peers also plays a role in shaping children's food preferences and food selection patterns.

4.3.7 Effects of social interaction on food preference

For the young child, eating is typically an occasion for social interaction. Others present can include other eaters who serve as models, or caretakers who are attempting to control the quantity and variety of food the child eats. The research reviewed below includes studies investigating (i) social learning, particularly the effects of peer and adult models on children's food preferences and consumption patterns, and (ii) the effects of child feeding practices employed by caretakers to control the child's behavior on the acquisition of food acceptance patterns.

Several studies in the literature have examined the relationship between children's food preferences and those of their parents. Parents could influence the food preferences of their children in three ways: by serving as models for their children, by controlling the food available to children, and through genetic transmission of predispositions. Three studies investigating this relationship have produced somewhat contradictory results (Birch, 1980b; Pliner, 1983; Pliner and Pelchat, 1986; Rozin *et al.*, 1984). The first study by Birch (1980b) explored this relationship among 128 preschool children and their parents. Preferences for a variety of foods, including fruits, vegetables, sandwiches and snack foods, were obtained from children and their parents. When the preference orders obtained from the children were correlated with those of their parents, only 10% of the mother–child and 6% of the father–child correlations were significant. Correlations obtained when children were randomly paired with other children's parents were significant in 8% of the cases. Parental preferences were no more strongly related to their children's preferences than were those of other parents in the same subcultural group. In both cases, the majority of the correlations were weak but positive, suggesting a commonality of preference among the subcultural group. This pattern of results is consistent with the idea that parents influence children's preferences by limiting the set of foods to which the children are exposed; our data indicated that among our rather homogeneous sample, parents were making similar sets of foods available to the children.

In contrast to the lack of parent–child similarity in preference reported for preschoolers and their parents, two studies investigating family resemblances in food preferences in families with older, college age children report parent–child preference correlations that are significantly higher than those of unrelated adult–child pairs in the sample (Pliner, 1983; Rozin *et al.*, 1984), although the unrelated adult–child correlations also tended to be positive. Rozin *et al.* (1984) reported that parent–child pairs were in closest agreement in response to items on disgust and contamination sensitivity. Again, both modeling and commonality of exposure could be operating to produce these family resemblances, and the pattern does not provide information regarding the relative importance of the two processes. In the case of the families with college age children, there would have been much more time and hence many more opportunities for both types of influence to occur than in the case of families with preschoolers. In addition, developmental differences in preferences and food rejections would serve to reduce the magnitude of the correlations among preschoolers and their parents to a greater degree than in the families with college age children.

To examine the impact of social factors on children's food preferences, Duncker (1938) investigated

> 'objective reasons why, in a social group, certain likes and dislikes rather than others become dominant and traditional, and . . . the psychological mechanisms by which such likes and dislikes are instilled into the individual member of the group' (p. 489).

To explore these social-psychological questions, Duncker attempted to modify children's food preferences through 'social suggestion.' He felt that children's food preferences were 'especially fit for investigation.' In one experiment, he had preschoolers observe other children who were making food choices. Based on previously obtained information about their preferences in the absence of peers, the children observed others making food choices that were different from their own. Duncker noted the food choices made by the observers following their exposure to the peers' choices and compared these to those made prior to their exposure to the peer models. He then examined whether the degree of social influence was related to (i) the difference in age among the observers and observed children, (ii) the absolute ages of the children, and (iii) the relationship among the children. Duncker also included an adult model and fictional predecessors.

Duncker reported clear evidence for social influence; the children's choices changed in the direction of the models' choices. The effects were greater for younger children, down to a lower limit of about 2 years, and were greater when the children were friends or when a more powerful child was the model. Surprisingly, he found that an adult stranger was not effective in influencing the children's choices among the familiar foods. Fictional heroes were effective agents of social transmission. Duncker concluded that (i) social influence was dependent upon the relationship among the individuals and individual characteristics, (ii) influence tended to survive beyond the original situation and (iii) the influence was effective because it 'came to affect the sensory qualities of the food.'

There is no direct support in his data for the last statement, and this issue should receive attention. When a food preference is acquired or modified as a result of experience in a social context, does this change in preference reflect a change in the meaning of the object for the individual or the discovery of favorable qualities or characteristics in the food itself? This issue returns us to a question that arose in the context of associative conditioning of food preferences and aversions: should the conditioned stimulus be viewed as: (i) a signal predicting the unconditioned stiumulus, or as (ii) having acquired some of the qualities or characteristics of the unconditioned stimulus? As indicated earlier, research on conditioned aversions in humans is more consistent with the second interpretation.

We extended Duncker's approach (Birch, 1980a) in a study on the effects of peer models on children's food preferences and consumption patterns. To do this, we constructed social influence situations within a preschool lunch program. The preschoolers' preferences for familiar vegetables were assessed and seating at lunches was arranged based on these preferences. A target child who strongly preferred vegetable 'A' to vegetable 'B' was seated at lunch with three or four other children who preferred 'B' to 'A'. Children were served these pairs of vegetables in a series of four consecutive lunches and were asked to choose a vegetable. On the first day, the 'target' child chose first, followed by the peers, giving us a check on the validity of the preference data as a predictor of choices in the mealtime context. On the following three days, the target child chose last, following the peers' selections. Results indicated that the target children showed a significant shift from choosing their preferred vegetable on day 1 to choosing their nonpreferred food by day 4; consumption data corroborated these results. In post influence assessments, the target children showed an increased preference for the initially nonpreferred vegetable, while their peers did not. These data provide clear evidence that food preference can be altered by social learning.

In another experiment on the effects of social learning on children's food acceptance, Harper and Sanders (1975) investigated the effectiveness of familiar and unfamiliar adult models in inducing young children to eat novel foods. They offered novel foods to two age groups of children: 14- to 20-month olds and 42- to 48-month olds. Two conditions were included; either the adult offered the food to the child or the adult ate the food while offering it to the child. More children ate the novel food in the 'adult eating' condition than in the 'offer only' condition, and the younger children were more affected by the procedures than the older ones. In analyzing the form of observational learning that occurred, the authors attribute the results to simple social facilitation, the most basic form of social learning, in which the actions of the model arouse a similar action in the observer.

This form of observational learning does not require that a new behavior is acquired, but rather produces a change in the stimulus control of the behavior. Social facilitation provides a mechanism for initiating ingestion of the novel food. Once the child ingests the food, feedback on palatability and associative conditioning of food cues to the social contexts and physiological consequences of eating can come into play to influence whether or not the food becomes an accepted part of the diet. Galef (1976) has argued that direct social interaction with others of the group can provide a means of giving the young organism an initial experience with a novel food, then allowing mechanisms that produce affective change with exposure to operate.

We have continued to investigate the effects of social experience with food on food acceptance (Birch, 1981; Birch *et al.*, 1980b, 1982, 1984b), and have begun to look more closely at how the nature of the social interactions surrounding child feeding can influence the acquisition and modification of children's food preferences. The research described in the preceding paragraphs can be viewed as examples of social interaction providing a mechanism for initiating the child's contact with new foods. In addition to this initiating function, the results of research conducted in our laboratory indicate that the specific type and affective tone of the social interactions that surround feeding influence the acquisition of food preferences. In order to investigate this issue, we began by examining the literature on child feeding practices currently used in our culture. We then attempted to reproduce these feeding practices in the laboratory and, by monitoring the children's preferences and consumption patterns before, during, and after the imposition of the feeding practices, have assessed their impact on food preference.

Survey data on child feeding practices indicate that parents of preschool children frequently underestimate the adequacy of their young children's diets, and are concerned that their children are eating too little (Beal, 1957; Dierks and Morse, 1965; Kram and Owen, 1972; Stanek *et al.*, 1990). In their concern regarding this perceived inadequacy, parents often employ contingencies in order to effect increases in consumption (Kram and Owen, 1972; Stanek *et al.*, 1990). For example, in order to increase the child's consumption of peas above very low baseline levels, eating peas becomes the instrumental activity in a contingency, e.g. 'Eat your peas and then you can . . . (engage in some attractive activity, such as playing outside, watching TV, or eating chocolate ice cream).' We investigated whether the use of such contingencies had effects on the acceptability of food consumed instrumentally (Birch *et al.*, 1982, 1984a,b). To obtain initial data on this question, 12 preschool children's preferences for fruit juices were obtained before and after the imposition of a series of contingency schedules in which the children drank one of the juices in order to gain access to a play activity. Results indicated that 9 of the 12 children decreased their preference for the instrumentally consumed juice; the other three children showed no change. For the total sample, the negative shift in preference was significant, indicating that the repeated association of the food with the instrumental eating context had decreased the food's acceptability.

A second experiment on the effects of instrumental eating was conducted to obtain information on the psychological processes contributing to changing preferences (Birch *et al.*, 1984a,b). Each child was randomly assigned to one fo four contingency groups or one of two control groups, matched for age, sex and beverage preference. The four contingency conditions were generated by crossing two variables, each of which had two levels: type of reward (tangible or positive verbal feedback); and amount consumed (baseline or baseline plus an additional increment). The selection of type of reward as a variable was based on the extrinsic motivation literature (*cf.* Lepper and Greene, 1978), which indicates that while providing tangible rewards for the performance of activities reduces the attractiveness of those activities, providing feedback on the quality of performance does not. Positive verbal feedback has the effect of focusing the child on the activity that the adult wants the child to perform; in contrast, the use of salient tangible rewards focuses the child on the reward rather than on the instrumental activity. The other variable (amount consumed) was included to determine whether negative effects on preference noted previously were attributable, not to the contingency *per se*, but to having the children consume more of a food than their ideal, preferred amount. We hypothesized that this overconsumption of the food might, in and of itself, be aversive and account for the negative shifts in preference noted in the previous research. Children participated in a series of eight contingency or control sessions.

Results indicated significant negative shifts in acceptability for foods eaten to obtain rewards, in contrast to slight increases in acceptability for foods in the control groups. These changes in acceptability can be viewed as a result of associative conditioning in which food cues become associated with perceived social context cues present during eating. Depending on the affective tone of the social context, preference for the food can either increase or decline. Subsequently, these findings have been replicated by other investigators (Newman and Taylor, 1992). There is direct evidence that children see these instrumental contexts in a negative light. Lepper *et al.* (1982) reported that when children were told stories involving the eating of two imaginary foods (called hupe and hule), in which one food was in the instrumental component and the other in the reward component of a mealtime contingency, the children indicated that they would prefer the food used as a reward over the food eaten instrumentally.

Additional evidence on the central role of associative conditioning to social contexts in determining food preference during early childhood also comes from our work on the use of foods as rewards. We repeatedly presented one food to each child in one of four social contexts: as a reward for prosocial behavior; noncontingently paired with positive adult attention; at snacktime, along with other foods to control for exposure effects, or in a nonsocial context (Birch *et al.*, 1980ab). The series of 20 presentations was integrated into the ongoing preschool programs over a six-week period; teachers were trained to present the foods to the children. Results showed a significant *increase* in preference for the foods presented as rewards or paired with positive adult attention, and these increases in acceptability persisted for at least six weeks after the presentations ended. The basic findings were replicated and the changes in preference resulting from such associative conditioning generalized to foods perceived as similar to the presented food (Birch, 1981).

These results provide evidence for associative conditioning of food acceptability during early childhood. The results indicating that the use of foods in contingencies reduces liking suggests that these child feeding practices have effects on preference quite opposite to those intended by parents who use them. In the case of instrumental eating, parents employ this strategy to increase the child's consumption of a nutritionally desirable food. Our data indicate that instrumental eating results in declines in preference, making it less likely that the food will be consumed when the contingency is removed. In contrast, increases in preference result from the use of foods as rewards. It is worth noting that there is a confounding of food and function in these contingencies, with highly preferred foods serving as rewards, and non-preferred foods eaten instrumentally. These child feeding practices work against the establishment of healthy eating practices by altering preferences for specific foods used in contingencies: while preference for nutritionally desirable foods is reduced, preferences for palatable, sweet foods used as rewards is enhanced.

As stated in the discussions of factors influencing meal size, the use of external control by parents to regulate the child's eating may convey information to the child regarding the cues that are relevant to the control of intake. Children may be learning that internal cues such as hunger and satiety, or the changes in preference that occur from hunger to satiety are not relevant in the control of intake. Rather external cues such as time of day, or the amount of food remaining on the plate come to control food intake. The use of foods as rewards and as pacifiers and treats also communicates to children that there are reasons other than hunger for eating; through associative conditioning, children could learn to use foods to fulfill emotional as well as physiological needs.

4.3.8 The impact of food preferences on food selection and intake

Evidence presented in the prior discussions of meal size reveals that children are responsive to the energy density of foods, and that energy density can serve to control meal size and meal timing. Via early learning and experience, energy density also influences food *selection* in two ways. First, the energy density of foods consumed at a first course in a meal can influence which foods are selected in subsequent courses (Birch *et al.*, 1989a). Second, as reviewed above, children can learn to prefer energy dense foods via associative conditioning of foods' flavors with the consequences of ingesting those foods (Birch, 1992).

With respect to food selection, we have shown that the energy density of foods consumed at one meal can influence the size of the next meal: the consumption of a high energy meal leads to reduced consumption at the next eating occasion. Examination of how children adjust their food intake in the second meal reveals the importance of food preferences. The reduced intake following energy dense meals occurs because children reduce the variety of foods eaten following energy dense foods: *which* foods children consume in the next meal are affected by the energy density of foods consumed in the previous meal. After consuming a high energy in the first course, the children consumed fewer foods *ad libitum*. When the intake data were analyzed taking the children's food preferences into account, we noted that relative to intake following the low energy first course, the children reduced their intake by eliminating the less preferred foods from meals that followed the more energy dense first course, while they continued to consume their preferred foods. In fact, their preferred foods tended to be relatively energy dense ones with high levels of sugar and fat, and those not eaten tended to be good sources of micronutrients, generally lower in energy density (Birch *et al.*, 1989a, 1993b). The consumption of energy dense foods at one meal can limit dietary variety by reducing the variety of foods consumed at subsequent meals. Because foods eliminated tended to be nonpreferred foods, such as vegetables, lower in sugar and fat content but high in micronutrients, the resulting reduced variety may have negative effects on dietary quality.

Although children's food preferences can be powerful predictors of their consumption patterns, discussions of pediatric nutrition have typically ignored this fact and focused exclusively on the caregivers' role in providing children with nutritionally adequate diets. The impact of children's food selections on their dietary intake is frequently overlooked. Recent research (Fisher and Birch, in press) provides additional evidence regarding the importance of children's preferences as a control of their food intake and the overall macronutrient composition and quality of children's diets. Preschool children's ($n = 18$) food intake, including their consumption of high-fat foods and total dietary fat intake, were measured during six 30-hour periods of observation. In addition to children's fat preferences and intake, we obtained measurements of children's and parents' adiposity. Although the same menus were offered to all children, children's food selection patterns produced variability in the quality of their diets. The percent of energy from fat in the menus offered was 33%, but across individual children, the percent of energy from dietary fat ranged from 25% to 41%. Children's preferences for high fat foods were significantly related to their selection and consumption of high fat foods, total dietary fat intake and to parental adiposity. Children indicating strong preferences for high-fat foods had the greatest triceps skinfolds, a measurement of relative adiposity. Children who preferred high-fat foods had higher fat intakes, consuming a large percentage of their energy from high-fat foods, and had the heaviest parents. This research indicates that while the nutritional adequacy of young children's diets is constrained by the foods provided to children, the children's choices from among those foods exerts a considerable influence on the overall quality of their diets. Providing children with a variety of healthful foods from which to select their diets is necessary but not sufficient to ensure nutritional adequacy. Finally, the fact that children's fat preferences, consumption of high-fat foods, and total dietary fat intake were related to parental adiposity highlights the central role of familial factors on the food selection and the controls of food intake.

4.4 Summary and implications for child feeding

Infants enter the world with a set of reflexive behaviors and predispositions that shape the development of food habits. In infancy, there are relationships between meal size and meal interval that reflect an emerging regulatory process. Children are sensitive to the energy content of food and regulate meal size and food selection. These early regulatory processes are modified by learning and experience. With respect to the timing of meals, initially, demand feeding is desirable, allowing the infant a high degree of control over meal timing. When infants are fed on demand, there is evidence that via experience and learning, meal size, meal timing and the relationships between the timing and size of meals change with experience and learning. Finally, with respect to food selection, children come into the world predisposed to learn about food: to learn what to eat and what to like; to prefer some foods and reject others. This learning is based on repeated experience with food and eating, and via associations formed between food's sensory cues and the effect generated by the social contexts and physiological consequences of eating.

We have presented evidence that infants and young children are capable of assuming a high degree of self control over meal timing, meal size, and within the constraints imposed by parents on what foods are offered, over food selection. Children differ in their responsiveness to internal cues of hunger and satiety in regulating meal size and total daily energy intake, and in the extent to which they have learned to use other cues to control the timing and size of meals. These other cues include the presence of food, time of day, and the social and environmental contexts previously associated with food and eating. Research has begun to reveal that individual differences in the control of food intake emerge during childhood, and differences among children are related to parenting style in child feeding, especially differences in the balance of parent–child control in child feeding. By adulthood, based on years of learning and experience, the controls of food intake have become complex. A variety of learned controls of food intake may operate either in addition to, or instead of, internally generated cues signalling hunger and satiety. The increasingly high incidence of obesity and eating disorders suggests that for many in our society, hunger and satiety signals are not functioning as controls of food intake to maintain energy balance: chronic dieters consistently ignore hunger cues, and binge eaters continue eating well beyond normal satiety.

What does research on the developing controls of food intake in children imply regarding child feeding practices? What practices should parents follow to facilitate the development of styles of intake control that maintain energy balance and result in healthy diets? We encourage parents to focus on the long term goal of developing healthy self controls of feeding in children, and to look beyond their immediate concerns regarding composition and quantity of foods children consume. One goal is having the child accept a wide variety of foods. Children should be given substantial control over food intake, especially meal size. While allowing children control over how much is eaten, parents should work towards shaping the timing of the child's meals to the adult pattern of the culture by imposing some control over meal intervals. Because children's food preferences and food selection are tightly linked to the familiarity of foods, parents can have a powerful positive influence on the nutritional adequacy of children's diets via the array of foods that they offer the child (Satter, 1986, 1987, 1990). Research on associative conditioning of food preferences suggests that parents should forgo the temptation to control children's eating by imposing contingencies and coercive practices. Although rewarding children for eating, coercing them to eat, or using foods as rewards can give parents control over children's eating in the short run, the learned food preferences that result tend to be antithetical to the selection of healthy diets. Such child feeding practices can be harmful because in addition to limiting opportunities for self control, children are learning to dislike the foods that are 'good for them' and to love the 'junk' foods that should be consumed only in moderation.

To facilitate healthy self control of eating, children should have repeated opportunities to sample healthful foods in noncoercive, positive contexts, so that via associative learning processes, some of the foods offered will become preferred and accepted. Although evidence is limited, we also caution that parents should not severely restrict access to highly palatable 'junk' foods, foods high in sugar, salt and fat, because we suspect that such restricted access may make these forbidden foods even more attractive. Parental influence should be focused not on controlling intake at individual meals, but should operate to allow children to develop preferences and food selection patterns consistent with a healthy diet. We do not imply that children should be given total control over food selection, as advocated in one recent book on child feeding (Hirschmann and Zaphiropoulos, 1985). This position is not consistent with the literature on the development of children's food preferences and the controls of food intake. First, research does not support the view that specific hungers and the 'wisdom of the body' will lead the child to select an adequate diet. In addition, children have an unlearned preference for sweet and salty tastes (Cowart and Beauchamp, 1986). Children are neophobic, and reject new foods (especially those that are not sweet or salty). Finally, children have a propensity to learn to prefer energy dense foods. Given these propensities, and the array of foods available in supermarkets today, consider the likely outcome of the following thought experiment. Using Clara Davis' dietary self selection protocol, at weaning, we give infants and toddlers complete freedom to select their diets by allowing them to select from among all the foods available in US supermarkets today. Similar to Davis' infants, our participants begin the study at weaning, so all foods are initially novel to them. However, in striking contrast to Davis' findings, we would predict that children's 'supermarket' diets would not be nutritionally adequate; from among the foods available, they would select foods high in sugar, salt, and fat, and it is unlikely that the children would ever learn to like and consume other novel, less inherently palatable foods. As Davis pointed out, the 'trick' of her successful dietary self selection studies was in what the children were offered.

To increase the likelihood that children will select foods that will comprise an adequate diet, we must provide parents with information about how children learn to accept new foods. Armed with such information, parents would not expect that new foods (especially those low in sugar, salt and fat) will be accepted the first time they are offered, and will be prepared to offer new foods repeatedly, with the expectation that acceptance can emerge gradually. In addition, once parents appreciate the role of social context in shaping children's food likes and dislikes, they may be motivated to provide opportunities for children to eat foods in positive social contexts, in the absence of coercion. Many healthful foods, such as vegetables, that we want children to consume are not inherently palatable. Especially for foods that are not sweet or salty, associative learning can be pivotal in determining whether these foods will be accepted or rejected.

Finally, providing accurate information regarding appropriate portion sizes for young children can reduce parental anxiety about the adequacy of their children's diets. In our informal discussions with parents of young children, we find that many parents frequently overestimate the amounts of food that young children need to eat. The discrepancy between what parents think children should eat and the child's actual intake can make parents very anxious, and serve to underscore parental convictions that the child is not capable of controlling her own food intake. This can lead parents to attempt to assume greater control over feeding, particularly over how much the child eats. Certainly parents with misconceptions about portion sizes are not well prepared to assume responsibility for controlling children's meal size. We have found that providing parents with information about more appropriate portion sizes for their young children can help to reduce parental anxiety about whether their child is getting enough to eat, allowing parents to relinquish some control over meal size.

Let us return to Clara Davis' advice, given more than 60 years ago: during the transition to the adult diet, parents should restrict themselves to assuming responsibility for what foods are made available to children. The current research findings support the view that given experience with an adequate variety of healthful foods in a noncoercive feeding environment, the normally developing child will learn to accept and consume a variety of foods in sufficient quantities to comprise a nutritionally adequate diet, and will sustain growth and health. These same views are reflected in Satter's more contemporary advice to parents (1986, 1987, 1990). With respect to meal size, all the evidence suggests that children have some capacity to regulate energy intake by adjusting meal size. If presented with a healthy array of foods, the child will obtain adequate nutrients; parents should not assume control of meal size, or resort to coercing or controlling tactics to induce the child to eat. Preliminary findings suggest that parental control can have adverse effects on children's responsiveness to the energy content of food, which can serve as an effective control of food intake. In summary, parents should provide children with a variety of healthful foods, set limits to shape the child toward adult meal patterns, but allow the child control of whether and how much to eat.

References

Agras, W.S., Berkowitz, R.I., Hammer, L.C. and Kraemer, H.C. (1988) Relationships between the eating behaviors of parents and their 18-month-old children: A laboratory study. *International Journal of Eating Disorders*, **7**, 461–468.

Ainsworth, M.S. and Bell, S.M. (1969) Some contemporary patterns of mother–infant interaction in the feeding situation. In A. Ambrose (ed.), *Stimulation in early infancy* (pp. 133–162). New York: Academic Press.

Angyal, A. (1941) Disgust and related aversions. *Journal of Abnormal and Social Psychology*, **36**, 393–412.

Anliker, J.A., Bartoshuk, L., Ferris, A.M. and Hooks, L.D. (1991) Children's food preferences and genetic sensitivity to the bitter taste of 6-n-propylthiouracil (PROP). *American Journal of Clinical Nutrition*, **54**, 316–320.

Barker, L.M., Best, M.R. and Domjan (eds) (1977) *Learning mechanisms in food selection.* Houston, TX: Baylor University.

Bartoshuk, L.M. (1979) Influence of chemoreception and physiologic state on food selection. Paper presented at the Workshop on Nutrition Behavior and the Life Cycle. National Institutes of Health, Bethesda, MD.

Bartoshuk, L.M. and Beauchamp, G.K. (1994) Chemical senses. *Annual Review of Psychology*, **45**, 414–449.

Beal, V. (1957) On the acceptance of solid foods and other food patterns of infants and children. *Pediatrics* **61**, 448–457.

Beauchamp, G. (1980, May) *Ontogenesis of taste preferences.* Paper presented at the International Organization for the Study of Human Development, Campione, Italy.

Beauchamp, G.K., Cowart, B.J. and Moran, M. (1986) Developmental changes in salt acceptability in human infants. *Developmental Psychobiology*, **19**, 17–25.

Beauchamp, G.K., Cowart, B.J., Mennella, J. and Marsh, R.R. (1994) Infant salt taste: Developmental, methodological, and contextual factors. *Developmental Psychobiology*, **27**, 353–365.

Bernstein, I.L. (1978) Learned taste aversions in children receiving chemotherapy. *Science*, **200**, 1302–1303.

Bernstein, I. (1981) Meal patterns in 'free running humans'. *Physiology & Behavior*, **27**, 621–624.

Bernstein, I. and Webster, M.M. (1980) Learned taste aversions in humans. *Physiology & Behavior*, **25**, 363–366.

Birch, L.L. (1979a) Dimensions of preschool children's food preferences. *Journal of Nutrition Education*, **11**, 189–192.

Birch, L.L. (1979b) Preschool children's food preferences and consumption patterns. *Journal of Nutrition Education*, **11**, 77–80.

Birch, L.L. (1980a) Effects of peer models' food choices and eating behaviors on preschooler's food preferences. *Child Development*, **51**, 489–496.

Birch, L.L. (1980b) The relationship between children's food preferences and those of their parents. *Journal of Nutrition Education*, **12**, 14–18.

Birch, L.L. (1981) Generalization of a modified food preference. *Child Development*, **52**, 755–758.

Birch, L.L. (1992) Children's preferences for high-fat foods. *Nutrition Reviews*, **50**, 249–255.

Birch, L.L. and Deysher, M. (1985) Conditioned and unconditioned caloric compensation: Evidence for self-regulation of food intake by young children. *Learning and Motivation*, **16**, 341–355.

Birch, L.L. and Deysher, M. (1986) Caloric compensation and sensory specific satiety: Evidence for self-regulation of food intake by young children. *Appetite*, **7**, 323–331.

Birch, L.L. and Marlin, D.W. (1982) I don't like it; I never tried it: Effects of exposure to food on two-year-old children's food preferences. *Appetite*, **4**, 353–360.

Birch, L.L., Marlin, D., Kramer, L. and Peyer, C. (1980a) Mother–child interaction patterns and the degree of fatness in children. *Journal of Nutrition Education*, **13**, 17–21.

Birch, L.L., Zimmerman, S. and Hind, H. (1980b) The influence of social-affective context on preschool children's food preferences. *Child Development*, **51**, 856–861.

Birch, L.L., Birch, D., Marlin, D. and Kramer, L. (1982) Effects of instrumental eating on children's food preferences. *Appetite*, **3**, 125–134.

Birch, L.L., Billman, J. and Richards, S. (1984a) Time of day influences food acceptability. *Appetite*, **5**, 109–112.

Birch, L.L., Marlin, D.W. and Rotter, J. (1984b) Eating as the 'means' activity in a contingency: Effects on young children's food preference. *Child Development*, **55**, 432–439.

Birch, L.L., McPhee, L., Shoba, B.C., Steinberg, L. and Krehbiel, R. (1987a) Clean up your plate: Effects of child feeding practices on the conditioning of meal size. *Learning and Motivation*, **18**, 301–317.

Birch, L.L., McPhee, L., Shoba, B.C., Pirok, E. and Steinberg, L. (1987b) What kind of exposure reduces children's food neophobia? *Appetite*, **9**, 171–178.

Birch, L.L., McPhee, L. and Sullivan, S. (1989a) Children's food intake following drinks sweetened with sucrose or aspartame: Time course effects. *Physiology and Behavior*, **45**, 387–396.

Birch, L.L., McPhee, L., Sullivan, S. and Johnson, S. (1989b) Conditioned meal initiation in young children. *Appetite*, **13**, 105–113.

Birch, L.L., McPhee, L., Steinberg, L. and Sullivan, S. (1990) Conditioned flavor preferences in young children. *Physiology and Behvior*, **47**, 501–505.

Birch, L.L., Johnson, S.L., Andresen, G., Petersen, J.C. and Schulte, M.C. (1991) The variability of young children's energy intake. *The New England Journal of Medicine*, **324**, 232–235.

Birch, L.L., Johnson, S.L., Jones, M.B. and Peters, J.C. (1993a) Effects of a non-energy fat substitute on children's energy and macronutrient intake. *The American Journal of Clinical Nutrition*, **58**, 326–333.

Birch, L.L., McPhee, L.S., Bryant, J.L. and Johnson, S.L. (1993b) Children's lunch intake: Effects of midmorning snacks varying in energy density and fat content. *Appetite*, **20**, 83–94.

Booth, D.A. (1985) Food conditioned eating preferences and aversions with interoceptive elements: Conditioned appetites and satieties. *Annals of the New York Academy of Sciences*, **443**, 22–41.

Booth, D.A., Lee, M. and McAleavey, C. (1976) Acquired sensory control of satiation in man. *British Journal of Psychology*, **67**, 137–147.

Booth, D.A., Mather, P. and Fuller, J. (1982) Starch content of ordinary foods associatively conditions human appetite and satiation, indexed by intake and eating pleasantness of starch-paired flavors. *Appetite*, **3**, 163–184.

Bruch, H. (1961) Transformation of oral impulses in eating disorders: A conceptual approach. *Psychiatric Quarterly*, **35**, 458–480.

Bryan, M.S. and Lowenberg, M.E. (1958) The father's influence on young children's food preferences. *Journal of the American Dietetic Association*, **34**, 30–35.

Cardello, A. (1996) In H.L. Meiselman and H.J.H. MacFie (eds), *Food choice, acceptance and consumption*. Glasgow, Scotland: Blackie A&P.

Cashdan, E. (1994) A sensitive period for learning about food. *Human Nature*, **5**, 279–291.

Chiva, M. (1982) Taste, facial expression and mother-infant interaction in early development. *Baroda Journal of Nutrition*, **9**, 99–102.

Cowart, B. (1981) Development of taste perception in humans: Sensitivity and preference throughout the life span. *Psychological Bulletin*, **90**, 43–73.

Cowart, B.J. and Beauchamp, G.K. (1986) Factors affecting acceptance of salt by human infants and children. In M.R. Kare and J.G. Brand (eds), *Interaction of the chemical senses with nutrition* (pp. 25–44). New York: Academic Press.

Crook, C.K. (1978) Taste perception in the newborn infant. *Infant Behavior and Development*, **1**, 52–69.

Davis, C.M. (1928) Self-selection of diet by newly weaned infants. *American Journal of Diseases of Children*, **36**, 651–679.

Davis, C.M. (1939) Results of the self selection of diets by young children. *The Canadian Medical Association Journal*, **41**, 257–261.

DeSnoo, K. (1937) The drinking child in the uterus. *Journal for Obstetrics and Gynecology*, **105**, 88–97.

Desor, J.A., Maller, O. and Turner, R. (1973) Taste in acceptance of sugars by human infants. *Journal of Comparative and Physiological Psychology*, **84**, 496–501.

Desor, J.A., Maller, O. and Andrews, K. (1975) Ingestive responses of human newborns to salty, sour, and bitter stimuli. *Journal of Comparative and Physiological Psychology*, **89**, 966–970.

Dierks, E.C. and Morse, L.M. (1965) Food habits and nutrient intakes of preschool children. *Journal of the American Dietetic Association*, **47**, 292–296.

Dudley, D., Moore, M. and Sunderlin, E. (1960) Children's attitudes toward food. *Journal of Home Economics*, **52**, 678–681.

Duncker, K. (1938) Experimental modification of children's food preferences through social suggestion. *Journal of Abnormal Social Psychology*, **33**, 489–507.

Dunshee, M.E. (1931) A study of factors affecting the amount and kind of food eaten by nursery school children. *Child Development*, **1**, 163–183.

Engen, T. (1982) *The perception of odors*. New York: Academic Press.

Fallon, A.E. and Rozin, P. (1984) The psychological bases of food rejections by humans. *Ecology of Food and Nutrition*, **13**, 5–26.

Fallon, A.E., Rozin, P. and Pliner, P. (1984) The child's conception of food: The development of food rejection, with special reference to disgust and contamination sensitivity. *Child Development*, **55**, 566–575.

Finberg, L., Kiley, J. and Luttrell, C.N. (1963) Mass accidental salt poisoning in infancy. *Journal of the American Medical Association*, **184**, 121–124.

Fischler, C. and Chiva, M. (1985, May) *Food likes, dislikes and some of their correlates in a sample of French children and young adults*. Paper presented at the Euro-Nutrition Meeting on Food Preferences and Food Habits, University of Giessen, Giessen, Germany.

Fisher, J.A. and Birch, L.L. (in press) 3–5 year-old children's fat preferences and fat consumption are related to parental adiposity. *Journal of The American Dietetic Association*.

Fomon, S.J. (1993) *Nutrition of normal infants*. St. Louis, MO: Mosby-Year Book, Inc.

Galef, B. (1976) Social transmission of acquired behavior. In J.S. Rosenblatt, R.A. Hinde, E. Shaw and C. Beer (eds), *Advances in the study of behavior* (Vol. 6, pp. 77–97). New York: Academic Press.

Galef, B.G., Jr. (1991) A contrarian view of the wisdom of the body as it relates to dietary self-selection. *Psychology Review*, **98**, 218–223.
Galst, J.P. (1980) Television food commercials and pro-nutritional public service announcements as determinants of young children's snack choices. *Child Development*, **51**, 935–938.
Galst, J.P. and White, M.A. (1976) The unhealthy persuader: The reinforcing value of television and children's purchase-influencing attempts at the supermarket. *Child Development*, **47**, 1089–1096.
Garb, J.L. and Stunkard, A.J. (1974) Taste aversions in man. *American Journal of Psychiatry*, **131**, 1204–1207.
Garcia, J., Hankins and Rusiniak, K. (1974) Behavioral regulation of the milieu intern in man and rat. *Science*, **185**, 824–831.
Garcia, S.E., Kaiser, L.L. and Dewey, K.G. (1990a) Self-regulation of food intake among rural Mexican preschool children. *European Journal of Clinical Nutrition*, **44**, 371–380.
Garcia, S.E., Kaiser, L.L. and Dewey, K.G. (1990b) The relationship of eating frequency and caloric density to energy intake among rural Mexican preschool children. *European Journal of Clinical Nutrition*, **44**, 381–387.
Glaser, A. (1964) Nursery school can influence food's acceptance. *Journal of Home Economics*, **56**, 680–683.
Goldberg, M.E., Gorn, G.J. and Gibson, W. (1978) T.V. messages for snack and breakfast foods: Do they influence children's preferences? *Journal of Consumer Research*, **5**, 73–81.
Hall, W.G. and Rosenblatt, J.S. (1977) Suckling behavior and intake control in the developing rat pup. *Journal of Comparative and Physiological Psychology*, **91**, 1232–1247.
Hammer, L.D. (1992) The development of eating behavior in childhood. *Pediatric Clinics of North America*, **39**, 379–394.
Harper, L.V. and Sanders, K.M. (1975) The effect of adult's eating on young children's acceptance of unfamiliar foods. *Journal of Experimental Child Psychology*, **20**, 206–214.
Hirschmann, J.R. and Zaphiropoulos, L. (1985) *Solving your child's eating problems*. New York: Ballantine Books.
Itani, J. (1958) In the acquisition and propagation of a new food habit in the natural group of Japanese monkeys at Taka Suki Yama. *Journal of Primatology*, **1**, 84–98.
Johnson, S.L. and Birch, L.L. (1994) Parents' and children's adiposity and eating style. *Pediatrics*, **94**, 653–661.
Johnson, S.L., McPhee, L. and Birch, L.L. (1991) Conditioned preferences: young children prefer flavors associated with high dietary fat. *Physiology & Behavior*, **50**, 1245–1251.
Kern, D.L., McPhee, L., Fisher, J., Johnson, S. and Birch, L.L. (1993) The postingestive consequences of fat condition preferences for flavors associated with high dietary fat. *Physiology & Behavior*, **54**, 71–76.
Kissileff, H.R. (1996) In H.L. Meiselman and H.J.H. MacFie (eds), *Food choice, acceptance and consumption*. Glasgow, Scotland: Blackie A&P.
Klesges, R.J., Coates, T.J., Brown, G., Sturgeon-Tillish, J., Modenhauer-Klesges, L.M., Holzer, B., Woolfrey, J. and Vollmer, J. (1983) Parental influences on children's eating behavior and relative weight. *Journal of Applied Behavioral Analysis*, **16**, 371–378.
Klesges, R.J., Malott, J.M., Boschee, P.F. and Weber, J.M. (1986) The effects of parental influence on children's food intake, physical activity, and relative weight. *International Journal of Eating Disorders*, **5**, 335–346.
Koivisto, U.K., Fellenius, J. and Sjoden, P. (1994) Relations between parental mealtime practices and children's food intake. *Appetite*, **22**, 245–258.
Kram, F.M. and Owen, G.M. (1972) Nutritional studies on United States preschool children: Dietary intakes and practices of food procurement, preparation and consumption. In S.J. Fomon and T.A. Anderson (eds), *Practices of low-income families in feeding infants and small children with particular attention to cultural subgroups* (pp. 3–18). Department HEW Publication No. 72-5605. Washington, DC: US Government Printing Office.
Lamb, M. and Ling, B.C. (1946) Analysis of food consumption and preferences of nursery school children. *Child Development*, **17**, 187–217.
Le Magnen, J. (1977) Sweet preference and the sensory control of calorie intake. In J.M. Weiffenbach (ed.), *Taste and development: The genesis of sweet preference* (pp. 355–362). Bethesda, MD: United States Department of Health, Education and Welfare.
Le Magnen, J. (1985) *Hunger*. New York: Cambridge University Press.

Lepper, M. and Greene, D. (eds) (1978) *The hidden costs of reward: New perspectives on the psychology of human motivation*. New Jersey: Lawrence Erlbaum Associates.
Lepper, M., Sagotsky, G., Dafoe, J.L. and Greene, D. (1982) Consequences of superfluous social constraints: Effects on young children's social inferences and subsequent intrinsic interest. *Journal of Personality and Social Psychology*, **42**, 51–65.
Lipsitt, L.P. (1977) Taste in human neonates: Its effect on sucking and heart rate. In J.M. Weiffenbach (ed.), *Taste and development: The genesis of sweet preference* (pp. 125–141) DHEW Publication No. NIH 77-1068. Washington, DC: US Government Printing Office.
Logue, A.W., Ophir, I. and Strauss, K. (1981) The acquisition of taste aversions in humans. *Behavior Research and Therapy*, **19**, 319–333.
Martin, I. and Levey, A.B. (1978) Evaluative conditioning. *Advances in Behavior Research and Therapy*, **1**, 57–102.
Matheny, R., Birch, L.L. and Picciano, M.F. (1990) Control of intake by human milk fed infants: Relationships between feeding size and interval. *Developmental Psychobiology*, **23**, 511–518.
McCarthy, D. (1935) Children's feeding problems in relation to the food aversions of the family. *Child Development*, **6**, 277–284.
Mela, D.J. (ed.) (1992) *Dietary fats*. Essex, England: Elsevier Science Publishers Ltd.
Mennella, J.A. and Beauchamp, G.K. (1991a) Maternal diet alters the sensory qualities of human milk and the nursling's behavior. *Pediatrics*, **88**, 737–744.
Mennella, J.A. and Beauchamp, G.K. (1991b) The transfer of alcohol to human milk: Effects on flavor and the infant's behavior. *New England Journal of Medicine*, **325**, 981–985.
Mennella, J.A. and Beauchamp, G.K. (1993) The effects of repeated exposure to garlic-flavored milk on the nursling's behavior. *Pediatric Research*, **34**, 805–808.
Mennella, J.A. and Beauchamp, G.K. (in press) The human infants' response to vanilla flavors in mother's milk and formula. *Infant Behavior and Development*.
Newman, J. and Taylor, A. (1992) Effect of a means-end contingency on young children's food preferences. *Journal of Experimental Child Psychology*, **64**, 200–216.
Nisbett, R.E. and Gurwitz, S. (1970) Weight, sex and the eating behavior of human newborns. *Journal of Comparative and Physiological Psychology*, **73**, 245–253.
Nowlis, G.H. (1973) Taste elicited tongue movements in human newborn infants: An approach to palatability. In J.F. Bosma (ed.), *Fourth symposium on oral sensation and perception: Development of the fetus and infant*. DHEW Publication No. NIH 73-546. Washington, DC: US Government Printing Office.
Otis, L. (1984) Factors influencing the willingness to taste unusual foods. *Psychological Reports*, **54**, 739–745.
Parker, L., Failor, A. and Weidman, K. (1973) Conditioned preferences in the rat with an unnatural need state: Morphine withdrawal. *Journal of Comparative and Physiological Psychology*, **82**, 294–300.
Peiper, A. (1963) *Cerebral function in infancy and childhood*. New York: Consultant's Bureau.
Peryam, D.R. (1963, June) The acceptance of novel foods. *Food Technology*, 711–717.
Phillips, B.K. and Kolasa, K. (1980) Vegetable preferences of preschoolers in day care. *Journal of Nutrition Education*, **12**, 192–195.
Pinilla, T. and Birch, L.L. (1993) Help me make it through the night: Behavioral entrainment of breast-fed infants' sleep patterns. *Pediatrics*, **91**, 436–444.
Pliner, P. (1982) The effects of mere exposure on liking for edible substances. *Appetite*, **3**, 283–290.
Pliner, P. (1983) Family resemblances in food preferences. *Journal of Nutrition Education*, **15**, 137–140.
Pliner, P. (1994) Development of measures of food neophobia in children. *Appetite*, **23**, 147–164.
Pliner, P. and Hobden, K. (1992) Development of a scale to measure the trait of food neophobia in humans. *Appetite*, **19**, 105–120.
Pliner, P. and Pelchat, M.L. (1986) Similarities in food preferences between children and their siblings and parents. *Appetite*, **7**, 333–342.
Pliner, P., Pelchat, M. and Grabski, M. (1993) Reduction of neophobia in humans by exposure to novel foods. *Appetite*, **20**, 111–123.
Rolls, B.J. (1986) Sensory-specific satiety. *Nutrition Reviews*, **44**, 93–101.

Rozin, P. (1976) The selection of foods by rats, humans, and other animals. *Advances in the study of behavior, Volume 6* (pp. 21–67). New York: Academic Press.

Rozin, P. (1977) The use of characteristic flavorings in human culinary practice. In C.M. Apt (ed.), *Flavor: Its chemical, behavioral, and commercial aspects.* Boulder, CO: Westview Press.

Rozin, P. (1984) The acquisition of food habits and preferences. In J.D. Matarazzo, S.M. Weiss, J.A. Herd and N.E. Miller (eds), *Behavioral health: A handbook of health enhancement and disease prevention* (pp. 590–607). New York: Wiley.

Rozin, P. and Fallon, A.E. (1980) The psychological categorization of foods and non foods: A preliminary taxonomy of food rejections. *Appetite*, **1**, 193–201.

Rozin, P. and Fallon, A.E. (1981) The acquisition of likes and dislikes for foods. In J. Sohms and R.L. Hall (eds), *Criteria of food acceptance: How man chooses what he eats* (pp. 35–48). Zurich: Forster Verlag.

Rozin, P. and Zellner, D. (1985) The role of Pavlovian conditioning in the acquisition of food likes and dislikes. In N. Braverman and P. Bronstein (eds), *Experimental assessments and clinical applications of conditioned food aversions* (pp. 189–202). New York: New York Academy of Sciences.

Rozin, P., Fallon, A. and Mandell, R. (1984) Family resemblance in attitudes to food. *Developmental Psychology*, **20**, 309–314.

Sanjur, D. and Scoma, A.D. (1971) Food habits of low-income children in northern New York. *Journal of Nutrition Education*, **3**, 85–94.

Satter, E. (1986) *Child of mine.* Palo Alto, CA: Bull Publishing Co.

Satter, E. (1987) *How to get your kid to eat . . . but not too much.* Palo Alto, CA: Bull Publishing Co.

Satter, E. (1990) The feeding relationship: Problems and interventions. *The Journal of Pediatrics*, **117**, S181–S189.

Sclafani, A. (1990) Nutritionally based learned flavor preferences in rats. In E. Capaldi and T. Powley (eds), *Taste, experience, and feeding* (pp. 139–156). Washington, DC: American Psychological Association.

Smith, G.P. and Greenberg, D. (1992) The investigation of orosensory stimuli in the intake and preference of oils in the rat. In D. Mela (ed.), *Dietary fats* (pp. 167–178). Essex, England: Elsevier Science Publishers Ltd.

Stanek, K., Abbott, D. and Cramer, S. (1990) Diet quality and eating environment of preschool children. *Journal of the American Dietetic Association*, **90**, 1582–1584.

Steiner, J.E. (1974) Discussion paper: Innate discriminative human facial expressions to taste and smell stimulation. *Annals of the New York Academy of Sciences*, **237**, 229–233.

Steiner, J.E. (1977) Facial expressions of the neonate infant indicating the hedonics of food related chemical stimuli. In J.M. Weiffenbach (ed.), *Taste and development: The genesis of sweet preference* (pp. 173–18) DHEW Publication No. NIH 77–1068. Washington, DC: US Government Printing Office.

Stunkard, A. (1975) Satiety is a conditioned reflex. *Psychosomatic Medicine*, **37**, 383–389.

Sullivan, S.A. and Birch, L.L. (1994) Infant dietary experience and acceptance of solid foods. *Pediatrics*, **93**, 271–277.

Waxman, M. and Stunkard, A.J. (1980) Caloric intake and expenditure in obese boys. *Journal of Pediatrics*, **96**, 187–193.

Weingarten, H.P. (1985) Stimulus control of eating: Implications for a two-factor theory of hunger. *Appetite*, **6**, 387–401.

Weiskrantz, L. and Cowey, A. (1963) The aetiology of food reward in monkeys. *Animal Behavior*, **XI**, 225–234.

Wenar, C. (1963) The reliability of developmental histories. *Psychosomatic Medicine*, **25**, 505–510.

Wright, P., Fawcett, J. and Crow, R. (1980) The development of differences in the feeding behaviour of bottle and breast fed human infants from birth to two months. *Behavior Processes*, **5**, 1–20.

Zahorik, D.M., Maier, S.F. and Pies, R.W. (1974) Preferences for tastes paired with recovery from thiamine deficiency in rats: Appetitive conditioning or learned safety. *Journal of Comparative and Physiological Psychology*, **87**, 1083–1091.

5 What does abnormal eating tell us about normal eating?

C. PETER HERMAN and JANET POLIVY

5.1 Introduction

Because we are intrigued with understanding normal eating, we are eager to address and answer the question posed by the editors of this volume. Compared with normal eating, abnormal eating has received the lion's share of research and professional attention, owing to its dramatic manifestations and consequences. (The major categories of abnormal eating are anorexia nervosa and bulimia nervosa, obesity having receded as an eating disorder because of professional uncertainty as to whether eating in the obese is truly disordered.) We are not sure that we understand the etiology and dynamics of anorexia and bulimia, and it seems clear that they are probably not the only abnormalities of eating worth attending to, but progress is being made. Certainly tremendous resources of time, effort and money have been dedicated to delineating and explaining these disorders, and to eliminating them either before or, more often, after they appear. Normal eating is the goal of therapy, and if perfectly normal eating is too much to hope for, then we will settle for eating that at least more closely approximates normality.

But what is normal eating? Less flamboyant than abnormal eating, it is nevertheless equally elusive. Little attention is paid, professionally or otherwise, to normal eating. No-one has a very good idea of what exactly it is. (This latter claim must be qualified by noting that almost everyone thinks that they know what normal eating is. You can disturb people's complacent assumption that they know what normal eating is by asking them to specify more precisely what they mean by 'normal eating' and forbidding them to respond glibly with a circular response involving 'the absence of abnormal eating.' Regrettably, this qualification must itself be qualified: most people ought to be disturbed by their inability to describe normal eating adequately, but more often than not, the demonstration of their ignorance and confusion does not disturb them.) The description and analysis of normal eating, then, remains woefully incomplete, and deserving of our concerted efforts.

How can we better understand normal eating? Presumably, the best thing to do would be to identify it and study it as it occurs in nature or society. But the problem here, of course, is identifying it. If we don't know what it is, how can we identify it, define it, isolate it—even begin to study it? Maybe the glib circular approach deserves another look. If we actually know something about abnormal eating—and psychology and psychiatry have certainly invested enough effort toward accumulating such knowledge—then maybe we can develop a better notion of normal eating by regarding it as what abnormal eating is not.

In principle, then, perhaps by describing the characteristics of abnormal eating and then engaging in a process of negation, we can delineate the characteristics of normal eating. This approach can escape the charge of glibness, though, only if it embraces some content: we are required to develop the picture of abnormal eating, so that normality may be seen in its negative image. Of course, when it comes to eating, this simple negation approach may not work too well. For instance, the difference between abnormal and normal may be a matter of degree. Consider the speed of eating. Normal eating may proceed at a certain pace, whereas abnormal eating may be quicker (as in binge eating), or slower (as in anorexia, where each bite seems to take forever). If eating is fast in bulimia, then what does that tell us about normal eating? That normal eating is not fast (but probably not too slow). But what exactly, or even roughly, is the pace of normal eating? There's no way of telling from studying the speed of abnormal eating (assuming that we could study the speed of eating in bulimia, a disorder about which there are plenty of anecdotal reports but few hard facts). In fact, there is no way of knowing that abnormal eating is fast, except insofar as we assume that the speed of eating we see in abnormal eaters is fast compared to some implicit normal rate. This exercise quickly becomes complicated, and perhaps impossible. We may not even be able to recognize abnormal eating unless and until we have a firm idea of normal eating.

Of course, the key element here may be 'the idea of normal eating.' What we consider abnormal about abnormal eating reveals what we assume is normal about normal eating. In other words, our very description of abnormal eating contains within it clues to what we mean by normal eating. Characterizing abnormal eating, then, uncovers our implicit theories of normal eating. Whether these theories are correct or not is, of course, an empirical question. But regardless of whether these theories are correct, they are useful as a window on the common assumptions that we hold about normal eating. These assumptions, in turn, provide an interesting springboard for research; assumptions can be tested. In the case of assumptions about normal eating, we will frequently find that these assumptions are simply incorrect. And as these incorrect assumptions are replaced by facts, we may in turn be forced to challenge our assumptions about what is abnormal in abnormal eating. After all, if we adopt new notions about what constitutes normal eating, we may well be forced to adopt correspondingly new notions about what constitutes abnormal eating. Conceivably, aspects of abnormal eating that are considered to be part of what makes abnormal eating abnormal may turn out to be perfectly normal. Let us then proceed to attempt to describe abnormal eating, bearing in mind of course the implications for what we are to conclude about normal eating.

How do we go about describing abnormal eating, though? Without wishing to drive the reader to distraction, we are obliged to point out that describing eating (abnormal or otherwise) is not an obvious or straightforward procedure. The problem is: what are the dimensions or features that we must identify—the parameters of eating that might in some way or another be abnormal? We mentioned above the feature of pace or speed of eating. Clearly, that is an example of a feature that might take on abnormal values in certain individuals or classes of individuals. But thinking of an easy example does not solve the more general problem of identifying the full catalog of dimensions on which people might display normality or abnormality in their eating. How does one go about identifying such features? The answer, quite clearly, must be 'by referring to the literature'. Although it is possible that the literature has overlooked certain features that might be abnormal—or normal features that might take on abnormal values—we are in no position to fill in these latent gaps. We must defer to the work that precedes us, and discuss those features that others have discussed already, if only because in the absence of published discussion of these features, we cannot say much about them. Of course, there is no way that in a chapter of this scope, or of any reasonable breadth, we could do justice to the entire literature on abnormal eating. The accumulation of data—especially if we include anecdotal, 'clinical' reports—is simply overwhelming. We are forced to sample, and to hope that our sampling technique is not overly biased. But enough delay—let us examine the literature. Maybe things will somehow become simpler and clearer.

5.2 Quantity

For want of a better place to start, let us consider amount eaten. We are all agreed that abnormal eating probably entails, among other things, eating an abnormal amount of food. People with eating disorders are notable for eating unusual quantities of food. What does the literature tell us about what is abnormal and what is normal when it comes to amount eaten?

Anorexia nervosa refers specifically to loss of appetite, which is reflected in a low intake. Bulimia nervosa, on the other hand, refers specifically to an inordinately large (ox-like) appetite, again reflected in intake. How little do anorexics eat? How much do bulimics eat? (And how much, somewhere in the middle, is normal?)

5.2.1 Anorexics

The definition of anorexia nervosa does not specify how little food must be eaten in order for one to be considered anorexic. The defining characteristics of anorexia include a low weight (or substantial weight loss), certain attitudes, body image distortions, and, where relevant, amenorrhea. Anorexics are notorious for eating sparsely (or parsley), but nowhere is it specified how little one must eat in order to become an anorexic. As a result, we cannot determine the dividing line between normal and abnormally low intake. This indeterminacy is exacerbated by the fact that most of the time, people who are indisputably normal are likewise not eating anything. It seems clear that determining a lower bound of normal intake must involve reference to some reasonably long period of time, given the highly skewed distribution of calories consumed per minute in all eaters, a distribution notable for the prevalence of zeros.

Another, possibly more fruitful approach would be to regard as abnormally low any level of intake insufficient to maintain body weight. (A normal intake would presumably suffice to maintain body weight.) One problem here is that different individuals require different intakes in order to lose or maintain weight. Owing to the slowing of metabolism that accompanies progressive weight loss, those who have already lost weight need fewer calories to maintain their lower weight; that is why it is so difficult for dieters to continue to lose weight. What might be an abnormally low (weight-losing) intake in Person A could well be a normal or even above-normal (weight-gaining) intake in Person B, if Person A has recently been gaining weight and Person B has been losing weight. A second problem is that maintaining a particular body weight (or losing or gaining weight) is not an exclusive function of intake; energy expenditure (including basal metabolic costs, post-prandial thermogenesis, and deliberate physical activity) plays a large role in weight maintenance and change. It would be imprudent to use weight stability as an index of normal intake when factors other than intake affect weight stability. It would seem, then, that specifying a particular level below which intake is abnormally low is a precarious undertaking. In any event, it has not been done. The result is that it is correspondingly impossible to specify the lower boundary of normal intake.

5.2.2 Bulimics

The very definition of bulimia nervosa (or of binge eating disorder, which appears to be bulimia nervosa without the accompanying purging) centers on

> 'consuming a large quantity of food in a short period of time.' (Johnson and Connors, 1987, p. 3)

Of course, we have to acknowledge—as Johnson and Connors do, and everyone else does, on reflection—that determining what counts as a large quantity of food is ultimately a question of rate of eating, not just amount *per se*. Eight thousand calories are not very many, if you take a week to consume them (just as zero calories are not preposterously few, if we consider brief time periods). DSM-IV (APA, 1994) defines a binge (in bulimia nervosa) as

> 'eating, in a discrete period of time (e.g., within any 2-hour period), an amount of food that is definitely larger than most people would eat during a similar period of time and under similar circumstances.'

This definition provides, albeit tentatively, a specific time period to consider, but ultimately does little to answer the question of how much one must eat to be considered a binger. This definition is particularly interesting in light of our present purpose, inasmuch as it refers to how much 'most people would eat.' The implication here is that we all know—or at least somebody knows—how much most people would eat in a particular period; in other words, a common understanding of what constitutes normal eating is implied. But as should be clear by now, we do not know how much eating is normal; indeed, we are looking at abnormal eating to find a clue about normal eating, but abnormal eating in this case is really nothing other than 'more than normal' eating, leaving us back where we started.

Some attempts have been made to prescribe the (minimum) quantity of a binge, but the variability in these specifications is disturbing. Chernyk (1981) requires 4000+ calories within two hours; Katzman and Wolchik (1984) require only 1200 calories for a binge, while Nevo (1985) requires only 1000 calories. We can attest to the fact that consuming 1000 calories in two hours does not require a Bunyanesque effort, and in fact this amount falls well within the boundaries of a standard gourmet meal at a good French restaurant (especially the kind that still uses rich sauces). Mitchell *et al.* (1985) report that their bingers averaged 3415 calories during binges, but the range extended from 1200 to 11 500 calories. This empirical approach is blurred by the fact that

> 'not all individuals are aware of calorie counts of various foods, and bulimics have been known to give very unreliable estimates of calories consumed during a binge.' (Johnson and Connors, 1987, p. 23)

Besides, learning what bingers actually consume does not help us determine precisely what (minimum) consumption is necessary to qualify as a binge. Without such a minimum, it would seem to be futile to try to establish the limits of normality.

Another complication derives from the fact that prodigious consumption is not confined to the eating disordered, and may well be part of everyday life for many people. Many people claim to 'binge' or 'pig out' as part of their normal recreational lives. You need not be a clinical case to participate in a pie-eating contest; in fact, many of us overindulge on an all-too-frequent basis, especially if we are at a party, and/or at a restaurant where there is a lot of food available, and/or if we have had a few drinks. De Castro has documented what he calls the social facilitation of eating (e.g. de Castro *et al.*, 1990), in which people eat more when they are in the presence of others than when they are alone (see Figure 5.1). Indeed, de Castro argues (de Castro and Brewer, 1992) that the amount consumed is a power function of the number of people present, which leads quickly to the possibility of very large individual intakes in large groups. Such groups, it should be emphasized, are not composed of eating disordered individuals; de Castro does not include such people in his studies, and eliminates even garden-variety dieters. (Later in this chapter, we will return to the topic of social influences on eating, and learn that neither anorexics nor bulimics react to the presence of others in the same way that most people do.)

The overindulgence induced by the group—through a mechanism that remains a mystery—is in some ways perfectly normal; de Castro regards it as a fundamental part of normal eating. This line of research raises the question of whether periodic overindulgence—as defined, perhaps, by subsequent regrets—is to be considered normal or abnormal. Almost everyone does it, so in that sense it is normal. By the same token, however, almost every normal person does abnormal things on occasion; perhaps socially-induced overeating is one of those things.

Consideration of the social facilitation of eating leaves us in a bit of a quandary. If this excessive consumption is considered normal, then it becomes clear that the upper bound of normal eating exceeds the lower bound of binge eating. There is no obvious way of demarcating where one leaves

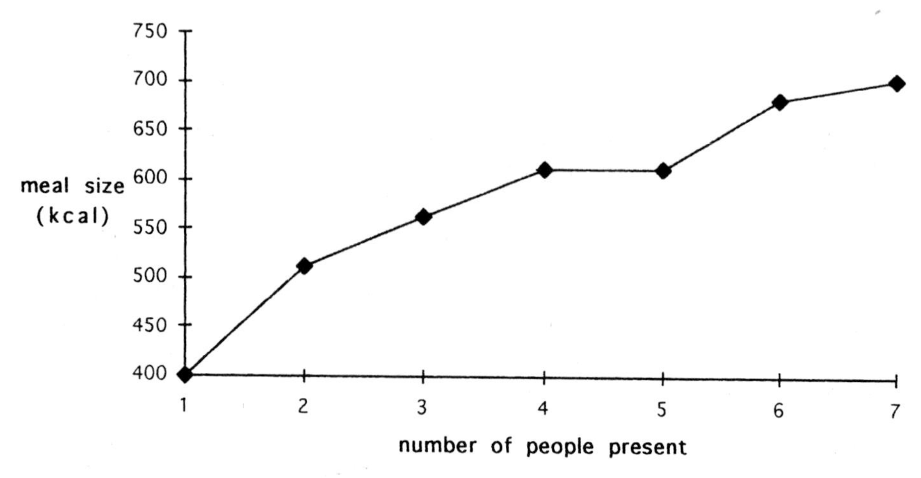

Figure 5.1 People present/meal size correlation. Adapted from de Castro and Brewer (1992).

off and the other begins. If, on the other hand, we decide that socially facilitated eating is abnormal, then it becomes clear that we cannot look to particular individuals to provide us with clear cases of abnormal eating. If virtually everyone behaves abnormally on occasion, then we cannot define normality in terms of who does behave that way and who does not. Normal people sometimes behave abnormally, and presumably abnormal people sometimes behave normally. Looking at diagnosed cases is not a surefire source of information on abnormality, since not everything that they do is abnormal. Mitchell *et al.* (1985) report that only 21% of their patients ate 2 normal meals per day; perhaps that is a low percentage, but it does indicate that bulimic patients do occasionally eat normal meals. Beumont (1995) notes that, like us, pathological bingers may consider as a binge any eating that is excessive; but pathological bingers may have a much lower subjective threshold for excess than we do. That is, what we would consider a normal amount of food might in their eyes be excessive, given their distorted view of what they 'should' be eating. This consideration suggests all the more forcefully that a binge may not contain as many calories as the more florid accounts of bingeing might lead us to expect.* Rosen and Leitenberg (1988) conclude from their own previous research (Rosen *et al.*, 1986) that 'most of the binge episodes fell within the same calorie range as the nonbinges' and that 'in most instances eating binges are much smaller than is commonly believed' (p. 165).

A final complication pertaining to quantity arises when we consider the binger who is only a couple of minutes into the binge. Or consider the binger who is interrupted during the binge. (Most bingers simply stop when their behavior is subjected to public scrutiny.) Is a binge that has not run its full course not a binge? DSM-IV (APA, 1994) defines a binge not only in (vague) terms of quantity but also requires that the episode be characterized by

> 'a sense of lack of control over eating during the episode (e.g., a feeling that one cannot stop eating or control what or how much one is eating).'

This opens up the possibility that out-of-control eating might be considered a binge even if the amount consumed is not excessive. It would seem to be unarguable that a binge has a beginning, a middle and an end. And if we consider just the beginning portion of the binge, in isolation, we may ask, Is the eating in that portion of the binge not bingeing? For normal eating to become bingeing, must a certain amount of food be consumed? We would argue that an interrupted binge—even if it is interrupted after only a few bites—is nevertheless a binge, however truncated it may be. It does not seem advisable to consider as binges only those binges that run their com-

*Beumont, in the same paragraph (p. 156), himself provides such a florid account when he claims that 'some patients may take as much as 30 times the recommended daily allowance of calories in one binge.'

plete natural course. However, Garfinkel (1995, p. 129) points out that this approach 'can lead to the strange situation of some patients "bingeing" on two cookies.' Indeed, there would seem to be almost no logical lower limit to binge intake, if the binge is defined in terms of the secondary aspect of 'lack of control.' Garfinkel (1995, p. 129) recommends that we

> 'define excessive quantity by objective standards. What is most relevant is that the person is eating more than is normal, given what other people eat, and the time and social circumstances of the person's last meal.'

This is close to the DSM-IV criterion, which, as we have seen, is more elusive than it seems (or must have seemed to those who articulated it).

5.2.3 The obese

From time immemorial, there have been allegations that the obese eat too much. Indeed, most laymen and probably even most professionals believe that the reason that people become obese is that they eat too much. (Even those who favor a genetic approach to the etiology of obesity may acknowledge that one route by which 'fat genes' produce a fat physique is by prompting the individual to overeat.) Empirical investigations of overeating in the obese, however, have not supported the widespread assumption of overindulgence. More precisely, the overeating displayed by the obese is sporadic: some, indeed most, fat people overeat only occasionally, and some do not overeat at all. When the obese overeat, moreover, it is usually in response to some particular set of conditions (Schachter and Rodin, 1974). In short, the obese do not appear to be much different than anyone else, and may in fact not be a homogeneous category, at the behavioral level. Doubts about whether the obese as a category of people are abnormal in their eating or in anything other than physique led to the elimination of obesity from DSM as a definite psychiatric condition. Later in this chapter, we will refer to the obese in the context of their alleged overreponsiveness to external cues that control eating; the literature on external control was derived principally from studies of the obese. It must be remembered, though, that we no longer 'officially' consider the obese to eat abnormally, so for purposes of this chapter, they must yield the floor to the anorexics and bingers.

5.3 Speed

Our discussion of abnormal and normal eating in terms of quantity consumed could not avoid the issue of speed or rate of eating. On close examination, an abnormal quantity of food is abnormal only when one considers how long it takes to eat it. An extremely large amount suddenly becomes a normal amount if an extremely long period of time is allowed for its consumption; by the same token, minimal consumption during a minimal time period cannot be regarded as particularly bizarre. In short, speed and quantity seem to be inextricably linked for purposes of defining abnormalities of eating. We should not expect, therefore, to learn a great deal more about normal eating from the literature on abnormal rates of intake than we learned from the literature on abnormal quantities.

5.3.1 Anorexics

Because they are so averse to food in the first place, anorexics understandably do not show a great deal of eagerness to eat, if eagerness is measured by speed of consumption. Anorexics 'eat painfully slowly', according to Beumont (1995, p. 153). Garfinkel and Garner (1982, p. 11) claim that anorexics often 'spend increasingly lengthy periods of time at a meal, at times barely completing lunch before supper.' We are likely to conjure up an image of the anorexic rearranging the food on her plate without actually eating any of it. Our assumption that motivation may be inferred from the speed with which the object is consumed—an assumption that goes beyond eating to all realms of appetitive behavior—would seem to be reasonable, but the following counterproposals should perhaps also be borne in mind.

First, if eating is so aversive to the anorexic, then perhaps we ought to expect her* to be eager to get it over with. If she 'does not deserve to eat', as she claims, then perhaps she should minimize the pleasure of eating, and we normally regard the drawing out of an activity as a way of enhancing its pleasures, not minimizing them. We suspect, though, that the painfully slow eating shown by the anorexic is a public performance; she drags out her eating when others are watching. If she eats her first bite quickly, she is likely to be encouraged to get on to her second bite. So by delaying the first bite, she postpones the struggle over the second bite. Given that she is required to spend a certain minimal amount of time at table, the best strategy to minimize consumption is to proceed as slowly as possible. Getting it over early by eating one's fill in a single quick bite does not work because those who would exert pressure on the anorexic to eat more have all the more time left over to exert more pressure.

A second consideration stems from our conviction that the anorexic, despite or because of her minimal intake, is extremely hungry. Bruch (1973, p. 45), while acknowledging that the anorexic usually does not or will not admit to her chronic hunger, never doubts that this hunger exists, and notes that the anorexic is occasionally 'overpowered by urges for food' and binges. Interestingly, Garfinkel and Garner (1982, p. 5) claim that 'most

*The use of the female pronoun in this paragraph, and that below, is not gratuitous. The principal eating disorders are much more prevalent in women.

patients report normal awareness of hunger' and that they experience 'intense hunger.' There is some consensus, then, that the anorexic is hungry, but little consensus on whether the anorexic realizes or admits to this hunger. The anorexic's hunger ought perhaps to be manifested in rapid intake, again on the assumption that heightened motivation is reflected in heightened speed. What we might expect is that the anorexic is loathe to eat, but that when she is forced to, or when her hunger finally overcomes her resolve to abstain, we should see a brief display of that underlying hunger. Because the hunger is so intense, it should produce a corresponding zeal for food. Admittedly, this display of hunger may not last long; after all, anorexics do not eat much; but while it lasts, perhaps it should be binge-like in its ferocity.

Certainly there are anorexics who binge (and then purge). Debate continues as to whether they ought to be considered anorexics who display the key behavioral signs of bulimia or as bulimics who manage to achieve the weight restrictions at the heart of anorexia (Garner and Fairburn, 1988). In either case, when they binge they do not dawdle (see §5.3.2 for a discussion of speed in bulimia); but it is clear that notwithstanding whatever hunger they may be experiencing, true anorexics (anorexic restrictors) are notoriously slow eaters. The only real question is whether this sluggishness is an essential characteristic of their eating, or whether it is created or accentuated by public eating performances, where the anorexic is expected to devote a certain seemly period of time to eating, so that dawdling allows her to eat up the time without eating up the food.

What can we conclude about the normal pace of eating from the pace of eating displayed by anorexics? As we expected, only that anorexics eat more slowly than normal. How much more slowly? To even begin to answer this question, we require some metric of speed, such as calories per minute. Of course, such calculations could be made, for anorexics or for normals—but they haven't been. No-one seems concerned enough about quantifying rate of eating to bother with these calculations. And had the calculations been made, we would no doubt not be much further ahead. The rate or speed of eating would have to be calculated as calories consumed per minute during a meal, and if we were to try to compare anorexics and normals (however defined) in this way, we would immediately bump up against the fact that the anorexic 'meal' is itself so bizarre as to preclude easy comparison with normal meals—and not only in terms of quantity, as discussed already. How can we hope to compare the speed of eating of two meals that are so radically different in both content and size? To compare speed realistically, we would have to hold meal size and content constant, and such constancy is nowhere to be found in the literature.

As a final twist on the question of abnormally slow eating, let us reprise the image of the anorexic rearranging the food on her plate, or hiding it away for later disposal, or surreptitiously dropping it for the dog. These tactics serve to avoid actually eating unwanted food, but on reflection, these tactics are not confined to anorexics. Children confronted with food that they do not like engage in precisely the same ruses, and even adults, who usually have more control over what ends up on their plate, will resort to these subterfuges when somehow they are expected to eat something particularly nasty. They will dawdle, stall, and slow their actual eating down to an anorexic pace. Is this abnormal, or is it a normal response when confronted with a demand to eat what one doesn't want to eat? The only difference between anorexics and the rest of us would seem to be that anorexics don't want to eat anything, whereas we reserve our distaste for a more limited range of foods. A slow pace of eating, then, may not be a true discriminator of normal and abnormal eating, but only a natural reflection of a normal/abnormal difference in what sorts of food are desirable (see §5.5).

5.3.2 Bulimics

The DSM-III definition of a binge referred explicitly to speed of eating, requiring that the food be consumed rapidly in order to qualify. Aside from the fact (see below) that this criterion has been weakened recently, the rapid rate requirement poses difficulties when examined closely. Must the rapidity of consumption (as measured in calories per minute) apply to the meal (or binge) as a whole or to each and every segment of the binge? It is certainly conceivable that normal eating could consist of rapid bursts of eating—maybe just as rapid, in a microscopic way, as binge eating—but in normal eating these bursts are separated by intervals of conversation (or just breathing) that are shorter or absent in bingeing. Or maybe normal eating, no matter how brief a segment of it we consider, never achieves the speed that typifies binge eating. Global considerations of how much a binge eater and a non-binge eater consume in, say, fifteen minutes will never answer this question. Thinking about abnormal eating certainly raises interesting questions about normal eating, but it does not provide obvious answers to those questions.

We referred earlier to the fact that rapid eating has been dropped from the DSM criteria for bulimia nervosa. The reason is simple: although rapid eating is one of the more flamboyant aspects of binges, it is not a reliable one. 'Many patients report a slow or moderate rate of eating,' according to Garfinkel (1995, pp. 128–129). Beumont (1995, p. 156) notes that

> 'people with bulimia nervosa have a tendency to eat rapidly during a binge, stuffing in a large amount of food within minutes. Chronic patients, however, will binge eat at a slower rate, particularly if there is little risk of discovery. Others report "picking" behavior, taking in small quantities at a time (i.e. a teaspoon of ice cream, a small piece of cake, or a portion of cheese) but continuing this "picking" for hours until they have ingested thousands of calories.'

After all, what's the rush? It has long been assumed that rapid eating—and the 'rush' that it creates—is an essential aspect of the binge, and that the same amount of food consumed in a leisurely fashion would not have the same reinforcing power. (An analogous situation applies to nicotine: Why would you absorb nicotine slowly into your bloodstream by way of a patch when you could experience a blast of nicotine by taking it in through your lungs?) Apparently, however, bulimics—at least some bulimics—are quite satisfied to adopt a leisurely pace. Perhaps it allows them to enjoy the food more. Beumont refers to the 'risk of discovery'; if it is low, a slower pace may ensue. We may infer that at least some of the rapid eating that characterizes bulimia is forced on the bulimic by the limited amount of time that she has available for satisfying her craving, or by her fear that she may be interrupted before she has properly finished. This view is very much at odds with the preceding assumption that rapid consumption was intrinsically (or psychodynamically, or metabolically) rewarding. Perhaps we must entertain the possibility that there are two sorts of binges, serving different purposes; in one case, speed is of the essence, whereas in the other it is only quantity that counts, and if time allows, there is no need for haste. To return to our original question: To whatever extent a leisurely pace characterizes binge eating, we are not likely to be able to distinguish it from normal eating on the dimension of speed.

Later in this chapter, we will address more explicitly the social context of eating, and will conclude that normal and abnormal eaters respond to the presence or absence of other people in quite different ways. For the moment, suffice it to say that our discussion of speed has revealed a social influence for both anorexics and bulimics. The anorexic, we argued, drags out the meal only when observed by others; she is socially compelled to engage in eating for a certain seemly length of time, and wastes as much of that time as possible. The bulimic, on the other hand, uses her time as efficiently as possible: if she is likely to be discovered, she crams her eating into as brief a period of time as its caloric expanse will allow, but if privacy is guaranteed, the pace may slow to normal or even a dawdle. In short, the extremes of speed that we associate with abnormal eating may be a function of the fact that the presence of others makes it difficult for both the anorexic and the bulimic to eat the quantity that they would prefer—nothing and everything, respectively—and this difficulty leaks out into the pace of eating. This social-induction approach to pace of eating is intriguing, but we should probably be careful before entirely abandoning the long-standing notion that the rapidity of binge eating may add psychologically or physiologically to whatever function binge eating serves.

In another contribution to the intersection of social factors and speed of eating, Rosenthal and McSweeney (1979) found that people eat more quickly while in the presence of others; this finding may be related to (or an alternative expression of) the social facilitation effect discussed above in

§5.2. Rosenthal and McSweeney's observations, of course, were obtained from normal subjects (and obese subjects, who did not differ in the social acceleration of eating). The import of this study would seem to be that the normal speed of eating is variable, responding to variations in social circumstances. (One is tempted to imagine an experiment in which an experimental confederate, eating alongside the subject, is instructed to vary the speed of eating deliberately, in order to determine whether subjects will keep pace.) The variability of speed in normal eating—and it is important, we think, to acknowledge that eating alone and in the presence of others must both be considered normal variants of the basic eating situation—means that there is no single pace that is absolutely normal; as with amount, we must tolerate a range of normality. But (a) we still do not have much of a sense of what the upper and lower bounds of that range might be, and (b) even if we did, our knowledge would be derived from a closer examination of normal eating, not abnormal eating.

The likelihood that there is a range of normality—that there is no single value for normal speed (or quantity or probably anything else)—is worth a bit more contemplation. By proposing a range of normality, we are not simply suggesting that some normal people eat at one (normal) rate while other normal people eat at another (normal) rate, although that is no doubt the case; some otherwise normal people are notorious for wolfing down their food whereas others seem to take forever to eat. (This is precisely the sort of thing that marital compatibility tests should include.) Neither the fast eater nor the slow eater may be regarded as abnormal; more likely, they will be regarded as normal variants. Beyond chronic differences in speed, though, we must acknowledge that the same individual will show different rates of consumption, depending on circumstances. As a trivial example, the speed of eating tends to vary with the degree of hunger or prior deprivation. Within the same meal, we are likely to start fast and finish slow, so that our speed decelerates over the course of the meal (Kissileff *et al.*, 1982). In short, we cannot specify what is normal, but we know that when it is specified, it will include large inter-individual and intra-individual variation. This conclusion will no doubt apply to all the other parameters of eating, not just speed.

5.4 Frequency

Everyone knows that three (square) meals a day is normal. The problem is that nobody seems to subscribe to this normal pattern. It may be normal prescriptively, but not descriptively. When it comes to meal frequency, reference to the behavior of explicitly abnormal populations does not seem to be a likely source of illumination. How many meals do anorexics eat? Arguably, they do not eat any. Of course, this argument leads us in the direction of defining a meal, an undertaking not for the faint of heart. Meals are typically defined as the consumption of a certain minimum number of calories, separated in time by a certain minimum interval from other consumption. If we adhere to the minimum calorie requirement, the anorexic may be excluded from meal-eating altogether. But granting that anorexics eat infrequently, as infrequently as they can manage, what does that tell us about the lower boundary of normal meal frequency? One meal a day? Two? Dieters and the obese are notorious for skipping breakfast, and other meals as well. And even people who are perfectly normal eaters in every respect will occasionally skip a meal, not in the interests of weight-reduction perhaps, but simply because they become absorbed in some other activity during the time usually allotted for eating. That people sometimes skip a meal inadvertently makes it clear that what is normal in the realm of meal-taking is not a matter of nature but of cultural norms: If one fails to eat at the accustomed time, there is no logical reason why one should not eat as soon as possible after the distracting activity has been completed. However, we are likely to find that the meal-misser will simply wait for the next scheduled meal, rather than satisfy his hunger at the next available opportunity, an opportunity that he himself could probably arrange at his early convenience. So we may conclude that the daily routine—culture, if you will—exerts a profound effect on meal-taking. At the same time however, the fact that our hypothetical meal-misser usually eats on a fixed daily schedule means that his hunger is probably temporally conditioned; that is, once the usual mealtime has passed, his hunger may well abate even if no food has been ingested, and his hunger may not re-emerge until the next accustomed mealtime. The conditioning of hunger (Weingarten, 1985) may typically converge with nutritional needs, so that usual mealtimes (and meal sizes) are adjusted to accommodate caloric requirements. This adaptation of physiology to culture and vice versa means that culture is not the only determinant of meal frequency; physiology will determine how much is eaten at meals and may even affect the setting of mealtimes in the first place. Culture need not develop arbitrarily; indeed, it usually develops in a way that accommodates physiology.

To return to the question of normal frequency of eating, we are probably safe in concluding that less than two meals per day is abnormally low. Still, we must be careful to remember that defining a meal is difficult, and perhaps more important, our original question was not 'What is the frequency of normal/abnormal meal-taking?' but 'What is the frequency of normal/abnormal eating?' Not all eating occurs at meals, as is evident to those of us who decry the apparently ever-expanding epidemic of snacking. Snacking typically (if vaguely) refers to eating that does not fall within the definition of a meal. This distinction, however, does not withstand much scrutiny, inasmuch as snacks often exceed in caloric volume any reasonable minimum definition of a meal. The precise nature of a snack, like everything else in this area, is elusive, but it would seem to have three crucial parameters. First, the snack should be small (but often as not it is not). Second, the snack usually escapes the 'well-rounded' ideal of a true meal, and may be devoted entirely to quasi-foods such as potato chips that could never by themselves be considered a meal. (At the same time, it must be conceded that normal meals no longer—did they ever?—live up to the 'basic food group representation' ideal; just this morning, our teenage daughter ate a breakfast consisting entirely of the remains of a container of ice cream. Perhaps it was ever thus.) The third defining aspect of a snack is that it occurs at a time outside normal mealtime. On reflection, it seems that snacks stand in relation to normal meals much as weeds stand in relation to normal plants. A weed is simply an unwanted plant; snacks, while not exactly unwanted, are simply meals that we refuse to dignify with the term 'meal', or prefer not to dignify with the term 'meal' so as to excuse our indulgence in eating that would not be acceptable as a meal.

Grazing is snacking by another name. Unlike snacking, which is generally frowned upon, presumably because it often consists of so-called junk food (food which falls short of some nutritional ideal, but perhaps not much more so than the rest of our eating, including what we call meals), grazing is often presented as an ideal approach to eating. Eating *ad lib* throughout the day is thought to represent a return to an earlier, healthier, almost Edenic eating style, before the industrial revolution alienated us from nature, including our own bodies. Grazing is also assumed, owing to its naturalness, to somehow spontaneously involve healthy food choices; the grazer is more likely to select vegetables in their natural state rather than as potato chips. Grazing is alleged to prevent the development of obesity; by freeing intake from artificial constraints, intake reverts to a nutritional balance of both content and quantity, and the digestive system is not forced to cope with the traumatic consequences of large meals, including storing the excess as fat (Woods, 1991). Whether grazing is actually different from or healthier than snacking is a difficult question, and perhaps simply a matter of semantics. 'Grazing' as a colloquial term seems to have captured the nutritional high ground. One imagines that if a grazer were found head first in a large bag of potato chips, the grazer might be recategorized as a snacker.

Whether or not grazing or snacking is normal obviously depends on your perspective. Snacking is certainly normal statistically. (One of us is regarded as somewhat pathological in his refusal to eat between meals, and he probably is.) Grazing—if we distinguish it from snacking in terms of the nutritional value of the food eaten—is less prevalent, probably owing to snacking's universal appeal as an opportunity to eat bad-for-you, good-tasting food in a positively sanctioned context ('Everyone does it!' and even serious dieters are expected to snack, if the availability of 'diet snacks' is any indication.)

This extended discussion of snacking and grazing brings us back to the question of the normal frequency of eating. One can hardly imagine what 'normal' might be, given the complexities that we have uncovered so far. In one of the few cases where anorexics and bulimics both appear to differ from normal in the same direction, Davis *et al.* (1985) found that bulimics consumed fewer meals than did normal controls. (Perhaps this is not as surprising as it might appear, when one considers that the prodigious consumption displayed by bulimics does not occur at meals, but during binges, which we may perhaps regard as pathological snacks; Davis *et al.* note that bulimics consume more snacks than normal.) So we may conclude that in terms of meals, there is a plausible number (say, two per day) any less than which we may regard as abnormal. Two meals per day, then, becomes the lower bound of normal frequency—but only if we think in terms of meals. The inclusion of snacks upsets our calculations of what may be regarded as the lower bound of normal eating frequency: two times per day is undoubtedly too low a threshold for normality if we include snacks.

At the upper bound of normality we are perhaps even more unsure of ourselves, especially since our abnormal eaters tend to eat with uncommon infrequency. It would appear to be impossible to specify how frequently one might eat and still be normal. If we subscribe to the grazing philosophy of natural eating, then arguably almost any conceivable frequency of eating might be normal. Certainly the normal range, however it might be defined, is well beyond three meals per day, if snacks are included. If we were to arrive at an accurate upper limit on the normal number of eating bouts per day, or per any unit of time, it would not be by examining abnormal eaters (bulimics or anorexics). Even the obese tend to eat less often (Schachter and Rodin, 1974) than normal, and so provide no help in determining how often is too often.

5.5 Type of food

Quantity, speed and frequency of eating, three relatively quantitative dimensions, provide hints that we might use abnormal eating to learn about normal eating, but these hints did not quite materialize into firm conclusions. If anything, it became clear that close consideration of the issues raised doubts about the ability of abnormal eating to teach us anything about normal eating. Of course, one hopeful possibility is that we were looking in the wrong place. Perhaps these more quantitative parameters are not the ones that clearly distinguish normal and abnormal eating. Let us look elsewhere.

One obvious place to look is the type of food consumed. Regardless of amount, pace or frequency, there is an assumption, sometimes implicit but often explicit, that abnormal eaters eat abnormal types of food, or at least an abnormal balance of foods. In short, their diet is aberrant. Can we learn what a normal diet looks like from examining the aberrant food choices of pathological eaters?

5.5.1 Anorexics

Anorexics, of course, are more notable for what they do not eat than for what they do eat. On the positive side, they are stereotyped as eating 'rabbit food'—carrots, celery, and other low calorie foods. But we cannot hope to identify the constituents of a normal diet by exclusion. After all, the foods that anorexics do eat are also part of a normal diet. We all eat 'rabbit food', and those of us who do not, recognize that we should.

The foods that anorexics reject are not quite identical with the complement of what they choose. Between the chosen and the rejected there is a gray area containing foods that the anorexic might eat but usually does not. Notorious for her all-or-none, black-and-white categorization of food, the anorexic in reality saves her positive and negative evaluations for the most extreme food choices, and ignores much in the middle. Moreover, the foods chosen and rejected are subject to periodic re-evaluation, according to the latest fad.

> 'The food choices of anorexia patients are determined by unhealthy attitudes and misconceptions acquired from dubious sources of information such as popular women's magazines. As the fads reported in these periodicals have changed over the years, so have the foods rejected by anorexia patients. Previous generations selectively avoided simple sugars and other carbohydrates (sweets and potatoes). Today, fatty foods and red meat are considered "unhealthy," and vegetarianism has become the most common dietary perversion. Great reliance is placed on energy-reduced dietary products, on foods with a high fiber content, and on supplementary vitamins.' (Beumont, 1995, p. 153)

To summarize, much of what the anorexic rejects is almost certainly part of a normal diet, however we might choose to define it; but we cannot simply define a normal diet as what the anorexic rejects, just as we cannot exclude from a normal diet what the anorexic accepts. The anorexic's food choices are simply too bizarre—or maybe in some ways not bizarre enough—to help us understand normality.

5.5.2 Bulimics

Like anorexics, bulimics are thought to engage in unusual food choices. Much of the time, of course, bulimics are in a restrained mode, eating minimally and/or eating 'permissible' foods. When the bulimic binges, however, she abandons not only her concern with quantity, but also her concern about restricting herself to specific food groups. Debate rages in the literature, however, over exactly what it is that bingers like to binge on. Some argue that bingers gravitate toward certain special types of foods, whereas others have argued that bingers are virtually oblivious of what they eat.

Johnson and Connors (1987, p. 42) assert that 'most binges appear to consist of sweets or salty carbohydrates. . . . These are foods that the patients generally regard as "forbidden".' Sweets (desserts) and salty carbohydrates (junk snacks) are precisely the sorts of foods that bulimics—like everyone else who is trying to eat healthily or keep weight off—ordinarily deny themselves. Most analyses of binge eating retain a somewhat psychodynamic flavor, emphasizing the notion that forbidden foods are especially attractive, presumably because their forbiddenness adds allure or incentive to them. (Another way to arrive at the same conclusion is to argue that as part of Nature's fundamental perversity, particularly alluring foods—sweet or salty carbohydrates—tend to be bad for your health and/or your weight, so that the conscientious eater must avoid them.) In either case, it seems that concerns with health or weight are forgotten temporarily during the binge, and the basic allure of certain foods is permitted free expression. There is some measure of consensus in the literature (e.g. Hamburger, 1951; Rosen *et al.*, 1986) about the preferred food of bingers, but a possible complication is introduced by the realization that the sweet or salty carbohydrates that are so popular are not only tasty and forbidden but also very easy to consume in quantity. They go down easily and require little or no preparation beyond ripping open the packaging. The accessibility of these types of foods and the rapidity with which they can be consumed brings us back to our earlier discussion of quantity and pace; perhaps bingers prefer these foods not because—or not only because—they taste good, but because they lend themselves to rapid and prodigious intake.

The proposal that 'binge foods' are selected because they are easy to eat meshes with some claims in the literature suggesting that bingers, in their avidity to eat rapidly and extensively, actually do not pay much attention to what they eat. Going back at least as far as the famous case of Ellen West, we discover that during binges she threw herself 'indiscriminately . . . on any foods which happen to be at hand' (Binswanger, 1958, p. 252). (Binswanger complicates things for us by mentioning that West would often consume large quantities of tomatoes and oranges, which hardly correspond to the sort of forbidden junk that bingers supposedly prefer; perhaps in West's day, fruit was more readily available than were nacho chips.) Kaye *et al.* (1986), in one of the few empirical studies in this area, concluded that bingers do not consume a disproportion of carbohydrates in binges (i.e. they consume carbohydrates during binges in the same proportion, relative to fats and proteins, as they do when not bingeing). We have noted elsewhere that there is extensive clinical evidence suggesting that bingers ignore palatability and become virtually oblivious to the quality of the food that they are devouring (Herman and Polivy, 1988a). Indeed, the authors went so far as to suggest (p. 34) that part of the guilt that the bingers experience after the binge is attributable not simply to their having overeaten but to the fact that their overeating was 'wasted' on foods that they did not particularly care for. One way to reconcile the claim that bingers preferentially consume forbidden, tasty carbohydrates with the competing claim that they eat indiscriminately is to propose that bingers, like everyone else, eat what they prefer when it is possible, but settle for just about anything when they have to. When the binge begins, are the preferred carbohydrates available? If so, there will be no hesitation about what to eat. The interesting question is, What happens if the preferred carbohydrates run out before the binge is properly completed? The binge could simply stop, of course; or equally possible, the binge could continue, using whatever other food was available, in descending order of preference. Whether or not the binge would continue would seem to depend on what is motivating the binge. Is it that the desire to consume attractive, forbidden food temporarily overcomes the customary inhibitions against eating for pleasure? In that case, the absence of the desired food would eliminate the incentive for continued bingeing. On the other hand, perhaps the purpose or goal of the binge is not to indulge in forbidden food but to achieve some sort of emotional satisfaction through the extensive, rapid consumption of food. All things equal, preferred foods will be selected for the binge, but in their absence, less desirable foods will do. Other scenarios are possible; for instance, some bingers plan their binges, stocking up on binge treats in order to avoid running out. In this case, things are ambiguous; the binge will end when the preferred food runs out, but only because the food has been designed to run out as the binge terminates.

Whether the binge is driven by the opportunity to eat attractive food, or whether it is driven by a desire to eat that will persist even in the absence of attractive food is a question that at the moment has no definite answer. The fact that the previous literature contains hints of two opposed alternative answers may well mean that both interpretations are correct. Some bingers may binge for one reason, while others binge for the other, with correspondingly different reactions to the end of the preferred food supply. The very same binger may binge for different reasons on different occasions. Some binges are planned; some are spontaneous. Some binges are designed to produce a positive emotional state; others are reactions to a negative emotional state. It is probably unwise to generalize about the purpose of a binge, and correspondingly unwise to generalize about the role of different types of food in serving the purpose of a binge.

Regardless of how we might decide this issue, it is worth remembering that the type of food that bingers prefer for their binges is more or less exactly the same type of food that we all prefer for our snacks. What distinguishes a binge from normal eating cannot be the type of food that is preferred. Perhaps it is the type of food that is eaten when the preferred food is unavailable; when normal eaters finish a box of cookies, they do not go on to eat the box itself. But even normal eaters will eat the inedible if they are hungry enough; hunger alters our threshold of acceptability. The binger who eats 'inedible' food acts as if she too is famished. Given her chronic deprivation (between binges), perhaps she is. Or perhaps she is simply famished 'emotionally.' It does not seem, in the final analysis, that food type is likely to provide the secret to illuminating the difference between abnormal and normal eating.

5.6 Internal and external cues

Our discussion of the 'purpose' of the binge, of what drives the binge, leads to a consideration of another dimension of eating along which normal/abnormal differences have been alleged. There has long been an assumption in the eating literature that it is normal to eat in response to 'internal' cues (i.e. physiological signals—either peripheral or central—of hunger and satiety) and inappropriate to have one's eating regulated by 'external' cues (i.e. food-related cues in the immediate environment). We have inherited from physiological psychologists the notion that (normal) feeding is an example of a negative feedback system, in which food deprivation over time arouses internal hunger signals which prompt eating, which in turn eliminates hunger signals and produces satiety signals, which in turn terminate eating (until deprivation once again arouses hunger signals). Departures from this well-regulated system characterize and perhaps even define pathologies of eating.

One version of such pathology occurs when internal signals that ought to terminate eating instead stimulate further eating. For instance, in our own lab, we have explored the phenomenon of 'counter-regulation', in which dieters respond to a high-calorie preload not by reducing their subsequent ad lib intake, as do nondieters, but by eating even more than if they had not been preloaded at all (Herman and Polivy, 1988b). This response of overeating after overeating is also prevalent in bulimics (Johnson and Connors, 1987). There is little doubt that preloading—especially if the preload is rich in calories—can incite further eating, but a question remains as to the nature of the internal signal that provides the trigger. Booth (1988) has argued that bulimics may become conditioned (through purging-based reinforcement) to interpret satiety cues as cues for further eating. For the bulimic, the more food that is consumed, and the more uncomfortable satiety pressures become, the more pleasurable it is when these pressures (and the corresponding fear of caloric overindulgence) are removed by vomiting or other forms of purging. Thus, ironically, the availability of purging may lead the bulimic to eat as much as possible, and may make satiety signals into signals to eat even more.

We have argued that the signal to overeat following a rich preload is not a physiological signal of the same type that prompts eating in hungry people, but rather a cognitive trigger based on the perception of having overeaten and thereby broken one's diet, rendering further dieting useless. This conclusion is bolstered by the observation (e.g. Polivy, 1976; Spencer and Fremouw, 1979) that the perceived calorie content of the preload is more important than its actual calorie content in triggering counter-regulatory eating. In any case, the response of bingers to a rich preload is certainly abnormal, and highlights the normal response to a rich preload—caloric compensation. The fact that the normal response is opposite to the abnormal response provides our first clear instance of abnormality clarifying or 'telling us something' about normality. Unfortunately, while we may learn something about normal eating from abnormal eating in response to preloads, it must be conceded that what we have learned is something that we already knew. It had been assumed all along that caloric compensation was the norm; the paradoxical response to a preload shown by dieters/bingers was something that was 'discovered' and explored later. Moreover, other than the general trend of normal eaters to compensate for caloric preloads, we cannot claim a great deal of knowledge about the normal response. For instance, how precise is the caloric compensation to a preload? Since Spiegel's (1973) review, nothing much has happened to alter the conclusion that caloric regulation in normal eaters is imprecise. The authors (Herman *et al.*, unpublished results) have demonstrated, for instance, that the extent to which the ad lib food bears a sensory resemblance to the preload food will affect the extent to which precise ad lib caloric compensation occurs; the typical response is far from perfect (one-to-one) compensation for the preload; the reduction in ad lib intake does not correspond to the caloric value of the preload. As an example, people eat less following a preload when the sensory properties of the ad lib food closely resemble those of the preload than when they are radically different.

The fact that weight remains remarkably stable from year to year, despite evident variations in daily or meal-to-meal intake, is often taken as evidence for the exquisite operation of the negative feedback hunger/satiety system, at least over the long run. It must be remembered, though, that our weight (and its stability) is a function of more than our intake (and its stability); metabolic and other energy expenditure factors are capable, in principle, of accounting for long-term weight stability without necessarily invoking precise intake compensation.

In summary, the abnormal pattern of counter-regulation highlights the normal compensatory response to a preload, but (a) we cannot really claim that we learned of this compensatory pattern by examining the counter-regulatory pattern, and (b) the details of the normal compensatory pattern remain obscure. More generally, responsiveness to internal cues of hunger and satiety, the basis of the negative feedback model, would seem to be a prerequisite for normal eating. Bruch (1973) has argued that far from being a purely natural ability, our capacity to attend and respond appropriately to internal hunger and satiety signals is a learned phenomenon, and like most learning, is vulnerable to error. Eating disorders and obesity, in her view, stem from (among other things) poor learning experiences and a consequent ineptness at recognizing and/or responding to the physiological cues that form the substrate of hunger and satiety. Again, we may assume that normality is distinguished by appropriate learning experiences, but at the same time, the learning approach emphasizes gradations of success, so that among normal eaters, we cannot expect to find perfect responsiveness or regulation. The extent to which normal eaters fall short of perfection in this respect is virtually unexplored.

As for the role of external cue responsiveness, most discussions begin with Schachter's (1968) contention that, unlike normal weight eaters (whom he assumed to be normal eaters in every respect), the obese were especially and inappropriately responsive to external food cues; this responsiveness led them to eat when there was no biological reason to eat. Subsequent research (see Schachter and Rodin, 1974, for a review) has confirmed the sensitivity of the obese to external cues, but the past two decades of research has mostly served to remind us of what we knew all along: namely, that normal eaters are also quite responsive to external cues. Normal eaters eat more preferred food than nonpreferred food, even if there is no immediate physiological support for that behavior. The authors (Herman *et al.*, 1983) found, for instance, that displaying an attractive dessert increased the likelihood that people would select it; this effect was greater among the overweight subjects than among the normal weight, but the normal weight were by no means unaffected.

The responsiveness of normal eaters to external cues can be quite dramatic. When people are hungry (after 24 h of food deprivation), they tend to show greater responsiveness to the taste of their food—eating more good-tasting food but less bad-tasting food—than when they are not hungry (Kauffman *et al.*, 1995). The explanation for this phenomenon is still uncertain, but what is not uncertain is that normal individuals do respond, sometimes quite strongly, to external food cues. (The responsiveness of normal eaters to social facilitation of eating [see above and below] may also be interpreted as an example of external cue responsiveness.) By the same token, the responsiveness of abnormal eaters to external cues is far from ubiquitous. As we discussed above, binge eaters have been alleged to be oblivious to what they are eating. The conclusion of almost three decades of research on the control of eating by external cues must be that what distinguishes normal from abnormal eating is not that normality involves inattention to external cues whereas abnormality involves overattention to external cues. Rather, it appears that the major difference is that the responsiveness of abnormal eaters to external cues is unconstrained by the operation of internal cues, whereas for normal eaters, internal cues put some limits on the influence of external cues. The normal eater approaching satiety is likely to slow down or stop eating, even if the available food is highly attractive; the abnormal eater may persist, despite the discomfort of satiety. Because responsiveness to external cues, like responsiveness to internal cues, appears to be a matter of degree, the distinction between normal and abnormal behavior is subject to the same sorts of complications that we encountered in our discussion of other quantitative distinctions (e.g. quantity, pace and frequency of eating); accordingly, we should not expect abnormal eating to provide us with very much useful information about normal eating.

5.7 Social influence

We mentioned above de Castro's research (e.g. de Castro *et al.*, 1990) on the social facilitation of eating. Before, we were concerned with the issue of exactly how much normal and abnormal eaters ate. We now return to this research in order to consider it from a different angle—the effect of the presence of others on eating. We saw before, in the case of response to preloads, that the clearest lessons about normal eating to be learned from abnormal eating were those that occurred when the normal and abnormal eating patterns were not merely different in degree but different in direction. In the case of preloading, the lesson about normal eating was one that we probably already knew; in the case of social influence, we may learn something about normal eating that is not so well-known. The fact that people eat more when the people around them are eating more (de Luca and Spigelman, 1979; Nisbett and Storms, 1974; Polivy *et al.*, 1979; Rosenthal and McSweeney, 1979) may come as no surprise, but the fact that normal individuals are just as likely to display this modelling effect as anyone else came as a surprise to those who, following Schachter's analysis of obese/normal differences, expected the obese (or otherwise abnormal) eater to be more susceptible to social influence. Recall that the fundamental assumption about normal eating is that it is governed by internal cues representing hunger and satiety; if we take this assumption seriously, there is no room for social influences in the control of eating. But quite clearly, social influences do operate in the control of normal eating, and it would be ludicrous to try to escape this conclusion by, say, arguing that to the extent that eating is governed by social influences it is not normal.

The presence of others who eat a lot stimulates normal eating, as we have seen in the modelling studies. De Castro's social facilitation studies focus on the mere presence of others, irrespective of whether or not they are eating. We may safely assume that in de Castro's eating situations (which are recorded in diary form by the subjects in his studies), the other people present are often eating a lot; but no doubt they are often not eating a lot, and presumably social facilitation occurs regardless.

What about abnormal eaters? As we saw earlier, anorexics are often forced to eat by others, whose cajoling or threatening presence constitutes a powerful social influence to eat more. Nevertheless, it is not appropriate to consider this 'overeating' by anorexics a case of social induction of eating in the same sense as the normal modelling and social facilitation studies that we have discussed. For one thing, anorexics eat more in the presence of others only in a strictly arithmetic sense—more than they might otherwise eat, but not more than any rational criterion of minimal eating. Moreover, this social induction of eating is much more direct;, normal eaters do not have to be told or begged to eat more in the presence of others. The interesting thing about the social induction of eating in normals is that it occurs despite the absence of any direct requests. In short, it is fair to say that the presence of others does not have the same effect on anorexics as it does on normals; although both types of eaters may appear to eat more, the dynamics—not to mention the amounts consumed—are quite different.

Bulimics differ quite dramatically in their response to the presence of others. Pyle *et al.* (1981) and Schlundt and Johnson (1990) claim that some bingers will tolerate the presence of others; that is, they will continue their binge even though others (preferably other bingers, one would imagine) are present. The vast majority of authorities, however, are quite insistent that the presence of others is a sure terminator of a binge. Indeed, it may be that introducing others into the binge environment is the only reliable way to stop the binge. Johnson and Connors (1987, p. 42) cite several studies converging on the conclusion that binges occur when the binger is alone.* Bingeing (along with purging) is a private behavior, something that the binger is concerned to hide from others. Although bingers often report that they have little or no control over the binge, either when it begins or once it gets going, the fact remains that if they are interrupted during the binge by the intrusion of another person, they typically manage to bring their out-of-control binge to a sudden halt. This ability to stop eating, even eating that is allegedly unstoppable, raises fascinating questions about our ability to exert self-control (Herman, 1996a).

Do people other than bulimics eat less when confronted with other people? Although obese people occasionally eat more in the presence of others—Schlundt *et al.* (1985) found that whereas bulimics are more likely to binge when alone, the obese are more likely to binge when in company—it is certainly not the case that the obese invariably eat more with others. Maykovich (1978) found that the obese ate a lot when alone or with other obese diners, but much less when with normal weight diners. Similarly,

*Parenthetically, Johnson and Larson (1982) found that not only are bingers more likely to binge when alone, but that they are more likely to be alone in the first place.

de Luca and Spigelman (1979) found that obese subjects ate more when paired with an obese confederate (who ate a lot) but considerably less when paired with a normal weight confederate (who ate a lot). For the obese, then, it depends on the sort of person in whose presence they are eating; normal weight subjects inhibit the obese eater's intake.

An inhibitory response to the presence of others is also found in dieters. (Whether dieters are to be considered abnormal or not is open to debate, but the very fact that their behavior departs from that of nondieting and otherwise nonpathological eaters suggests that at least in some sense they are abnormal; see Polivy and Herman (1987) for a discussion.) The presence of others eliminates the 'bingeing' to which dieters are prone after consuming a large preload (Herman *et al.*, 1979); even the implied presence of others, as occurs when the experimental subjects know that others will eventually be made aware of how much they have eaten, will prevent counter-regulation (Polivy *et al.*, 1986). In Figure 5.2, we see that following a preload, dieters (restrained eaters) eat much less when they are made more aware of how much they are eating (self attention) or led to believe that others are or will be aware of how much they are eating (public attention) than when they eat without any selfconsciousness, social or personal (control). Eliminating overeating, however, is not quite the same as inhibiting eating altogether. In the Herman *et al.* (1979) study, at least, dieters who were not preloaded (and who would otherwise have eaten minimally) actually ate more than expected when an observer was present, suggesting that the effect of the observer was to 'normalize' eating. In Figure 5.3, we see that dieters (restrained eaters) eat in the same fashion as nondieters (unrestrained eaters) when they are observed. In the absence of an observer (not shown), the dieters eat more following the larger preload than following the smaller preload.

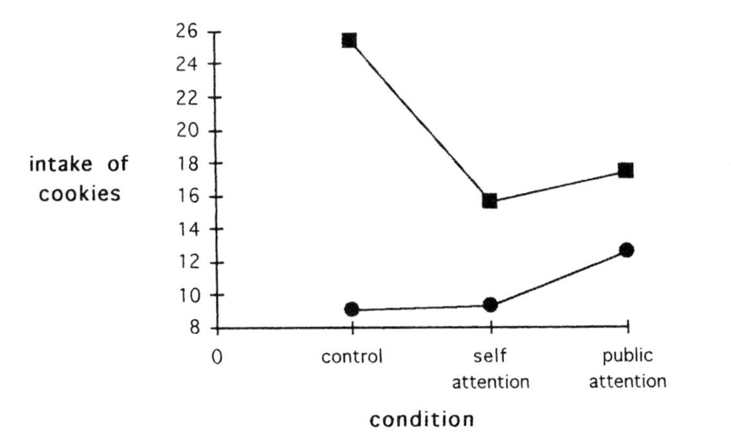

Figure 5.2 Study 2 preload. ●, Unrestrained; ■, restrained. Adapted from Polivy *et al.* (1986).

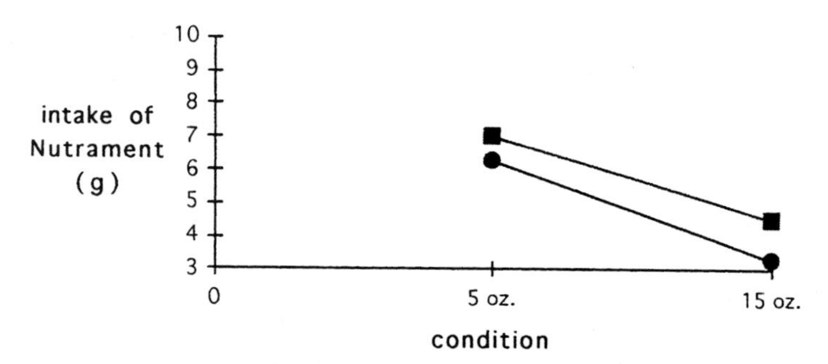

Figure 5.3 Effects of an observer on eating behavior. ●, Unrestrained; ■, restrained. Adapted from Herman *et al.* (1979).

With some qualifications, then, we may conclude that the presence of others is more likely to inhibit eating than to promote it in abnormal eaters. Like the response to a preload, the reaction of normal and abnormal eaters to the presence of other people appears to be diametrically opposed. What does this tell us about normal eating? To answer this question adequately requires that we delve a little deeper and ask why bulimics and dieters and the obese often eat less when others are watching. The answer would seem to be that these abnormal eaters are all somewhat ashamed of their eating, or at least their overeating. If someone watches them when they might engage in this shameful overeating, then they simply do not do it, tempted as they might be.

Normal eaters, by contrast, are not ashamed of their overeating. For one thing, they usually do not overeat, and in some sense cannot overeat, since overeating presupposes that there is some limit or quota of consumption that must not be surpassed; normal eaters usually do not think of their eating in such terms. The presence of others, particularly others who are eating, leads normal eaters to eat more, even to 'binge', but there is nothing particularly shameful about such binges, particularly since shame requires the disapproval of others, and the others are participating in the binge as well.

As in the case of preloading, where normal and abnormal eaters showed opposite reactions, the effect of social influence—its differential effect on normal and abnormal eaters—depends crucially on the fact that the abnormal eater imposes on herself a limit of intake which she must not exceed. No such arbitrary limit exists for the normal eater. Preloading disinhibits eating in the abnormal eater because it exceeds the limit, and leaves the individual with no reason to refrain from eating. The presence of others inhibits eating in abnormal eaters, but only when they have exceeded their limit and begun a binge. Lacking such self-imposed limits, the normal eater has no reason to binge when preloaded, and no reason not to 'binge' in the presence of others. The lesson of abnormal eating is to remind us that normal eating proceeds without arbitrary, diet-imposed limits on consumption; without such limits, and without a pervading concern about breaking them, eating may proceed 'normally'.

5.8 Precipitants

One final aspect of eating that may be worth looking at for our purpose is the cause or trigger of eating, which may discriminate normal from abnormal eaters. Abraham and Beumont (1982), in discussing binge eating, cite several precipitants, as reported by patients: tension, eating something, and being alone were the top three. We have already discussed eating something (preloading) and being alone (social influence); these aspects were notable inasmuch as both of them involved circumstances in which normal and abnormal eaters reacted oppositely. The third factor mentioned by Abraham and Beumont (1982), tension, likewise shows a directionally different pattern in normal and abnormal eaters. Tension (or arousal, or stress, anxiety, or general dysphoria) all tend to suppress eating in normal eaters, presumably because the physiological concomitants of distress (especially adrenalin/glycogen release) have an anorexic effect. In one of the first experimental studies of eating, Schachter *et al.* (1968) demonstrated that the induction of fear significantly reduced eating in normal subjects (although not in the obese); subsequent research (especially Slochower, 1983) confirmed the findings of Schachter *et al.*

The suppressive effect of distress does not extend to abnormal populations. The obese, dieters, and bingers—three groups that overlap in this regard—all tend to eat more when they are upset. Indeed, binge eaters report distress to be the single most reliable precipitant of a binge (Abraham and Beumont, 1982; Mitchell *et al.*, 1985). Drinking alcohol is another binge precipitant that suppresses eating in normal eaters, owing to the high caloric density of alcohol (Polivy and Herman, 1976). Like preloading and social influence, both distress and alcohol consumption, in our view, operate differently in abnormal eaters because of their impact on the maintenance of dietary restraint—or more accurately, owing to their disruption of dietary restraint. In Figure 5.4, we see that dieters (restrained eaters) eat more when they are anxious (high anxiety) than when they are calm (low anxiety). Nondieters (unrestrained eaters) show the opposite pattern, a pattern that makes more physiological sense. These data pertain to people who have not recently eaten ('deprived'); in preload subjects (not shown), the clearly opposed effects of anxiety on dieters and nondieters are not evident. In Figure 5.5, we see that whereas alcohol suppresses eating in nondieters (unrestrained eaters), it induces more eating in dieters (restrained eaters). This pattern, paralleling the effect of anxiety shown in Figure 5.4, appears only when people are aware that they have consumed alcohol ('told alcohol'). The obese, dieters, and bingers all share a concern

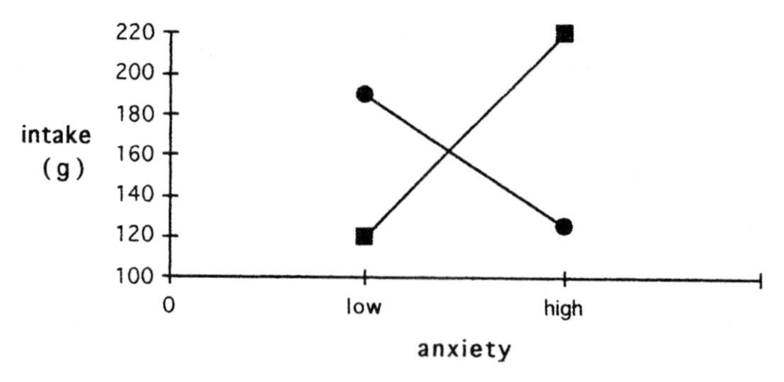

Figure 5.4 Anxiety, hunger and eating behavior. ●, Unrestrained; ■, restrained. Adapted from Herman *et al.* (1987).

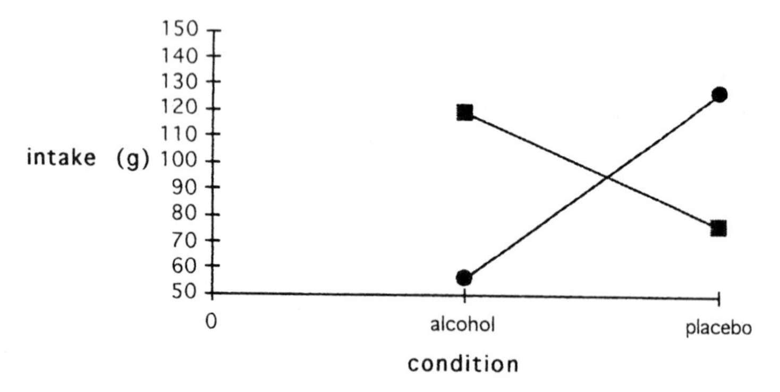

Figure 5.5 Effects of alcohol on eating behavior. ●, Unrestrained; ■, restrained. Adapted from Polivy and Herman (1976).

with restraining their intake; distress and intoxication both serve to (temporarily) undermine that concern, and unleash customarily suppressed eating in a binge-like manner. Normal eaters do not have customarily suppressed eating to unleash; their response to these factors is more straightforward, based on their primary physiological, satiating effects.

5.9 Conclusions

What has our cursory survey of parameters of abnormal eating told us about normal eating? Disorders of memory are assumed to be of interest because of, among other things, their ability to tell us about normal memory. Likewise for disorders of personality and normal personality. Admittedly, the notion that how something goes wrong can tell us about how it normally works is appealing. Lesions are inflicted on the brain in order to identify normal brain functioning, by negation. But does the analogy work with pathologies of eating and normal eating? Ironically, the fact that we know so little about normal eating, while thinking that we know so much (Herman, 1996b), weakens the analogy. Our ignorance about normal eating, combined with our complacent assumption that we understand it, puts us in the uncomfortable situation of pretending to know what is abnormal about abnormal eating (and what is normal about normal eating) when we are often guessing. The charming notion that normal eating is governed exclusively by internal hunger and satiety cues is no longer tenable, even though the object of much eating disorder treatment is to 'return' eating to this idyllic state.

As we have seen, when the difference between abnormal and normal eating is a matter of degree, as it appears to be for most of the quantitative parameters, we cannot learn much about normality from abnormality. In the more interesting cases where the abnormal and normal eater react oppositely to the same factor, we may be better able to appreciate fundamental differences between normal and abnormal eating, the most prominent difference being that abnormal eating is 'pathologized' by the attempt to regulate one's eating by reference to permissible limits of consumption, limits that often are exceeded or overcome, with damaging results for the exercise of self-control. Our survey of abnormal eating indicates that the major differences between abnormal and normal eating—the 'directional' differences where abnormal eaters do one thing and normal eaters do precisely the opposite—are in each case attributable to the fact that the abnormal eater attempts to impose an artificial constraint on her eating. The effects of such self-imposed constraints are complex (see Herman and Polivy, 1984, for a discussion), but in each case they lead to eating that does not 'make sense', or makes sense only when one takes the artificial constraint into account. What is truly abnormal about abnormal eating, then, is the attempt to impose artificial limits (usually derived from an arbitrary diet agenda) on what one eats. Normal eating, on reflection, is eating that is unconstrained, or constrained only by the influence of natural factors (not only physiological factors, but social and other 'external' factors as well).

Making connections between abnormal and normal eating, as we have seen, is not a simple or straightforward task. Our ignorance about both makes drawing conclusions a precarious undertaking. Still, we offer the preceding conclusion about the pathological role of dietary constraints in the hope that it will clarify thinking in this area. We must balance this hopeful thought, however, with the more ominous recognition that abnormal eating in the sense we have described it may, in our culture at least, have achieved statistical normality: Dietary constraints of one sort or another appear to have become the norm in our society (Polivy and Herman, 1987).

References

Abraham, S. and Beumont, P.J.V. (1982) How patients describe bulimia or binge eating. *Psychological Medicine*, **12**, 625–635.

American Psychological Association (1994) *Diagnostic and Statistical Manual of Mental Disorders* (4th edn) Washington, DC: APA Press.

Beumont, P.J.V. (1995) The clinical presentation of anorexia and bulimia nervosa. In K.D. Brownell and C.G. Fairburn (eds), *Eating Disorders and Obesity: A Comprehensive Handbook*. New York: Guilford, pp. 151–158.

Binswanger, L. (1958) The case of Ellen West. In R. May, E. Angel and H. Ellenberger (eds), *Existence*. New York: Basic Books, pp. 237–364.

Booth, D.A. (1988) Culturally corralled into food abuse: The eating disorders as physiologically reinforced excessive appetites. In K.M. Pirke, W. Vandereycken and D. Ploog (eds), *The Psychobiology of Bulimia Nervosa*. Berlin: Springer-Verlag, pp. 18–32.

Bruch, H. (1973) *Eating Disorders*. New York: Basic Books.

Chernyk, B. (1981) Sex differences in binge eating and related habit patterns in a college student population. Paper presented at the meeting of the Association for the Advancement of Behavior Therapy, Toronto.

Davis, R., Freeman, R. and Solyom, L. (1985) Mood and food: An analysis of bulimic episodes. *Journal of Psychiatric Research*, **19**, 331–335.

de Castro, J.M. and Brewer, E.M. (1992) The amount eaten in meals by humans is a power function of the number of people present. *Physiology & Behavior*, **51**, 121–126.

de Castro, J.M., Brewer, E.M., Elmore, D.K. and Orozco, S. (1990) Social facilitation of the spontaneous meal size of humans occurs regardless of time, place, alcohol, or snacks. *Appetite*, **15**, 89–101.

de Luca, R.V. and Spigelman, M.N. (1979) Effects of models on food intake of obese and non-obese female college students. *Canadian Journal of Behavioural Science*, **11**, 124–129.

Garfinkel, P.E. (1995) Classification and diagnosis of eating disorders. In K.D. Brownell and C.G. Fairburn (eds), *Eating Disorders and Obesity: A Comprehensive Handbook*. New York: Guilford, pp. 125–134.

Garfinkel, P.E. and Garner, D.M. (1982) *Anorexia Nervosa: A Multidimensional Perspective*. New York: Brunner/Mazel.

Garner, D.M. and Fairburn, C.G. (1988) Relationship between anorexia nervosa and bulimia nervosa: Diagnostic implications. In D.M. Garner and P.E. Garfinkel (eds), *Diagnostic Issues in Anorexia Nervosa and Bulimia Nervosa*. New York: Brunner/Mazel, pp. 56–79.

Hamburger, W.W. (1951) Emotional aspects of obesity. *Medical Clinics of North America*, **35**, 483–499.

Herman, C.P. (1996a) Thoughts of a veteran of self-regulatory failure. *Psychological Inquiry*, **7**, 46–50.

Herman, C.P. (1996b) Human eating: Diagnosis and prognosis. *Neuroscience and Biobehavioral Reviews*, **20**, 107–111.

Herman, C.P. and Polivy, J. (1984) A boundary model for the regulation of eating. In A.J. Stunkard and E. Stellar (eds) *Eating and its Disorders*. New York: Raven, pp. 141–156.

Herman, C.P. and Polivy, J. (1988a) Restraint and excess in dieters and bulimics. In K.M. Pirke, W. Vandereycken and D. Ploog (eds), *The Psychobiology of Bulimia Nervosa*. Heidelberg: Springer, pp. 33–41.

Herman, C.P. and Polivy, J. (1988b) Studies of eating in normal dieters. In B.T. Walsh (ed.), *Eating Behavior in Eating Disorders*. Washington, DC: American Psychiatric Association Press, pp. 95–111.

Herman, C.P., Polivy, J. and Bradizza, C.M. (unpublished) Sensory specific satiety as a moderator of caloric compensation.

Herman, C.P., Polivy, J. and Silver, R. (1979) Effects of an observer on eating behavior: The induction of 'sensible' eating. *Journal of Personality*, **47**, 85–99.

Herman, C.P., Olmsted, M.P. and Polivy, J. (1983) Obesity, externality, and susceptibility to social influence: An integrated analysis. *Journal of Personality and Social Psychology*, **45**, 926–934.

Herman, C.P., Polivy, J., Lank, C. and Heatherton, T.F. (1987) Anxiety, hunger, and eating behavior. *Journal of Abnormal Psychology*, **96**, 264–269.
Johnson, C. and Connors, M.E. (1987) *The Etiology and Treatment of Bulimia Nervosa*. New York: Basic Books.
Johnson, C.L. and Larson, R. (1982) Bulimia: An analysis of moods and behavior. *Psychosomatic Behavior*, **44**, 333–345.
Katzman, M. and Wolchik, S. (1984) Bulimia and binge eating in college women: A comparison of personality and behavioral characteristics. *Journal of Consulting and Clinical Psychology*, **52**, 423–428.
Kauffman, N.A., Herman, C.P. and Polivy, J. (1995) Hunger-induced finickiness in humans. *Appetite*, **24**, 203–218.
Kaye, W.H., Gwirtsman, H.E., George, D.T., Weiss, S.R. and Jimerson, D.C. (1986) Relationship of mood alterations to bingeing behavior in bulimia. *British Journal of Psychiatry*, **149**, 479–485.
Kissileff, H.R., Thornton, J. and Becker, E. (1982) A quadratic equation adequately describes the cumulative intake curve in man. *Appetite*, **3**, 255–272.
Maykovich, M.K. (1978) Social constraints in eating patterns among the obese and overweight. *Social Problems*, **25**, 453–460.
Mitchell, J.E., Hatsukami, D., Eckert, E. and Pyle, R.L. (1985) Characteristics of 275 patients with bulimia. *American Journal of Psychiatry*, **142**, 482–485.
Nevo, S. (1985) Bulimic symptoms: Prevalence and ethnic differences among college women. *International Journal of Eating Disorders*, **4**, 151–168.
Nisbett, R.E. and Storms, M.D. (1974). Cognitive and social determinants of food intake. In H. London and R.E. Nisbett (eds), *Thought and Feeling: Cognitive Alteration of Feeling States*. Chicago: Aldine. pp. 190–208.
Polivy, J. (1976) Perception of calories and regulation of intake in restrained and unrestrained subjects. *Addictive Behaviors*, **1**, 237–243.
Polivy, J. and Herman, C.P. (1976) Effects of alcohol on eating behavior: Influences of mood and perceived intoxication. *Journal of Abnormal Psychology*, **85**, 601–606.
Polivy, J. and Herman, C.P. (1987) The diagnosis and treatment of normal eating. *Journal of Consulting and Clinical Psychology*, **55**, 635–644.
Polivy, J., Herman, C.P., Younger, J.C. and Erskine, B. (1979) Effects of a model on eating behavior: The induction of a restrained eating style. *Journal of Personality*, **47**, 100–112.
Polivy, J., Herman, C.P., Hackett, R. and Kuleshnyk, I. (1986) The effects of self-attention and public attention on eating in restrained and unrestrained subjects. *Journal of Personality and Social Psychology*, **50**, 1253–1260.
Pyle, R.L., Mitchell, J.E. and Eckert, E.D. (1981) Bulimia: A report of 34 cases. *Journal of Clinical Psychology*, **42**, 60–64.
Rosen, J.C. and Leitenberg, H. (1988) Eating behavior in bulimia nervosa. In B.T. Walsh (ed.), *Eating Behavior in Eating Disorders*. Washington, DC: American Psychiatric Association Press, pp. 163–173.
Rosen, J.C., Leitenberg, H., Fisher, C. and Khazan, C. (1986) Binge eating episodes in bulimia nervosa: The amount and type of food consumed. *International Journal of Eating Disorders*, **5**, 255–267.
Rosenthal, B. and McSweeney, F.K. (1979) Modeling influences on eating behavior. *Addictive Behaviors*, **4**, 205–214.
Schachter, S. (1968) Obesity and eating. *Science*, **161**, 751–756.
Schachter, S. and Rodin, J. (1974) *Obese Humans and Rats*. Potomac, MD: Lawrence Erlbaum Associates.
Schachter, S., Goldman, R. and Gordon, A. (1968) Effects of fear, food deprivation, and obesity on eating. *Journal of Personality and Social Psychology*, **10**, 91–97.
Schlundt, D.G. and Johnson, W.G. (1990) *Eating Disorders: Assessment and Treatment*. Boston: Allyn & Bacon.
Schlundt, D.G., Johnson, W.G. and Jarrel, M.P. (1985) A naturalistic functional analysis of eating behavior in bulimia and obesity. *Advances in behavior Research and Therapy*, **7**, 149–162.
Slochower, J. (1983) *Excessive Eating: The Role of Emotions and Environment*. New York: Human Sciences Press.

Spencer, J.A. and Fremouw, W.J. (1979) Binge eating as a function of restraint and weight classification. *Journal of Abnormal Psychology*, **38**, 262–267.

Spiegel, T.A. (1973) Caloric regulation of food intake in man. *Journal of Comparative and Physiological Psychology*, **84**, 24–37.

Weingarten, H.P. (1985) Stimulus control of eating: Implications for a two factor theory of hunger. *Appetite*, **6**, 387–401.

Woods, S.C. (1991) The eating paradox: How we tolerate food. *Psychological Review*, **98**, 488–505.

6 The contextual basis for food acceptance, food choice and food intake: the food, the situation and the individual

H.L. MEISELMAN

6.1 Introduction

Research on determinants of food acceptance and food choice has emphasized sensory and other properties of the food, as well as psychological and physiological aspects of the person. In recent years there has been increased attention to the eating environment or context (Rozin and Tuorila, 1993; Hirsch and Kramer, 1993; Meiselman and Kramer, 1993; Bell and Meiselman, 1995). Until recently, there has been little attempt to organize contextual variables affecting food acceptance and choice; Rozin and Tuorila (1993) have suggested three potential organizing principles for contextual variables. They suggested a distinction between variables that are simultaneous to eating and variables that are separated in time (both past and future), which they called temporal. Second, they also noted that a distinction by reference unit could be used, distinguishing a single food exposure (e.g., bite) from a meal and a pattern of eating. And third, they noted that contextual variables could be distinguished as food or nonfood. In their review of context, they used a hybrid of these three organizing principles, although they emphasized the simultaneous–temporal distinction. Their paper represents an early and important attempt to organize contextual research. Bell and Meiselman (1995) have also very recently organized contextual variables. They divided variables into those which are antecedent to food choice and those which are present at the food choice situation. The former ones bring to the eating environment, for example, expectations, prior experience, habits, etc. The latter exert their influence directly during the eating experience, for example, social facilitation and physical variables of the eating situation. Thus, Bell and Meiselman emphasize the temporal dimension, also emphasized by Rozin and Tuorila.

In the present review and organization of context, I will try to include the dimension of application of food context research. I suggest that the three main application areas for contextual research are food product development, food service or catering, and diet and health. Food product development deals with everything from simple fruit juice drinks to complex packaged meals with dozens of ingredients. Product development also includes packaging and labeling. A number of recent texts have dealt with product development methodology but have not emphasized contextual issues. Perhaps the closest is Moskowitz (1993), who moves from traditional sensory evaluation to include a broader consumer research perspective. Moskowitz also directly deals with issues of food and package and label, that is, the food and its immediate context.

Food service or catering includes the entire range of restaurant, factory and business feeding, as well as institutional feeding (hospital, military, etc.). Catering ranges from one time to long-term, three meals per day, and ranges from very simple to very luxurious. Food service or catering has often been omitted from food research, relative to product development and diet/health. Yet, catering represents the environments in which many foods are consumed. Meiselman (1992) called for greater research on real foods in real environments as one strategy for better prediction of food-related behavior. Meiselman, Bell, and colleagues have carried out several studies in catering environments, but there is generally little use of the catering environment for research.

Diet and health have been main concerns related to food for the past 20 years or more. Diet and health concerns relate to nutrition and its influence on the incidence of various diseases and to long-term patterns of eating rather than to individual foods or meals. The perspective in health and diet issues has moved from very brief snapshots of behavior to longer-term patterns of behavior. The 24 h recall has given way to longer observations of diet. By examining the influence of these longer-term effects on the relationship of eating context to diet, we can understand how one's pattern of eating is maintained. These include the physical and social eating environment.

The issues concerned with food product development, food service/catering, and diet/health can be related to the organizing continuum of bite/meal/pattern of eating suggested by Rozin and Tuorila as follows:

Product development	Bite
Food service/catering	Meal
Diet/health	Pattern of eating

Most product development usually deals with testing of bites or sips of products, rarely with entire multicomponent meals or longer-term patterns of eating. Food service/catering is concerned with meals, but not usually with patterns of meals, which is the interest of the diet/health area. In this chapter the author would like to suggest that an alternative organization for contextual research could be the following: the food, the situation, and the individual. Although no one organizing principle results in a completely clean separation of context variables and principles, it is important to explore a number of organizing principles while contextual research is relatively young. Such an organization could be useful for understanding relationships among variables and concepts and for stimulating further research in this important and growing area. Contextual research is important both for understanding the fundamentals of food-related behavior and for permitting more successful application of sensory science and food habits sciences. In the following sections of the chapter, selected contextual research dealing with the meal and the situation is reviewed and then the individual is considered in order to demonstrate this organization of context and to seek avenues for further contextual research and discussion.

Those favoring a market research perspective have also suggested a related organization of contextual material related to food acceptance. For example, Oude Ophuis (1993) suggests that the traditional marketing approach of place, product and person be applied as follows:

place—setting
product—labeling
person—attitudes

Throughout this chapter, use is made of the term *reference event* to refer to the food behavior under consideration. For example, the reference event might be a person eating an apple. Depending on one's perspective, the reader might be interested in the person's sensory response(s) to the apple, one's physiological state before, during, or after eating the apple, or one's perception of satiety before, during, or after eating the apple, etc. As Rozin and Tuorila stated:

> 'The context will be taken to mean that set of events and experiences that are not part of the reference event but have some relationship to it.'

Thus, the reference event includes the sensory, physiological and behavioral responses to the reference food; while the context will include those factors which surround the reference event and influence it.

6.2 The food

Whether food itself is part of the context obviously depends on what the reference event is. If one is considering a single food item, then other food items present are part of the context. If one is considering a meal, then other meals are part of the context. Thus, food itself can be considered part of the context for food.

What variables or events should be considered for the food part of context? Rozin and Tuorila mention the sip or bite, the meal, and the pattern of eating, which I will call diet. At the bite or sip level, they consider simultaneous chemosensory context, interactions of chemical senses with oro–nasal sensations, taste–smell confusions, and temporal contextual effects. For the basic sensory scientist, these bite/sip contexts have significant value for consideration. However, from the perspective of eating or of application of contextual research, I argue that the sensory experience of a food is seen as an integrated whole and that trying to analyze this flavor contextually is artificial. The purpose of promoting contextual or situational research, in general, is to promote fundamental understanding of food-related behavior and to make food research and application able to predict better and to mirror real life better. Context has been added to the food and individual variables, because we have come to realize that we cannot predict food-related outcomes from these alone. Therefore, I propose that context research begin at the level of the whole food and the whole sensory response. Sensory research at a more analytical level is not properly part of food context. It is possible, in fact, that context does not affect palatability as much as it affects food choice and intake. This will be discussed in later sections.

Bell and Meiselman (1995) do not deal with food *per se* for the reasons just discussed. Since their focus is the eating environment, they tend to deal with food as seen in the eating environment, i.e., in combination with other foods. Hence, they do discuss perception of variety and actual variety. Although many of the variables they discuss are directly influenced by food (exposure, neophobia, ethnicity, appropriateness, etc.), they do not deal with the food itself.

6.2.1 Effects of one food on another in a meal

Meals usually consist of more than one food, and even snacks might consist of beverage plus food (fruit, biscuit, sweet, etc.). Beverages and confections are probably the foods most frequently consumed singly.

When one food is served with other foods present, how do the foods interact? Or in terms of this paper, when a reference food is served, how do the accompanying or context foods affect its selection, acceptance and intake? There has not been a large amount of published research on this question. In addition, most of the research has been on fixed menus, often in nonrealistic settings. Meiselman (1992) has called for more research in actual eating situations in order to obtain this type of data.

A series of papers has dealt with food item compatibilities and menu planning. Understanding the rules for how foods combine is essential in menu planning and especially in attempts to automate menu planning. One of the earliest research papers to consider menu combinations was Eindhoven and Peryam (1959) who were extending their pioneering work with single item food preferences (Peryam and Pilgrim, 1957) to consider two-item combinations. They:

> 'hypothesized that when two foods are combined . . . , the individual components partially lose their identity in favor of a new identity, the combination, with its own unique associated preference.' (p. 379)

Eindhoven and Peryam prepared rating sheets where main dishes were combined with potato or vegetable and where potato–vegetable combinations were also rated. The authors demonstrated this 'unique associated preference' by showing that many food pairs had combination effects, that is, the preference for individual components were nonadditive. For example, the baked ham–sweet potato rating was higher than baked ham–french fries, although individually, french fries are preferred to sweet potatoes. Eindhoven and Peryam used ten main dishes, ten vegetables and five potatoes.

Moskowitz and Klarman (1977) used multidimensional scaling for two-item and three-item combinations selected from five main dishes, five potatoes and five vegetables. They developed a compatibility index for pairs of items and triples of items. To obtain an overall compatibility rating, they multiplied each item by its particular compatibility coefficient and then added the values for all the items.

A series of papers has attempted to model multicomponent meals. Rogozenski and Moskowitz (1982) used a questionnaire approach to ask people to rate various meal components and overall meal acceptance. Turner and Collison (1988) conducted their meal evaluations in a student training restaurant. Both studies found that the main dish contributed about one-half the weight of the meal. Turner and Collison noted that customers at the same tables tended to agree on ratings; this social effect will be covered later.

More recently, Hedderley and Meiselman (1993) have attempted meal modeling in a free choice eating situation in which students could select any meal (any combination of items) in their student cafeteria. This introduced many problems not encountered in the studies above, which used set meals either in questionnaires (Rogozenski and Moskowitz) or in actuality (Turner and Collison). The authors found that the student meals could be grouped into sandwich meals, pizza meals and main dish meals. When meals were so divided, the central item in each meal was even more dominant than in previous studies.

Thus, the acceptability of food items is partially determined by their food context, i.e., the foods with which they are served. Research has consistently shown the strong dominance of the main dish in an overall meal. This phenomenon could also be used to determine under what conditions this main dish dominance disappears or is reversed. Do sweet lovers place this same importance on dessert? Do others emphasize beverages? There is much to learn about inter-item food compatibility, which will help us understand and predict the consumer.

Food item compatibilities are an essential component of modeling overall acceptance and selection for multicomponent meals. Hence, the developer of prepared packaged meals needs to understand compatibility in order to make optimal individual item selections and substitutions. For example, is breaded fish best served with french fried potatoes (chips) or mashed potatoes? If french fried potato is not available (because of price, etc.), what is the best substitute? The metrics which determine the solutions to these questions will change with a number of demographic variables (age, gender, race, etc.) and might also change with cultural changes (North American, British, French, etc.). The questions themselves might have no validity under certain demographic conditions (baby food, certain geographic areas). In order to make compatibility modeling more useful in applied contexts, more model development is needed, along with the growth of appropriate databases.

6.2.2 *Food variety*

Variety can be short-term, within one meal, or longer-term, over many meals. Ideally, one would distinguish meal variety from dietary variety. Food variety (and the individual's response to it) is distinguished from general variety-seeking, which is covered in the section under the individual.

A number of laboratory approaches have studied variety. Spiegel and Stellar (1990) found that three flavors of small, bite-size sandwiches increased sandwich intake over a single flavor. Probably the briefest phenomenon of variety is exhibited in the laboratory phenomenon of sensory specific satiety (Rolls, 1986). When a variety of foods are shown to a subject in the laboratory, Rolls (1990) has suggested that the phenomenon of sensory specific satiety might be a key to understanding how variety is maintained within a meal. In sensory specific satiety the pleasantness of foods decreases with direct exposure (tasting, eating, etc.). Rolls argues that, in addition to cultural effects, the changes in palatability of foods as eating proceeds encourages variety in the meal to encourage continued palatability. This model appears to imply a sequential meal or, at least, a sequential eating of what is offered. Rolls suggests that sensory specific satiety could generalize to other similar foods.

There is very little data on food variety in adults in nonlaboratory situations. Rozin and Markwith (1991), who also note this lack of data, go on to demonstrate how consistent variety-seeking is in different food groups. Subjects indicated their annual variety within the food groups of soups, sodas, fruits and vegetables. All but one correlation between variety of different foods was significant, ranging from 0.60 for fruits × vegetables, down to 0.25 for soups × vegetables. Sodas × vegetables was not significantly correlated. Thus, Rozin and Markwith conclude that people may be classifiable on the basis of their variety behavior.

The role of variety in natural eating has received very little attention. While some animals have evolved very narrow food repertoires (low variety), most humans have evolved very wide food repertoires (high variety).

It would be interesting to try to link diet variety to the observed consistency of individual day-to-day intake variation. On a more methodological level, since some subjects consistently demonstrate higher dietary variety and some subjects consistently demonstrate higher day-to-day intake variation, it is interesting to question whether these subjects are the typical subjects of laboratory studies (Tarasuk and Beaton, 1992).

Dietary variety is enormously important in all applied aspects of food context. For food product developers, variety determines the limits of new product development, and variety-seeking (discussed in §6.4.3) is an important basis for the continued need for new products. In food service or catering, variety is the source of continued changes to menus. In both fast food restaurants and full service restaurants, customers demand variety. Finally, in the area of diet and health, variety in the diet is a concern for those seeking to provide adequate nutrition but not excessive intake. The lack of good variety data noted above is a problem in applying variety issues to all three application areas. At present, there does not seem to be much hope of the situation changing.

6.2.3 Food culinary context

Considering that ethnic foods are now very big business and considering the potential for additional business in ethnic foods, there has been very little research on understanding the basis for culturally identified foods. For example, Italian food is extremely popular in the USA; what characteristics identify a food as Italian? E. Rozin (1983, 1992) has defined cuisine in terms of basic foods, processing techniques and characteristic flavor principles. Bell, Meiselman and colleagues have recently tried a number of experimental approaches to identify determinants of cultural identification.

Bell and Paniesin (1992) developed chicken recipes from several ethnic cuisines and studied how sauce, spice and product name influenced perceived product ethnicity for American subjects. They concluded that sauce alone influenced ethnic identification for many of the cuisines.

Meiselman and Bell (1993) developed four pasta recipes and studied how sauce and product name influenced perceived product ethnicity for British subjects in a laboratory setting. In addition to categorizing dishes according to ethnic identity, Meiselman and Bell asked subjects to assign a rating on ethnicity scales. The presence of Italian name increased perceived Italian ethnicity, and adding meat and/or cheese increased British ethnicity. While naming had an influence on ethnic identification, it had less influence on palatability, which varied with different recipes.

Most recently, Bell *et al.* (1994) attempted to vary ethnic environment in a restaurant setting. Identical recipes were served for two days without specific ethnic theme decor and for two days with an Italian theme, including Italian menu names. The ethnic theme affected food selection but not palatability. Overall, meal ethnicity increased on Italian theme days. This is an example of how the eating situation, not the food, can influence culinary context.

Thus, we are just beginning to investigate cultural or ethnic food context influences on food selection, acceptance and intake. A product's sensory characteristics, its name and the environment can all have an impact on how the product is perceived. Much more work is needed to develop laboratory and field research methods to identify critical cultural food variables. Such methods will permit the identification of critical variables and the collection of appropriate databases for application in product development, catering and health areas. Perhaps no greater opportunity exists now than in Europe, with its large number of different cultures within a relatively small geographic area. The birth of the European Common Market, the growth of trading and the breakdown of solely regional markets all demand a greater understanding of food culinary context.

6.2.4 Package, name, label

From the consumer's perspective, a food is often associated with a name, and a package is associated with a label. Most processed foods are sold packaged. Most foods in restaurants are selected by name, not by sensory evaluation of the food; therefore, I include package, name and label as part of the general food context.

There is a large literature on use of brand names. However, how food products are named or labeled has a big influence even without the brand name identification. Much of the research on name or label has dealt with novel or unusual foods. Seaton and Gardiner (1959) found that knowledge of intended use enhanced ratings of liquified military pilot meals. Wolfson and Oshinsky (1966) found that commercial chocolate milk was rated the same with different labels, but the rating for a liquid diet for space use was enhanced when its label indicated its intended use. Cardello *et al.* (1985) found that identifying soups as regular soups or special medical products affected relative hedonic ratings.

Subjects' expectations are probably a key factor in name/label/package effects and will be dealt with under individual context. For example, Levin and Gaeth (1988) found that subjects expected 75% lean meat to taste better than 25% fat meat, although actual testing yielded no difference.

6.2.5 Summary: food context

Thus, a number of factors involving the food, including its name and package, contribute to eating context. Most apparent are the interactions among meal items, with the main dish being most prominent. Variety is a key factor in food context, although variety research has generally not yet considered long-term complex diets. Culinary context is one means of achieving variety and has become one of the main growth areas for product development and catering. Understanding what controls ethnic identity of foods will be critical in commercially expanding culinary products without destroying their distinctiveness. The package for food, with a label and name, also needs to be researched at a level more basic than the commercial level. All of this demonstrates that the most immediate context for food is other food, and all of the influences and interactions need to be understood. Food context has enormous application potential. The food business is very large, and understanding food context will give a cognitive edge to the food companies which understand it better.

6.3 The eating situation

When we think of eating context or situation, perhaps the most important area—and the most neglected—is the eating situation itself. The eating situation includes all of those variables in the physical/social eating environment. There has been very little published research on the eating situation with respect to product development, catering or health. As has been noted elsewhere (Hirsch and Kramer, 1993; Meiselman and Kramer, 1993), the issue of obesity stimulated research on situational factors in the 1970s, and the issue of consumption of military rations stimulated research on situational factors in the 1980s and 1990s.

One approach to social and physical aspects of different eating situations has been proposed by Schutz (1988) who suggested the use of an 'appropriateness scale'. Instead of asking for product ratings of liking, similarity, or another dimension, Schutz tried to capture situational differences in a rating scale of how appropriate a food is in different listed situations. Schutz used 48 uses, and most of these do not refer to different physical eating situations. Schutz is trying to predict use intention rather than where the use will occur.

While appropriateness scaling might have potential in identifying situational differences, the technique can be very burdensome. Since each food is rated for appropriateness in each situation, the resulting matrix can be very large. Fox example, while a list of 10 foods and 10 uses would require only 100 ratings, a list of 100 foods and 50 uses would require 5000 separate ratings. Further, when factor analysis is used in the data analysis it adds a qualitative element to understanding contextual differences, although it could be helpful in uncovering common features of different contexts.

6.3.1 Where the food is eaten

In general, there is not a body of food acceptance or choice data comparing food acceptance or food choice in different eating situations. I have noted previously the lack of published studies comparing food acceptance in the laboratory and in the field (Meiselman, 1992). Such data might exist in the proprietary files of food product and catering companies.

The US military research on eating situation originated with the observation that military troops in the field for 34 days ate less food and rated it differently than did a comparison group of students at a local university (Meiselman *et al.*, 1988; Hirsch and Kramer, 1993; Meiselman and Kramer, 1993). The difference in caloric intake between field and nonfield conditions was approximately 1000 kcal per day, certainly an impressive difference for 30 days' data. Also, the students rated the food one scale point lower in nonfield conditions than did the soldiers in the field. Soldiers rated the food higher but ate less. Of course, it was not clear how much the difference in populations between the two groups (military–students) influenced the outcome; both groups consisted of males of the same age group who were generally healthy. When Hirsch and Kramer (1993) compared soldiers in the field and out of the field to attempt to replicate the initial observation, they observed the same daily 1000 kcal difference. Thus, the military researchers came to realize that in addition to the food variables which they had studied for years, situational variables were critical in determining food selection.

Comparisons of eating in different eating situations have been studied in the context of obesity research (Stunkard and Kaplan, 1977; Coll *et al.*, 1979). Although the purpose of these and related papers was to compare the eating habits of overweight and normal weight individuals, the normal weight data are interesting by themselves. Stunkard and Kaplan concluded that food choice was an important measure in future studies; Meiselman *et al.* (1988) also concluded that such behavioral measures should be considered in addition to, or instead of, acceptability measures. Stunkard and Kaplan concluded that:

> 'The circumstances under which food is eaten are probably of far greater importance in determining how it is eaten than many of the other parameters which we have discussed.' (p. 98)

Coll *et al.* (1979) observed different meal sites and different snack sites and concluded:

> 'These findings are consistent with the six earlier studies of food choice, which concluded that the major influence on how much people choose to eat is where they eat. . . .' (p. 795)

The authors also noted that palatability did not clearly differentiate among the different sites. Considering these far-reaching conclusions about the importance of eating site, there has been very little recent research.

6.3.2 Physical environment

The preceding section covered different general eating situations, e.g., laboratory and field. This section deals with the physical attributes of different eating situations. The catering industry has a large stake (not steak!) in how the physical environment affects food acceptance and choice, but there is little published research in this area. Textbooks and articles present guidelines for the lighting, decorating and sound aspects of dining areas, but these are generally not based on published research. It is not known whether the large catering companies or fast food companies have proprietary research data in this area.

One line of published work on environment and behavior deals with social psychological effects (Russel and Pratt, 1980; Mehrabian, 1980). This approach has led to food-related research. Milliman (1986) presented either slow or fast music in a restaurant and observed that customers took more time to eat with the slow music. In addition, customers with slow music spent the same amount of money on food, but spent significantly more on drinks.

Another line of published food research deals with the effect of climate on food intake (Marriott, 1993). This work deals with the special needs of the military when eating outdoors, and it generally has a physiological orientation (caloric needs, etc.).

It was noted above that earlier studies on the eating environment were related to theories of obesity. Very few physical context variables have been systematically studied. One of these is effort to obtain food. In a series of studies (Nisbett, 1968; Meyers *et al.*, 1980; Levitz, 1975), obese and normal weight subjects were compared in eating situations which presented foods with different degrees of cue prominence and different degrees of effort. Both Meyers *et al.* and Levitz used hospital cafeterias and varied the accessibility of desserts. More effort was required to obtain certain desserts. In general, the accessibility/effort condition affected how many people selected dessert, but there was no difference between obese and normal subjects' selection. Further, Meyers *et al.* reported that people were more willing to expend the extra effort for a high calorie dessert than for a low calorie dessert. Thus, in this setting, effort affected food choice and acceptance.

Nisbett observed that overweight individuals took more sandwiches than did normal weight individuals when offered one sandwich in a laboratory setting. However, it is relevant in the present discussion to note that when one sandwich was offered, all groups (underweight, normal weight, overweight) took more from the refrigerator. But when three sandwiches were offered, no group averaged more than three; that is, on the average, no group took more from the refrigerator.

Engell *et al.* (1990) manipulated effort to obtain water from a pitcher for subjects at a lunch meal in the laboratory. Subjects drank approximately twice as much water when it was on the table in front of them than they did when the water was 30 feet away in the same room or 40 feet away in another room.

In two studies with colleagues at Bournemouth University in England, Meiselman (Meiselman *et al.*, 1994) varied effort to obtain a particular test food by moving that food out of a university cafeteria line and placing it in a separate line. This required students to stand in two lines and pay at two points to obtain the treatment item with their lunch. Food choice, food acceptance and intake measures were obtained. In the first study, chocolate confection was moved for one week, after a one-week baseline period. In the second study, a more complex design was used. Observations were made every Tuesday and Thursday at lunch for eight weeks. There was a two-week baseline period, a three-week effort treatment period, and finally, a three-week recovery period using the original baseline conditions.

The results of the two effort studies were identical: The effect of effort was specific to the item moved. In Study 1 chocolate confection selection rate droppped, and in Study 2 crisp (potato chip) selection rate dropped. Further, the drop in selection of the treatment food was accompanied by an increase in selection of one or several foods. In Study 1, subjects increased selection of dessert, fruit, and accessory food groups; in Study 2, subjects increased selection of starch items. Finally, in Study 2, crisp selection did not return to its original baseline level after three weeks of recovery conditions. Also of interest, there was no clear pattern of acceptance rating changes with the treatment and recovery conditions.

6.3.3 Social environment

Here, we consider the effects of social interaction on food choice and acceptance. Most people will agree that other people affect one's food choices and food appreciation, but until recently there has been very little published research.

deCastro and colleagues have published a series of papers on social facilitation of eating. Their basic method is to have adults complete seven-day diet diaries while they go about their normal daily routine. In addition to recording every item they eat or drink, subjects are asked to record a number of other pieces of information, including where they eat and the number of people eating with them. deCastro and deCastro (1989) reported that the number of people present was positively correlated with caloric content of the meal. The number of people present had a stronger correlation than any other single variable (hunger, premeal interval, estimated premeal stomach content). Even when those who ate alone were excluded from the analysis, there was a significant correlation between number of people present and meal size, suggesting that the overall effect was not due to eating alone vs. eating socially.

deCastro *et al.* (1990) reported further analysis of the same overall data set to determine whether the social facilitation observed above was due to an artifact which covaried with the social factor. The authors concluded that because strong correlations were found between meal size and number of people for all three daily meals and for snacks, for meals eaten in different places, and for meals eaten with or without alcohol, the correlations resulted from a true social facilitation. deCastro (1990a) further studied alcohol intake and concluded that the number of people present was a major contributor to the amount of alcohol ingested, with psychological and physiological factors being less important. deCastro (1990b) concluded that social facilitation in eating increases the size and duration of meals but not the rate of eating. deCastro posits a number of ways that social facilitation might increase meal size and duration.

Now that the correlation between intake and social influence seems well established, there will be new data undoubtedly limiting this social facilitation effect. Feunekes *et al.* (1995) collected food diaries from 50 free-living young adults. While the overall correlation between meal size and number of people present was statistically significant, the effect held for breakfast but not for lunch or dinner. In agreement with deCastro, the authors demonstrated the role of meal duration using a path analysis technique. The importance of meal duration was also implicated in the environmental variable of music; recall Milliman's finding that with slow music, people spent more time at the table and drank more.

In the preceding correlational studies, deCastro has been careful to point out that such studies do not prove causation between social number and amount eaten, although the weight of all the studies is compelling. A number of studies, however, have manipulated social influence as a variable. Edelman *et al.* (1986) used the cover of a routine taste test to compare food intake and acceptance ratings in social and isolated conditions. The latter were carried out in separate taste test booths, and the former with four or five participants around a table. Subjects ate considerably more of the sample lasagna in the social (*vs.* isolated) condition. For normal weight subjects there was no correlation between acceptance rating and consumption. However, in the social condition, intake and acceptance were significantly correlated. The experimenters observed that eaters in the social condition appeared to linger at the taste test and continue eating, informally supporting deCastro's observation of increased meal duration with social facilitation.

Engell *et al.* (1990; reported in Hirsch and Kramer, 1993) also used a taste test scenario to test social influence on soldiers. Soldiers and their sergeant were recruited to taste test lunches of military rations. In each group, the sergeant acted as a confederate, making positive or negative comments, and also eating almost all the meal or approximately 65% of the meal. When the sergeant was negative and ate less, soldiers ate less and rated their food lower; when the sergeant was positive and ate more, soldiers ate more and their food acceptance tended to be higher.

Thus, social context can facilitate or depress eating. However, the relationship between social context and food liking is less clear. Almost all laboratory taste testing is a social situation involving at least the subject and the experimenter and perhaps other subjects. In taste testing, we have assumed that social interaction is controlled. We need to understand better these possibly powerful effects.

6.3.4 Summary: situational context

Thus, the eating situation itself has emerged in the past decade as a major contributor to food-related behavior. Both the physical situation and the social situation have received attention, although research in both is really in its infancy compared to other aspects of food-related behavior and physiology.

Obviously, situational context is of tremendous importance to the food service or catering industry, which depend on context to stimulate business. Unfortunately, the catering industry is probably the least involved in research of all areas of application. Those interested in diet and health could also benefit from greater utilization of situational variables, rather than relying on food and individual interventions.

6.4 The individual

6.4.1 Is the individual part of the eating context?

If we consider the reference event to be an individual eating an apple, then the individual appears to be part of the reference event. If we take a momentary snapshot of the person eating the apple, then the person's sensory, physiological and psychological reactions to eating are part of the reference event. The taste, smell, and texture of the apple; the body's reaction to the intake of food; and the changes in the psychological variables of appetite, hunger, etc.—all of these and others are parts of the reference event.

Are the attitudes, opinions and habits which the individual brings to the eating situation part of the reference event or part of the context? Before the eater even sees the apple, he has prior views on a number of issues which will bear upon the apple-eating reference event. Does the eater like apples? If the eater has never eaten an apple, what is his reaction to novel foods? Rozin and Tuorila (1993) argued that 'a person's experiences and expectations/cognitions form a context for his perceptions.' The alternative would be to retain the attitudinal aspect of the individual within the individual, along with the sensory, physiological and psychological reactions to food. The individual could then interact with context. The position taken here is that the broad area of attitudes and habits is part of context, i.e., outside the reference event.

6.4.2 Food preferences and aversions

Food preference is a general predisposition for a particular food, independent of an eating situation. General liking for strawberries and general dislike for liver are examples of preferences. In a particular situation one might dislike strawberries (poor quality) or like liver (excellent preparation). The study of food preferences was one of the earliest aspects of food psychology utilizing the nine-point hedonic scale developed by the US Army (Peryam and Pilgrim, 1957), which has continued to track the preferences of military personnel (Meiselman, 1988).

Food preferences have been used as predictors of behavior in a wide variety of situations. In fact, the original use of the nine-point hedonic scale was as a predictor of food acceptance (Peryam and Girardot, 1952; Peryam and Pilgrim, 1957). Food preferences are used for menu planning (Balintfy *et al.*, 1975), for analyses of health issues such as obesity (see Dobbing, 1987), to predict product acceptability (Cardello and Maller, 1982), and to predict food intake (Kamen, 1962). However, food preferences reflect general dispositions toward food names; food preferences are not assessments of particular product quality or situational appropriateness. Hence, it should not be expected that food preferences should predict product acceptance or intake in specific situations.

Furthermore, it is known that preferences do not reflect an average product. Schutz and Kamenetzky (1958) showed that preference ratings correspond to high product quality, not average product quality. When people say they like strawberries, they are referring to good strawberries rather than average products.

Recently, Frank and van der Klaauw (1994) have tried to relate food preference to the underlying chemical senses of taste and smell. Subjects who liked many foods (likers) showed higher olfactory intensity and hedonic ratings and showed higher ideal taste intensities. While many questions remain about the methods used in this study, it is an important attempt to link food preferences and sensory factors. Using a different research paradigm, Meiselman analyzed food preferences of sweet foods and other foods to demonstrate significant differences among different demographic groups (Meiselman, 1977; Wyant and Meiselman, 1984).

One can consider food aversions to be the dislike end of the preference sclae, or one can consider food aversions to be qualitatively different from likes and dislikes. Rozin and Fallon (1987) suggest that food rejection results from three possible motivations: (i) perceived negative sensory properties, (ii) anticipation of social or physical harm, or (iii) knowledge of the origin or nature of food. These three factors result in four different types of food rejection: (i) distaste, primarily motivated by sensory factors; (ii) danger, anticipating harm; (iii) inappropriateness, because the item is not food; and (iv) disgust, because of origin or contamination. Pliner *et al.* (1992) found that dislike of a familiar food was a good predictor of willingness to eat; while dislike or the perception of danger of a novel food were both good predictors.

While there has been research on food preferences (food likes) and on aversions and disgusts, there has been very little work on the borderline area between likes and dislikes. Foods which are mildly liked or disliked probably represent more frequent everyday occurrences and probably are not rejected as clearly as food strongly disliked or completely avoided.

6.4.3 Variety seeking

One line of variety research has focused not on the food but on the individual's attitude toward and response to variety. Recently, Van Trijp and colleagues have begun to systematically and quantitatively consider variety seeking. Van Trijp and Steenkamp (1992) developed a questionnaire which measures variety seeking and which consists of eight items with five-point agree–disagree scales (Table 6.1). They have based variety seeking on the motivational concept of optimal stimulation level which can modify exploration for food.

The variety seeking scale has not yet been applied to the broad issue of variety in meals or diets. Van Trijp and Finnish colleagues (Van Trijp *et al.*, 1992) used questionnaires to study variety of cheeses and spreads in Finnish households. Differences in variety seeking were more related to cheeses

Table 6.1 The items constituting the VARSEEK-scale (translated from Dutch)

1. When I eat out, I like to try the most unusual items, even if I am not sure I would like them.
2. While preparing foods or snacks, I like to try out new recipes.
3. I think it is fun to try out food items one is not familiar with.
4. I am eager to know what kind of foods people from other countries eat.
5. I like to eat exotic foods
6. Items on the menu that I am unfamiliar with make me nervous.
7. I prefer to eat food products I am used to. (R)
8. I am curious about food products I am not familiar with.

Source: Van Trijp and Steenkamp (1992), Appendix A, p. 192.

than to spreads. Van Trijp *et al.* suggested that the cheeses had greater sensory differences than spreads, a requirement for satisfying a need for variety (see Rolls *et al.*, 1982).

6.4.4 Neophobia–neophilia

I noted above that some of the research on food labeling dealt with novel or unusual foods. Similarly, there has been substantial interest in attitudes toward novel or unusual foods. Until recently, there has been little ability to quantify this attitude concerning food novelty. Pliner and Hobden (1992) have developed a Food Neophobia Scale using standard scale construction techniques, although their population sample was mainly college undergraduates. The Food Neophobia Scale (Table 6.2) consists of ten items, five positively worded and five negatively worded. The potential range of scores on the scale is 10–70. Test–retest reliability ranged from $r = 0.82–0.91$. The authors also constructed a General Neophobia Scale to test reactions to novel situations in general. This scale has eight items, with a potential range of scores of 8 to 56. Pliner and Hobden (1992) sought to validate the Food Neophobia Scale in a number of ways. For example, they found that the scale predicted willingness to try different novel foods. Also, more highly neophobic subjects reported being less familiar with, and having eaten fewer times, the unfamiliar foods.

The development of a neophobia scale is an important contribution to food habits methods and will advance research on factors which initiate or restrict eating. Further research will need to extend the concept and

Table 6.2 Food neophobia scale

	Item
1.	I am constantly sampling new and different foods. (R)
2.	I don't trust new foods. (R)
3.	If I don't know what is in a food, I won't try it. (R)
4.	I like foods from different countries. (R)
5.	Ethnic food looks too weird to eat. (R)
6.	At dinner parties, I will try a new food. (R)
7.	I am afraid to eat things I have never had before. (R)
8.	I am very particular about the foods I will eat. (R)
9.	I will eat almost anything. (R)
10.	I like to try new ethnic restaurants. (R)

Source: Pliner and Hobden (1992), Table 1, p. 109.

measurement of reluctance to eat to foods other than foreign and unusual foods. Many people are reluctant to eat familiar foods, and this probably represents a greater dietary impact.

Variety seeking and neophobia probably tap some common individual attitudes concerning novelty. Van Trijp (1993) has characterized variety seeking as positive motivation and neophobia as negative motivation. He distinguished the motivation underlying the two, with neophobia being based in the extrinsic motivation of perceived wish and variety seeking being based in the intrinsic motivation of curiosity. Further, van Trijp noted that neophobia probably relates to absolute product novelty, whereas variety seeking relates to short-term or relative product novelty. Finally, van Trijp argues that high neophobics can be low variety seekers, but high variety seekers are not necessarily low neophobics. Neophobia and variety seeking are probably correlated, at least in Western cultures. The measurement instruments to quantify them appear to draw upon common issues relating to novelty. Further research will be able to determine the relationship between neophobia and variety seeking and the relevance of each to food acceptance and choice.

6.4.5 Restrained eating

Restrained eating is a psychological concept referring to the intention to restrict food intake. Hence, it belongs in this section on attitudes. Whereas food neophobia and variety seeking might exist universally, restrained eating is believed to exist mainly in Western Europe, North America and other areas within so-called contemporary Western culture. It is an interesting question whether the attitudes discussed in this section are universal. Do all people seek variety, avoid novelty and restrict intake? Are these all culturally determined, showing up in Western cultures but not elsewhere? Should broader concepts of food attitudes be universal? These are interesting questions which are beyond the scope of this chapter and beyond my expertise.

Restrained eating can be measured by a number of different scales, of which I will present three. Restraint measurement was introduced by Herman and Polivy (1975) as an alternative explanation for the differences in eating behavior between normal weight and obese subjects. Restraint was seen as the tendency of persons to restrict intake in order to control body weight. An earlier 11-item restraint scale was revised into a 10-item scale (Table 6.3) (Herman and Polivy, 1980) which became the standard of its type. Stunkard and Messick (1985) proposed a new questionnaire to overcome problems with Herman's restraint scale. The problems were

Table 6.3 Revised restraint scale

1. How often are you dieting? Never; rarely; sometimes; often; always. (Scored 0–4)
2. What is the maximum amount of weight (in pounds) that you have ever lost within one month? 0–4; 5–9; 10–14; 15–19; 20+. (Scored 0–4)
3. What is your maximum weight gain within a week? 0–1; 1.1–2; 2.1–3; 3.1–5; 5.1+. (Scored 0–4)
4. In a typical week, how much does your weight fluctuate? 0–1; 1.1–2; 2.1–3; 3.1–5; 5.1+. (Scored 0–4)
5. Would a weight fluctuation of 5 pounds affect the way you live your life? Not at all; slightly; moderately; very much. (Scored 0–3)
6. Do you eat sensibly in front of others and splurge alone? Never; rarely; often; always. (Scored 0–3)
7. Do you give too much time and thought to food? Never; rarely; often; always. (Scored 0–3)
8. Do you have feelings of guilt after overeating? Never; rarely; often; always. (Scored 0–3)
9. How conscious are you of what you are eating? Not at all; slightly; moderately; extremely. (Scored 0–3)
10. How many pounds over your desired weight were you at your maximum weight? 0–1; 1–5; 6–10; 11–20; 21+. (Scored 0–4)

Source: Herman and Polivy, in Stunkard, M., *Obesity* (1980), pp. 208–225.

failure to predict obese behavior and confounding of restraint with weight fluctation and social desirability. Stunkard and Messick developed a 51-item questionnaire to measure three dimensions of eating: cognitive restraint of eating (36 items), disinhibition, and hunger. Their restraint scale is shown in Table 6.4.

Van Strien *et al.* (1986) also developed an alternative to Herman's scale. The Dutch Eating Behaviour Questionnaire has 33 items, with scales for restrained eating (10 items), emotional eating (13 items), and external eating (10 items). The restrained eating scale (Table 6.5) is similar to Stunkard and Messick's (1985) 'cognitive restraint factor.'

To date, measurement of dietary restraint has mainly occurred within the context of obesity and other abnormal eating patterns. Restraint might turn out to be an important contextual determinant of eating in normal weight adults.

Table 6.4 The restraint scale I of the three-factor eating questionnaire

			Factor number
1. When I smell a sizzling steak or see a juicy piece of meat, I find it very difficult to keep from eating, even if I have just finished a meal.	**T**	F	2
2. I usually eat too much at social occasions, like parties and picnics.	**T**	F	2
3. I am usually so hungry that I eat more than three times a day.	**T**	F	3
4. When I have eaten my quota of calories, I am usually good about not eating any more.	**T**	F	1
5. Dieting is so hard for me because I just get too hungry.	**T**	F	3
6. I deliberately take small helpings as a means of controlling my weight.	**T**	F	1
7. Sometimes things just taste so good that I keep on eating even when I am no longer hungry.	**T**	F	2
8. Since I am often hungry, I sometimes wish that while I am eating, an expert would tell me that I have had enough or that I can have something more to eat.	**T**	F	3
9. When I feel anxious, I find myself eating.	**T**	F	2
10. Life is too short to worry about dieting.	T	**F**	1
11. Since my weight goes up and down, I have gone on reducing diets more than once.	**T**	F	2
12. I often feel so hungry that I just have to eat something.	**T**	F	3
13. When I am with someone who is overeating, I usually overeat too.	**T**	F	2
14. I have a pretty good idea of the number of calories in common food.	**T**	F	1
15. Sometimes when I start eating, I just can't seem to stop.	**T**	F	2
16. It is not difficult for me to leave something on my plate.	T	**F**	2
17. At certain times of the day, I get hungry because I have gotten used to eating then.	**T**	F	3
18. While on a diet, if I eat food that is not allowed, I consciously eat less for a period of time to make up for it.	**T**	F	1
19. Being with someone who is eating often makes me hungry enough to eat also.	**T**	F	3
20. When I feel blue, I often overeat.	**T**	F	2
21. I enjoy eating too much to spoil it by counting calories or watching my weight.	T	**F**	1
22. When I see a real delicacy, I often get so hungry that I have to eat right away.	**T**	F	3
23. I often stop eating when I am not really full as a conscious means of limiting the amount that I eat.	**T**	F	1
24. I get so hungry that my stomach often seems like a bottomless pit.	**T**	F	3
25. My weight has hardly changed at all in the last ten years.	T	**F**	2
26. I am always hungry so it is hard for me to stop eating before I finish the food on my plate.	**T**	F	3

Table 6.4 *Continued*

			Factor number
27. When I feel lonely, I console myself by eating.	T	F	2
28. I consciously hold back at meals in order not to gain weight.	T	F	1
29. I sometimes get very hungry late in the evening or at night.	T	F	3
30. I eat anything I want, any time I want.	T	F	1
31. Without even thinking about it, I take a long time to eat.	T	F	2
32. I count calories as a conscious means of controlling my weight.	T	F	1
33. I do not eat some foods because they make me fat.	T	F	1
34. I am always hungry enough to eat at any time.	T	F	3
35. I pay a great deal of attention to changes in my figure.	T	F	1
36. While on a diet, if I eat a food that is not allowed, I often then splurge and eat other high calorie foods.	T	F	2

Source: Stunkard and Messick (1985), Appendix, pp. 81–82.

Table 6.5 Restraint scale of the Dutch eating behaviour questionnaire

1. If you have put on weight, do you eat less than you usually do?
2. Do you try to eat less at mealtimes than you would like to eat?
3. How often do you refuse food or drink offered because you are concerned about your weight?
4. Do you watch exactly what you eat?
5. Do you deliberately eat foods that are slimming?
6. When you have eaten too much, do you eat less than usual the following day?
7. Do you deliberately eat less in order not to become heavier?
8. How often do you try not to eat between meals because you are watching your weight?
9. How often in the evening do you try not to eat because you are watching your weight?
10. Do you take into account your weight with what you eat?

Source: Strien *et al.* (1986), Table 2, p. 304.

6.4.6 *Expectations*

Cardello (1994) has presented a series of studies of consumer expectations and consumer acceptance and has also developed a model to account for how consumers evaluate products. Cardello has used both novel and more common products and has focused on when consumer expectations are not met (disconfirmation). He has proposed two alternative models of disconfirmation: a contrast model, when ratings move away from the expected rating, and an assimilation model, when ratings move toward the expected rating. Results have supported assimilation, which means that products with high expectations tend to actually be rated higher, and products with low expectations tend to be rated lower than their baseline values.

The reader is referred to the chapter by Cardello in this volume for more detail.

6.4.7 Summary: individual

The individual eater can be considered a part of the eating context and has received substantial research interest from the traditional fields of psychology and physiology. Study of preference behavior, and measurement of individual preferences, was an early component of food behavioral research, and dietary restraint has been of interest in the dietary area for decades. Recently, the area of aversions has been added. What remains is the broad middle area of weak preferences and weak aversions. A number of attitudinal dimensions have received special attention. Variety seeking was noted above under food variety; neophobia and neophilia can now be better quantified because of new techniques. The role of individual expectations is also emerging as a critical variable.

6.5 Summary

This chapter has sought to provide an organization for and selected review of the variables which form the context of eating. When eating is viewed in terms of a reference event, such as an individual eating an apple, the context is all those events and experiences (variables) which are not part of the reference event. An attempt has been made to break out the food context itself, that is, the effects of nonreference foods on the reference food. In addition, situational context, the effects of nonreference physical and social situational variables have been separated. Finally, the question has been asked whether the individual's attitudes which precede the reference event can form another aspect of context. Together, the food, the individual and the situation form the entire context of a reference eating event.

Although context research is a relatively new part of research on the spectrum of variables which affect food acceptance and food consumption, studies to date indicate that contextual variables can have profound effects. Future research will uncover new contextual variables and will also uncover how contextual variables interact with the other main classes of variables in the eating situation. These are the food variables, i.e., the psychology and physiology of the individual. Understanding the interaction of these different classes of variables is the challenge for the future. Rather than dealing with only one variable or only one class of variables, we will need to deal in research with complex interaction of variables, just as the individual does when eating. The research methods for this undertaking will include both laboratory and field research, both controlled measurement and uncontrolled observation. By opening up our field to contextual variables and by opening our field to a variety of methodological approaches, we will be able to fully understand the control of food acceptance and food consumption.

References

Balintfy, J.L., Sinha, P., Moskowitz, H.R. and Rogozenski, J.G. (1975) The time dependence of food preferences. *Food Product and Development*, Nov, 33–36, 96.
Bell, R. and Meiselman, H.L. (1995) The role of eating environments in determining food choice. In D. Marshall (ed.), *Food Choice and the Consumer*. Glasgow: Blackie A&P.
Bell, R., Meiselman, H.L., Pierson, B.J. and Reeve, W.G. (1994) Effects of adding an Italian theme to a restaurant on the perceived ethnicity, acceptability, and selection of foods. *Appetite*, **22**, 11–24.
Bell, R. and Paniesin, R. (1992) The influence of sauce, spice and name on the perceived ethnic origin of selected culture-specific foods. In L.S. Wu and A.D. Gelinas (eds), *Product Testing with Consumers for Research Guidance: Special Consumer Groups*, 2nd Vol. ASTM STP 1155. pp. 22–36. Philadelphia: ASTM.
Cardello, A.V. (1994) Consumer expectations and their role in food acceptance. In H.J.H. MacFie and D.M.H. Thomson (eds.), *Measurement of Food Preferences*. pp. 253–297. Glasgow: Blackie A&P.
Cardello, A.V. and Maller, O. (1982) Relationships between preferences and food acceptance ratings. *Journal of Food Science*, **47**, 1553–1557, 1561.
Cardello, A.V., Maller, O., Masor, H.B., DuBose, C. and Edelman, B. (1985) Role of consumer expectancies in the acceptance of novel foods. *Journal of Food Science*, **50**, 1707–1714, 1718.
Coll, M., Meyer, A. and Stunkard, A.J. (1979) Obesity and food choices in public places. *Archives of General Psychiatry*, **36**, 795–797.
deCastro, J.M. (1990a) Social, circadian, nutritional, and subjective correlates of the spontaneous pattern of moderate alcohol intake of normal humans. *Pharmacology, Biochemistry & Behavior*, **35**, 923–931.
deCastro, J.M. (1990b) Social facilitation of duration and size but not rate of the spontaneous meal intake of humans. *Physiology & Behavior*, **47**, 1129–1135.
deCastro, J.M. and deCastro, E.S. (1989) Spontaneous meal patterns of humans: influence of the presence of other people. *American Journal of Clinical Nutrition*, **50**, 237–247.
deCastro, J.M., Brewer, E.M., Elmore, D.K. and Orozco, S. (1990) Social facilitation of the spontaneous meal size of humans occurs regardless of time, place, alcohol or snacks. *Appetite*, **15**, 89–101.
Dobbing, J. (1987) *Sweetness*. London: Springer-Verlag.
Edelman, B., Engell, D., Bronstein, P. and Hirsch, E. (1986) Environmental effects on the intake of overweight and normal weight men. *Appetite*, **7**, 71–83.
Eindhoven, J. and Peryam, D.R. (1959) Measurement of preferences for food combinations. *Food Technology*, **XIII** (7), 379–382.
Engell, D., Mutter, S. and Kramer, F.M. (1990) Impact of effort required to obtain water on human consumption during a meal. Proceedings of Eastern Psychological Association Annual Meeting.
Feunekes, G.I.J., DeGraaf, C. and Van Staveren, W.A. (1995) Social facilitation of food intake is mediated by meal duration. *Physiology & Behavior*, in press.
Frank, R.A. and van der Klaauw, N.J. (1994) The contribution of chemosensory factors to individual differences in reported food preferences. *Appetite*, **22**, 101–124.
Hedderley, D. and Meiselman, H.L. Modelling meal acceptability in a free choice environment. Submitted to *Food Quality and Preference*.
Herman, C.P. and Polivy, J. (1975) Anxiety, restraint, and eating behavior. *Journal of Abnormal Psychology*, **84**, 666–672.
Herman, C.P. and Polivy, J. (1980) Restrained eating. In A.J. Stunkard (ed.), *Obesity*. pp. 208–225. Philadelphia, PA: W.B. Saunders.
Hirsch, E. and Kramer, F.M. (1993) Situational influences on food intake. *Nutritional Needs in Hot Environments*. pp. 215–244. Washington, DC: National Academy Press.
Kamen, J.M. (1962) Reasons for nonconsumption of food in the Army. *J. of the American Dietetic Association*, **41**, 437–442.
Levin, I.P. and Gaeth, G.J. (1988) How consumers are affected by the framing of attribute information before and after consuming the product. *Journal of Consumer Research*, **15**, 374–378.

Levitz, L.S. (1975) The susceptibility of human feeding behavior to external controls. In G.A. Bray (ed.), *Obesity in Perspective*. Washington, DC: US Government Printing Office (DHEW Publication No. NIH 75-708).

Marriott, B.M. (1993) *Nutritional Needs in Hot Environments*. Washington, DC: National Academy Press.

Mehrabian, A. (1980) *Basic Dimensions for a General Psychological Theory*. Cambridge, MA: Oelgeschlager, Gunn and Hair.

Meiselman, H.L. (1977) The role of sweetness in the food preference of young adults. In Weiffenbach, J.M. (ed.), *Taste and Development: The Genesis of Sweet Preference*. DHEW Publication No. 77–1068, National Institutes of Health, Maryland. pp. 269–281.

Meiselman, H.L. (1988) Consumer studies of food habits. In Piggott, J.R. (ed.), *Sensory Analysis of Foods*, 2nd edn London: Elsevier Applied Science. pp. 267–334.

Meiselman, H.L. (1992) Methodology and theory in human eating research. *Appetite*, **19**, 49–55.

Meiselman, H.L. and Bell, R. (1993) The effects of name and recipe on the perceived ethnicity and acceptability of selected Italian foods by British subjects. *Food Quality and Preference*, **3**, 209–214.

Meiselman, H.L. and Kramer, F.M. (1993) The role of context in behavioral effects of food. In National Academy of Sciences, An Evaluation of Potential Performance Enhancing Food Components for Operational Rations, 177–197.

Meiselman, H.L., Hirsch, E.S. and Popper, R.D. (1988) Sensory, hedonic and situational factors in food acceptance and consumption. In D.M.H. Thomson (ed.), *Food Acceptability*. pp. 77–87. London: Elsevier.

Meiselman, H.L., Hedderley, D., Staddon, S.L., Pierson, B.J. and Symonds, C.R. (1994) Effect of effort on meal selection and meal acceptability in a student cafeteria. *Appetite*, **23**, 43–55.

Meyers, A.W., Stunkard, A.J. and Coll, M. (1980) Food accessibility and food choice. *Archives of General Psychiatry*, **37**, 1133–1135.

Milliman, R.E. (1986) The influence of background music on the behavior of restaurant patrons. *Journal of Consumer Research*, **13**, 286–289.

Moskowitz, H.R. (1993) *Food Concepts and Products*. Trumball, CT: Food and Nutrition Press.

Moskowitz, H.R. and Klarman, L. (1977) Food compatibilities and menu planning. *Journal of the Institute of Canadian Science and Technology Alimenta*, **10** (4), 257–264.

Nisbett, R.E. (1968) Determinants of food intake in obesity. *Science*, **159**, 1254–1255.

Oude Ophuis, P. (1993) Better understanding of context effects through Means-End Chain Theory. In *Food Quality—Consumer Relevance*, p. 61 (Abstract). Proceedings of the 5th SENS Plenary Meeting, Reggio Emilia, Italy.

Peryam, D.R. and Pilgrim, F.J. (1957) Hedonic scale of measuring food preferences. *Food Technology*, **11**, 9–14.

Peryam, D.R. and Girardot, N.F. (1952) Advanced taste-test method. *Food Engineering*, **24** (7), 58–61, 194.

Pliner, P. and Hobden, K. (1992) Development of a scale to measure the trait of food neophobia in humans. *Appetite*, **19**, 105–120.

Pliner, P., Pelchat, M. and Grabski, M. (1993) Reduction of neophobia in a human by exposure to novel foods. *Appetite*, **20**, 111–123.

Rogozenski, J.G. and Moskowitz, H.R. (1982) A system for the preference evaluation of cyclic menus. *Journal of Foodservice Systems*, **2**, 139–161.

Rolls, B.J. (1986) Sensory specific satiety. *Nutrition Reviews*, **44**, 92–101.

Rolls, B.J. (1990) The role of sensory specific satiety in food intake and food selection. In E.D. Capaldi and T.L. Powley (eds), *Taste, Experience and Feeding*. pp. 197–209. Washington, DC: American Psychological Association.

Rolls, B.J., Rowe, E.A. and Rolls, E.T. (1982) How sensory properties of foods affect human feeding behavior. *Physiology and Behavior*, **29**, 409–417.

Rozin, E. (1983) *Ethnic Cuisine: The Flavor Principle Cookbook*. Stephen Greene Press.

Rozin, E. (1992) *Ethnic Cuisine: The Flavor Principle Cookbook*. New York: Penguin Books.

Rozin, P. and Fallon, A.E. (1987) A perspective on disgust. *Psychological Review*, **94**, 23–41.

Rozin, P. and Markwith, M. (1991) Cross-domain variety seeking in human food choice. *Appetite*, **16**, 57–59.

Rozin, P. and Tuorila, H. (1993) Simultaneous and temporal contextual influences on food acceptance. *Food Quality and Preference*, **4**, 11–20.
Russel, J.A. and Pratt, G. (1980) A description of the affective quality attributed to environments. *Journal of Personality & Social Psychology*, **38** (2), 311–322.
Schutz, H.G. (1988) Beyond preference: Appropriateness as a measure of contextual acceptance. In D.M.H. Thomson (ed.), *Food Acceptability*. pp. 115–134. London: Elsevier.
Schutz, H.G. and Kamenetzky, J. (1958) Response set in measurement of food preference. *Journal of Applied Psychology*, **43**, 175–177.
Seaton, R.W. and Gardiner, B.W. (1959) Acceptance measurement of unusual foods. *Food Research*, **24**, 271.
Spiegel, T.A. and Stellar, E. (1990) Effects of variety seeking of food intake of underweight, normal-weight, and overweight women. *Appetite*, **15**, 47–62.
Stunkard, A.J. and Kaplan, D. (1977) Eating in public places: a review of reports of the direct observation of eating behavior. *International Journal of Obesity*, **1**, 89–101.
Stunkard, A.J. and Messick, S. (1985) The three-factor eating questionnaire to measure dietary restraint, disinhibition, and hunger. *Journal of Psychosom Res.*, **29**, 71–83.
Tarasuk, V. and Beaton, G.H. (1992) Day-to-day variation in energy and nutrient intake: Evidence of individuality in eating behavior. *Appetite*, **18**, 43–54.
Turner, M. and Collison, R. (1988) Consumer acceptance of meals and meal components. *Food Quality and Preference*, **1**, 21–24.
Van Strien, T., Frijters, J.E.R., Bugers, G.P.A. and Defares, P.B. (1986) The Dutch eating behavior questionnaire (DEBQ) for assessment of restrained eating, emotional, and external eating behavior. *Int'l. J. of Eating Disorders*, **5**, 293–315.
Van Trijp, H.C.M. (1993) Contextual effects and variety seeking behavior for foods. In *Food Quality—Consumer Relevance*, p. 60 (Abstract). Proceedings of the 5th SENS Plenary Meeting in Reggio Emilia, Italy.
Van Trijp, H.C.M. and Steenkamp, J.-B.E.M. (1992) Consumers' variety seeking tendency with respect to foods: measurement and managerial implications. *European Review of Agricultural Economics*, **19**, 181–195.
Van Trijp, H.C.M., Lahteenmaki, L. and Tuorila, H. (1992) Variety seeking in the consumption of spread and cheese. *Appetite*, **18**, 155–164.
Wolfson, J. and Oshinsky, N.S. (1966) Food names and acceptability. *Journal of Advertising Research*, **6** (1), 21–23.
Wyant, K.W. and Meiselman, H.L. (1984) Sex and race differences in food preferences of military personnel. *Journal of the American Dietetic Association*, **84**, 169–175.

7 Marketing and consumer behaviour with respect to foods

HANS C.M. van TRIJP and
MATTHEW T.G. MEULENBERG

7.1 Introduction

Except for some food production for private consumption, most food products are purchased in the market. This is the domain of food marketing. Broadly speaking, marketing as a scientific discipline is concerned with the exchange processes that occur between a firm or organization that offers products or services to the market, and the target group(s) of potential buyers in its environment. Marketing focuses on policies and strategies which firms adopt to satisfy the needs and wants of the target groups more effectively and efficiently than competitors do. The basic assumption is that the organization that can do so most effectively and efficiently will be most successful in achieving its organizational objectives.

The first scientific contributions to the marketing discipline appeared in the beginning of this century (e.g. Shaw, 1912; Weld, 1917; Copeland, 1923). Since then, marketing has evolved from an emphasis on production to selling and from selling to a full fledged marketing operation. In the early days of marketing, most markets were characterized by a demand potential that exceeded product supplies. At the same time lack of money prevented many consumers from materializing their potential demands. In those circumstances many companies focused marketing operations on the physical product and on the efficiency of production and distribution. This approach is generally referred to as the 'production concept' in marketing. Marketing efficiency, in particular distribution efficiency, was an important marketing subject. The emphasis on the performance of marketing functions, the role of marketing institutions and the impact of product characteristics on the marketing process is reflected in early definitions of marketing as:

> 'all those activities involved in the distribution of goods from producers to consumers and in the transter of title thereto' (Bartels, 1970: 41).

According to Bartels (1970: 36) the key issues in the marketing discipline in the period 1910–1920 were: 'selling, buying, transporting and storing'. Product policy and promotional activity were not of central concern yet.

By the 1930s, in many Western markets basic consumers' needs were satisfied and shortages in basic food supply were solved. Food markets gradually developed into buyers' markets. As a result food suppliers had to put more efforts in stimulating demand and meeting competition. The struggle for the consumers' money became harder. Up until the 1950s this struggle was mainly fought by the 'selling' approach to marketing which Kotler (1994: 17) defines as:

> 'the selling concept holds that consumers, if left alone, will ordinarily not buy enough of the organization's products. The organization must therefore undertake an aggressive selling and promotion effort.'

Basically, the idea was that any product could be sold through aggressive and effective selling techniques. Neither identification of consumers' needs and wants, nor the product in terms of its quality and need satisfying value had a central position in this approach.

In the second part of this century the evolution from the 'selling concept' towards the true 'marketing concept', i.e. '. . . delivering the desired satisfactions more effectively and efficiently than competitors' (Kotler, 1994: 18) took place in many markets. This evolution was largely induced by changing market conditions. Purchasing power and consumer spending increased and product supply became more abundant in quantity and variety because of productivity increase, R&D efforts in agricultural and food industry, and price and income support to agriculture, such as the Common Agricultural Policy of the European Union. In fact, agriculture and food markets changed from a seller's to a buyer's market in which customers are critical and demanding. In addition, retail companies became big businesses with a strong impact on food marketing. They have developed specific marketing policies, are cost conscious and have a strong bargaining power *vis-à-vis* food producers (e.g. Heijbroek *et al.*, 1994). Adjustment of market supply to the needs and wants of potential buyers has become a necessary condition for market success in the light of demanding customers and of fierce competition between food suppliers.

As a result of these market developments and developments in marketing theory, marketing management has become the dominant stream in marketing theory (e.g. Howard, 1957; McCarthy, 1960; Borden, 1964). Marketing management is

> 'the process of planning and executing the conception, pricing, promotion and distribution of ideas, goods, and services to create exchanges that satisfy individual and organizational objectives' (Kotler, 1994: 13).

It rests on four main pillars: market focus, customer orientation, coordinated marketing and profitability. This approach requires a good insight into the market structure and customer behaviour. Important issues are amongst others:

> 'Which market segments can be distinguished? What are the needs and wants of potential buyers? How do potential buyers make their product choice decisions? How do potential buyers react to a specific combination of marketing instruments?'

In order to answer these questions and to evaluate their implications for marketing policies, organizations need to conduct marketing research. This marketing research is to a large extent based on theories of consumer behaviour. Through a co-ordinated use of marketing mix instruments: product, price, promotion and place (distribution), the company attempts to satisfy identified customer needs more effectively and efficiently than competitors do. This will increase the company's product's attractiveness relative to competing products, materializing in customer preferences and choice and ultimately in profitability. In light of the changing market opportunities, strategic marketing planning, aimed at the firm's long term profitability is becoming increasingly important.

The purpose of this chapter is to provide an introduction to marketing and consumer behaviour with respect to foods. We will focus on consumer marketing of processed and branded food products (e.g. Coca Cola, Nescafé) which conforms to the marketing approach taken in marketing management. Marketing of agricultural fresh produce (i.e. agricultural marketing) differs from general marketing theory due to specific characteristics of these products (e.g. its perishability, seasonality of production and consumption, and uncertainty about quality and quantity of production) and the market structure (small production units and specific marketing organization such as cooperatives, auctions, etc). For specific information about agricultural marketing the reader is referred to Kohls and Uhl (1990) and Meulenberg (1993). In section 7.2, we will introduce the general marketing concept from a marketing management perspective. We will discuss key elements of marketing systems and some of the marketing strategies at the corporate level. Marketing is executed at different levels within the company. Section 7.3 discusses marketing strategies at the business level, and section 7.4 discusses some basic considerations with respect to marketing mix decisions at the product level. As the marketing management approach critically depends on the understanding of consumer behaviour, relevant issues in consumer behaviour in general and their implications for food choice behaviour in affluent societies will be discussed in sections 7.5 and 7.6. Section 7.7 will discuss the consumer behaviour implications for food marketing. Section 7.8 presents an illustration of some of the key concepts in the context of consumer oriented product development, and section 7.9 contains the conclusions.

7.2 Marketing

A well known definition of marketing is

> 'Marketing is a social and managerial process by which individuals and groups obtain what they need and want through creating, offering and exchanging products of value with others' (Kotler, 1994: 6).

Marketing is in particular concerned with the efficiency and effectiveness of exchange processes. Somewhat simplified, exchange processes can be depicted as at least two parties being involved, each having specific needs and wants and something of value to offer that may contribute to the realization of the other party's objectives. Each party feels it is appropriate to engage in an exchange process with the other party, but is free to accept the other party's offering (see Houston and Gassenheimer, 1987). Both parties enter an exchange process with specific objectives (profit making and/or need satisfaction) and offer something of value to the achievement of the other parties' objective (food products *vs.* money). Thus, both parties may benefit from a particular exchange but also have access to alternative parties with whom an exchange process might be undertaken.

The marketing process can be understood by using some concepts of systems thinking (e.g. Churchman, 1968), such as system's objectives, environment, instruments, subsystems, planning and control. From a marketing point of view a company may be looked upon as a *marketing system*, with several *subsystems*. Many large companies are organized into a number of more or less autonomous Strategic Business Units (SBUs), in which the marketing activities for a set of related products are coordinated. SBUs may be defined in terms of customer groups that will be served, customers needs that will be met, and the technology that will satisfy the need (Abell, 1980). SBUs carry the strategic responsibility for a particular product assortment and reflect the different businesses a company is operating. SBUs consist of several individual products and/or brands for which concrete marketing plans are developed and executed. At each of these levels, specific marketing objectives may be discerned in terms of Return on Investment, sales or market share. Objectives will have to be realized within a *marketing environment*, consisting of the target group of customers, the competitive suppliers, the distribution structure, and government and lobby groups, such as consumers' unions. For example a producer of diet food will be focussing his product on a specific target group of consumers, will be trying to serve customers better than competitive suppliers, will offer services in particular logistical services to retail companies, and will abide by government laws and regulations, such as the food and drugs law, and will try to satisfy the policies and requirements of consumers' unions. The marketing system has *instruments* available to realize the marketing objective. At the level of individual products and brands these are the marketing mix instruments or the 4 Ps:

> Which benefits do I offer to the consumer (*Product*), at what costs (*Price*), by what information/persuasion (*Promotion*), at which place and time (*Place/distribution*)?

For instance, our diet food producer might develop a health food for diabetics, which is sold in regular food shops at a fair price, and which is promoted by advertising in special weeklies and leaflets to nutritionists and medical doctors. The development and implementation of a marketing plan requires a specific marketing organization (*subsystems*). Marketing activities of the company are reflected in *marketing plans*. Procedures are developed for monitoring, evaluating and *controlling* the marketing plans, which requires a systematic feed back from the environment about (dis)satisfaction of clients, both retailer and consumer.

Given the fierce competition and dynamics in the marketing environment in which food companies operate, strategic planning is of crucial importance for the achievement of the company's long term objectives. Strategic planning is

> 'the managerial process of developing and maintaining a viable fit between the organization's objectives, skills, and resources and its changing market opportunities. The aim of strategic planning is to shape and reshape the company's businesses and products so that they yield target profits and growths' (Kotler, 1994: 62).

Corporate headquarters are responsible for the corporate strategic plan to ensure a profitable future for the company. Corporate strategies have received considerable attention during the last two decades. Basic issues of the corporate marketing strategy are:

(a) *Definition of the company's mission.* This defines the organization's scope by answering fundamental questions such as: 'What is our business? Who are our customers? What kinds of value can we provide to these customers? and What should our business be in the future?' (Walker *et al.*, 1992).
(b) *Identification of company's strategic business units.* Most companies, even small ones operate several businesses. Ideally, a strategic business unit should be designed to incorporate a unique set of products aimed at a homogeneous set of markets. It should also have responsibility for its own strategic planning and profit performance and control over the resources that affect that performance (Walker *et al.*, 1992).
(c) *Analysis and evaluation of the current portfolio of businesses.* Based on their present and expected future performance, the corporate management will assign to the SBUs strategic planning goals and appropriate funding, to ensure future operation in profitable businesses.

		Relative Market Share	
		>1	<1
Market Growth Rate	> 10%	Stars	Question Marks
	< 10%	Cash Cows	Dogs

Figure 7.1 Boston Consulting Group's Growth–Share matrix.

Several *portfolio evaluation models* are available for the evaluation of SBUs, such as the Growth–Share matrix of the Boston Consulting Group, The General Electric Matrix and the Shell Directional Matrix (see e.g. Jain, 1993 for an overview). For ease of exposition, we will briefly illustrate portfolio analysis in terms of the Growth–Share matrix of the Boston Consulting Group which classifies the company's SBUs in terms of *market growth*, and *market share* relative to the largest competitor. On that basis SBUs can be classified as Stars, Cash Cows, Question Marks and Dogs (Figure 7.1).

The following strategies are suggested for the distinguished four cells. *Stars* and (if chances seem good) *Question Marks*: invest in order to increase future sales and share. Strong *Cash Cows*: maintain position or invest selectively. Weak *Cash Cows* and *Dogs*: increase cash flow regardless of long term effects. *Dogs*: withdraw so as to maximize cash flow.

Thus the prime responsibility at the corporate level is to define the corporate objectives and to develop strategic planning aimed at shaping and reshaping the companies SBUs so that the combined result of all SBUs generates satisfactory profits and growth. This may include divesting some businesses and development or acquirement of new businesses.

As the corporate long term objectives have to be realized in a competitive environment, competitive strategy at the corporate level has received considerable attention. In particular Porter (1980, 1985) has substantially contributed to the understanding of strategic aspects of competition. Porter (1980) argues that the degree of rivalry in an industry depends on Potential Entrants (threat of new entrants), on Buyers (bargaining power of customers), Suppliers (bargaining power of suppliers) and Substitutes (threats of substitute products or services). He has suggested three general competitive strategies: *Focus* on one or a limited number of narrow market segments; *Differentiate* by achieving superior performances in some important customer benefit area valued by the market as a whole, and *Overall Cost Leadership*, by achieving the lowest costs of production and distribution. In his book on competitive advantage Porter (1985) describes the characteristics of a good competitor amongst others as being realistic about the industry and its competitive position, knowing its costs and not unwittingly cross subsidizing product lines or understating its overhead.

7.3 Marketing strategies at the product and brand level

The mission statement of the company and the portfolio-analysis of business units constitute the framework within which marketing policies at the level of product groups and individual brands are developed and executed. This section focuses on marketing strategies, whereas the next section will focus on operational decisions with respect to the marketing mix.

Within the broader corporate mission, each SBU has a specific mission with associated marketing objectives. Based on an analysis of the strength and weaknesses of the firm in relation to the opportunities and threats in the external environment in which the business operates (the so-called SWOT-analysis), management will formulate goals and strategies to achieve these goals. At the level of products and brands, specific marketing programs will be planned for carrying out these strategies. Finally, procedures will be developed for organizing, implementing and controlling the marketing effort.

To a large extent, the business strategies aim at identifying specific market segments in which the company can deliver products which match with the needs and wants of the target group. As the specific choice of strategies is highly specific depending on the corporate mission, the competitive environment and resources of the company, there are no 'golden rules' for marketing strategy. Rather we will review some of the basic marketing strategies that may be adopted.

7.3.1 Market segmentation

Market segmentation builds on the idea that markets consist of customers that may differ in their specific needs and wants. Market segmentation aims at the identification of subgroups of customers (a market segment) that are internally homogeneous in some relevant aspects (e.g. their needs and wants) and different from other market segments. A company may develop a product and/or service which fits specifically to the needs and wants of a particular target group of customers. The segmentation strategy is familiar to many marketers of food products, since the affluent consumers of Western countries want to be served according to their specific wants and needs. Consequently there are not many opportunities for standard products to be served in the same way to every consumer. Water, salt and some other basic food products are exceptions. The ultimate type of market segmentation is customized production, which is more common in industrial markets, such as special machinery, than in food markets which are mostly mass markets.

7.3.2 Product differentiation

In highly competitive food markets producers try to differentiate their product supply from that of competitors. They want to create customer loyalty by doing better on one or more product characteristics and by being distinctive in the market from competitive suppliers. The ultimate form of product differentiation is a strong international brand, such as international food-brands of Kraft, Nestlé and Unilever. Being familiar and recognizable to consumers is a necessary condition for consumer loyalty. At present, branding is extensively used by retail companies (private labelling) in order to strengthen their image and/or to generate substantial gross margin at fair consumer price. Differentiation is not always worth while. Kotler (1994: 306) argues that differentiation is only meaningful if it satisfies the following criteria: The difference delivers a highly valued benefit to a sufficient number of buyers (i.e. *important*), it is not being offered by others or in a more distinct way (i.e. *distinctive*), it is *superior* to other ways to obtain the same benefit, it is visible to buyers (i.e. *communicable*), it cannot easily be copied (i.e. *preemptive*), consumers can afford to pay for the difference (i.e. *affordable*) and the company will find it *profitable* to introduce the difference. As a result, necessary conditions for effective product differentiation are high product standard, substantial market share, stable market supply, stable prices and substantial promotion.

Many authors use the term *positioning* for effective product differentiation, that is the act of placing a brand in that part of the market where it will receive a favourable reception compared to competing products.

> 'Positioning tells what the product stands for, what it is, and how customers should evaluate it' (Jain, 1993: 382).

It may be necessary to reposition a product, when the actual positioning does not generate sufficient sales. For example, one might change the position of milk in the product map of drinks from a healthy food being necessary for children to a modern tasty food that also suits youngsters.

7.3.3 Product development/innovation

Product development/innovation is extremely important for the company's long-term profits and growth. Many factors in the marketing environment, such as changing customer needs, competitive innovation activity and legislation threaten a product's long term success. This is reflected in the product life cycle describing the evolution of a product in the market from its introduction to market decline in terms of sales and profits. Theoretically product life cycles (PLC) are portrayed as an S-shaped curve, split up in the stages of introduction, growth, maturity and decline. By the end of the maturity stage, sales and profits are reducing. Apart from learning about generation of cash and profits, insight into the PLC is important because different marketing strategies may be appropriate at different stages of the PLC.

Product development and innovation is necessary to replace the products that are in the decline stage of the PLC. Product development and innovation strategies are classified as pro-active and re-active in nature, depending on whether the company takes an active role in initiating new product development or takes the role as a market follower (Urban and Hauser, 1993). Examples of re-active strategies are the 'me-too' strategy and the 'second-but-better' strategy. In these strategies, companies imitate new products introduced by innovating companies. They learn the flaws and weak points of a new product and try to introduce an improved product, without having to spend lots of money on basic research and marketing efforts in order to get a foothold in the market. The reverse side of the coin is that a reactive policy will have to carve out a place in the market in which the innovative companies have built up strong positions (the first-entry or pioneering advantage). Reactive policies can be blocked by patents or other unique expertise being only available with the first innovators. Nevertheless reactive strategies of imitating others have been practised by companies lacking the financial means for extensive R&D and marketing programs.

Product development and innovation may also be used in a more initiating way to realize growth (i.e. pro-active). According to Ansoff (1957) product development and market development are the fundamental means of realizing growth in sales for the company. Product development either for current or new markets (diversification) has become an intensively studied issue in marketing (e.g. Urban and Hauser, 1993), partly because of the high failure rate. It should be noted in this context that not all new products carry the same degree of newness. For example, Booz, Allen and Hamilton (1982) have classified new products according to their newness to the market and the company into six categories: New-to-the-world, New-product-lines, Additions to Existing Product Lines, Improvements in Revisions to Existing Products, Repositionings, and Cost Reductions. They conclude that only 10% of the new products are real innovations.

7.3.4 Competitive strategy

The competitive strategies discussed at the corporate level equally apply at the product and brand level, depending on the choice of one or more product market combinations. For example, some companies may try to create a strong position in the market by featuring low prices as the unique selling point to their customers. This strategy is based on low costs of production and logistics and on limited service and extras with the product (i.e. low cost strategy). A classic example of this strategy is a discount shop that competes in the market on the basis of low prices across its total assortment. Similarly, a company may focus on one specific segment (focus strategy) or attempt to achieve a competitive advantage in several or all market segments through a differentiation strategy.

7.4 Marketing tactics: organizing the marketing mix

Having formulated the basic marketing strategy for a product, the company has to plan specific marketing programs to carry out the strategies. Tactical decisions have to be made with respect to the four marketing mix elements: Product, Price, Promotion and Place. Marketing managers and product managers carry the responsibility for development and monitoring of specific marketing plans.

In this section some basic aspects to be considered in planning the marketing mix elements will be discussed. For more elaborate discussions on the marketing mix decisions, the reader is referred to the many textbooks on marketing (e.g. Kotler, 1994).

7.4.1 Product

Marketers think of products at different levels of inclusiveness, ranging from the *core product*, which determines what the product functionally does for the consumer, to the *product as marketed* which consists of the physical product augmented with all the marketing efforts associated with it (e.g. branding, packaging, additional services, guarantees etc.). In terms of product decisions, the product managers have to decide on the quality level of the physical product and how the desired quality image is realized by specific marketing mix decisions.

A key issue with respect to product decisions is the product quality level which will be offered to satisfy the needs and wants of the target group. Product quality and consumer perception of it have received considerable attention in the marketing literature (see e.g. Steenkamp, 1989). Insight into the quality attributes, the functional and psychological utility generating benefits that make up the consumer's quality perception, is important as these constitute the quality dimensions the product should compete on. In the food context, quality attributes include sensory and hygienic quality among other factors. Quality attributes may be delivered through intrinsic and extrinsic quality cues. Intrinsic quality cues are product characteristics that are intrinsic to the product and cannot be changed without altering the product. These cues find their basis in the physical product. On the other hand, quality image may also be influenced through other marketing efforts such as price, brand name, packaging, point of purchase etc. (extrinsic quality cues).

For many products packaging forms an integral part of the product mix. In addition to its protective function, it may contribute to quality attributes such as convenience. Packaging may also play an important role in the communication of information about the product and contribute to the product's quality image. In some cases, packaging may even become the product's unique selling point. For example, it is argued that Marlboro has become a world cigarette brand, *amongst others* by the introduction of a rugged box with a flip-up top. The real augmented product is the branded product. A brand is more that just a name or symbol and provides the consumer with a guarantee for specific product benefits. A brand has intrinsic value (brand equity) in that it may create brand awareness, stimulate brand loyalty to the familiar brand, serve as indicator for perceived quality and may evoke associations not explicitly communicated (Aaker, 1991). Development and maintenance of a positive brand image is extremely important, because much of consumer choice behaviour is based on routine and risk aversiveness. As the product's image may change under the influence of actual consumption experiences, it is extremely important that the product/brand lives up to its expectations (see also the chapter by Cardello). Therefore, product quality of the physical product is a prerequisite for long term success.

Strong brands may allow for brand extensions: i.e. new products introduced under the umbrella of the established brand. Obviously, a strong brand image can pave the way for a new product in the introduction stage of the life cycle, provided that the new product fits in with the brand's image. Consequently, if the image is based on general characteristics such as reliability, innovativeness, and high quality, successful brand extension will be easier to achieve than when image is based on specific characteristics. A particularly intriguing development is the increasing share of own brands of food retail shops. Own brands started out as an attractive proposition for retailers to substitute the low sales of a large number of less important brands, by one own brand with high turnover and being purchased in large quantities at a low price. Gradually own brands of many retail chains shift from price competition to quality competition.

7.4.2 Price

Price is the second marketing mix element that is under the marketer's control. From a financial perspective the price of a product should cover the 'out of pocket' costs (manufacturing and marketing costs), and on top of it generate a sufficient contribution to overheads and profits. Price setting thus depends on the expected quantities that can be sold at that price level. The determination of the 'break-even' point, the quantity that should be sold at a given price to cover fixed costs, is an important tool in this respect.

So, it is important for price setting to know the relationship between price and consumer demand. Consumers evaluate prices in relation to the product's perceived quality ('value for money') and the prices of competitive substitute products. In addition to considering price as a financial sacrifice to obtain the product, consumers may also use price as an indicator for quality (Gabor and Granger, 1966). Therefore too low prices may harm the company both financially and psychologically in terms of brand image.

Price setting is further complicated by the fact that consumer products are sold to final consumers through middlemen, both wholesalers and retailers. The price policy of a producer should suit these middlemen. Trade margins and discounts are important in this respect.

Apart from these long-term price-setting considerations, price promotions may be an important marketing instrument in the market to have new customers try the product. At the introduction stage of a new product, two common pricing tactics are the market-skimming pricing and penetration pricing. Market skimming entails relatively high introduction prices to skim a maximum amount of revenue from the market segments. When competition comes in, prices may be lowered as a defence. This tactic may be appropriate under market conditions of low price elasticity of demand, low levels of competition and modest financial resources of the firm. Market penetration pricing entails relatively low introduction prices in an attempt to create a substantial group of customers that may become brand loyal to the product in the long run. This tactic may be appropriate at high price elasticity of demand, strong potential competition and larger financial resources of the firm.

7.4.3 Promotion

In today's markets with abundant product supplies that can all satisfy the same basic needs, promotion, the communication with current and prospective customers, has become an important tool. Promotion plays an important role in creating brand awareness and brand recognition as well as in communicating the product's benefits (image development) and in stimulating demand.

Programming of promotion as a marketing instrument includes a number of interrelated decisions, such as identification of the target audience, determination of communication objectives, design of the message, selection of communication channels, allocation of total promotion budget, and decisions on the promotion mix. Of particular importance is the assessment of the effectiveness of promotional activities in relation to the communication objectives. Communication may be achieved through the promotional tools of advertising, personal selling, sales promotion, and public relations. One of the key decisions is the determination of the total promotional budget and its allocation across promotional tools and specific campaigns. To a large extent these decisions depend on the specific communication objectives the firm wants to achieve. These objectives should be translated into specific goals, e.g. an increase in awareness by a certain percentage of the target group or a specific number of new triers of the product. Advertising agencies will usually carry the responsibility of translating the objectives into a concrete form of communication in terms of message, format, timing etc. Marketing theory has developed a great many conceptual models and check lists for use in promotion programming. Also a great many analyses have been made about the impact of advertising on sales and communication objectives (e.g. Aaker and Myers, 1987; Blattberg and Neslin, 1993; Lilien *et al.*, 1992, chs. 6, 7).

7.4.4 *Place*

Distribution decisions are concerned with getting the product at a particular place at a specific time. Distribution activities are programmed within the strategic choice of a marketing channel, i.e. about when and where to supply products (e.g. through specialty shops, supermarkets, discounters etc.). The programming of distribution activities primarily consists of activities aimed at development and maintenance of the relationship with retailing organizations and physical distribution to ensure the product's availability at the point of purchase. Since physical distribution costs (transport, storage) account for a substantial share of total product costs they are crucial for overall profitability. Planning of transport and storage is improving because of better information exchange between producer and client on stocks, shippings and future sales. Progress in information technology (product scanning and Electronic Data Interchange) is revolutionizing distribution programming (see e.g. Chapman and Holtham, 1994). Physical distribution has become an element of *Logistics*, the total planning of physical product flows both of input, of materials in the company and final products to the customer. In logistics, models and concepts have been developed aiming at efficient and effective planning of product flows from producer to consumer (e.g. Bowersox *et al.*, 1986). One such concept is Just-In-Time (JIT):

> 'The basic philosophy of JIT is that inventory only exists to cover problems. By reducing inventories, problems in the manufacturing process are exposed. These problems must be solved before inventories can be further reduced.' (Bowersox *et al.*, 1986: 61)

Co-ordination of distribution both horizontally, with other activities in the company such as purchasing and production, and vertically, with distribution activities of other companies in the marketing channel, is very important for improving the efficiency and effectiveness of distribution.

7.4.5 *Coherence in marketing mix decisions*

The success of marketing programming to a large extent depends on the coherence in the decisions with respect to the individual marketing mix elements. The balanced choice of marketing mix decisions is reflected in the marketing plan that should realize the selected marketing strategies. For example, a product's positioning in the luxury segment, has implications for all marketing mix elements, such as the quality of the physical product and its packaging, the price (which should not be too low to avoid doubts about the product's quality), the choice of the distribution outlet (e.g. the luxury image may be supported by distribution through specialty shops), and communication which should support the luxury image.

7.5 Consumer orientation in marketing

The previous section emphasized the importance of customer orientation of the marketing management approach. In Figure 7.2, summarizing the marketing management approach in its environment, this is reflected in the central position of the customer.

Marketers attempt to adjust to these customers and their specific needs and wants by means of marketing programmes that are expressed in terms of the four controllable marketing mix elements: product, price, promotion and place. However, this process is embedded in an environmental context which is largely beyond the marketer's control. The environment includes such diverse factors as the economic environment, the cultural and social environment, the political and legal environment, the competitive environment, and the technological environment. Factors in each of these domains may exert an influence on the marketing management process either because they influence the structure of the customer's needs and wants or because they influence the marketing manager's freedom in his decisions with respect to the marketing mix elements. Given the dynamics in the uncontrollable environment and the uncertainties that arise from it, marketing and consumer behaviour research serve as an important input for

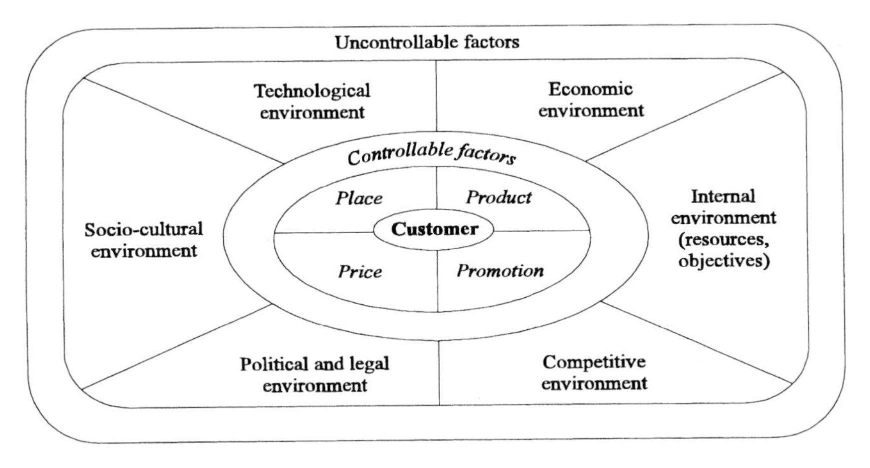

Figure 7.2 The marketing management approach in its environment (after McCarthy and Perreault, 1993).

marketing. To be effective and efficient marketing policy should be the mirror image of consumer behaviour and developments therein. In this section we will discuss some basic elements of consumer behaviour theory.

Consumer behaviour may be broadly defined as the study of

> 'the processes involved when individuals or groups select, purchase, use, or dispose of products, services, ideas, or experience to satisfy needs and desires' (Solomon, 1992).

This definition already highlights that consumer behaviour covers a very wide range of activities. It covers attainment and use of products as well as how consumers dispose of them. In addition to tangible products it also applies to non-tangible 'products' such as ideas and services. Finally, it emphasizes the (decision) processes consumers engage in.

The study of consumer behaviour has developed into a discipline in its own right. It takes a multi-disciplinary approach in that it integrates insights from different scientific fields. Psychology makes an important contribution through its concepts of motivation, perception, attitude, personality and learning theory. Sociology and social psychology are important sources of inspiration in terms of influences of groups, family structures, social class, reference groups etc. Theories about diffusion processes (i.e. the dissemination of a product from the moment of its introduction across different consumers and consumer groups in the population) also find their basis in sociology. Cultural anthropology provides insight into the ways societies function and in the values and norms that shape product consumption. Finally, economics more or less forms the basis from which consumer behaviour theory developed. From a system of axioms about the consumer, economic theory has largely focussed on how consumers react to changing prices and incomes.

The field of consumer behaviour has integrated many of these diverse insights, and has developed its own specific theories and models about consumer behaviour. Some of these early models of consumer behaviour attempted to capture an integral view on consumer behaviour. Examples include the models developed by Howard and Sheth (1969), Engel *et al.* (1986), Bettman (1979) and Howard (1989). Most of the current models, however, focus on specific elements of consumer behaviour like the gathering and processing of information and consumer reactions on specific marketing mix elements such as price and quality. Specific models have been developed for product quality perception (Steenkamp, 1989), service quality perception (Zeithaml *et al.*, 1990), brand loyalty (e.g. Wierenga, 1974), consumer response to price (Gabor and Granger, 1966) and variety seeking in product choice behaviour (e.g. Van Trijp, 1995). The broadness of the field of consumer behaviour renders it impossible to cover all its aspects. Rather we will focus on specific elements that have relevance for consumer behaviour with respect to foods. For more detail into consumer behaviour theory, the reader is referred to the numerous textbooks in this area such as Howard (1989), Schiffman and Kanuk (1991) and Solomon (1992).

Most comprehensive models for consumer behaviour focus on decision making at a relatively high level of elaboration. This type of consumer decision making is referred to as extended problem solving (Howard and Sheth, 1969). Although certainly not all consumer decisions are accompanied with such a high level of elaboration, these models provide a useful framework in which the various influential elements on consumer decision making can be classified. Most of these models recognize five stages in the consumer decision process: (i) problem recognition, (ii) search behaviour, (iii) evaluation of alternatives, (iv) purchase, and (v) evaluation of purchase outcomes. Each of these stages will be discussed below in some more detail.

7.5.1 Problem recognition

Problem recognition is the driving force behind consumer decision making and purchase behaviour. It results from a perceived discrepancy between the consumer's perception of the ideal state and the actual state he or she is experiencing. This discrepancy results in perceived tension and the desire to resolve the discrepancy, in the present context through purchase and consumption behaviour. The tension and the desire to resolve it is what we refer to as needs or motives (these two terms are often used interchangeably in the consumer behaviour context). A consumption motive is the driving force behind behaviour which gives direction and intensity to behaviour. Motives and needs constitute the 'why' of behaviour and for that reason take a central place in consumer behaviour theory. Motives and needs can be of different nature. In the psychological literature an important distinction is made between physiological motives and those of a psychological nature. Physiological (or primary) motives are innate and necessary to maintain a biological equilibrium (e.g. the needs for food, water, air, protection and sex). Psychological (or secondary) motives are acquired and relate to subjective psychological well being and the relationships with others (e.g. the need for luxury and the need for self-actualization). The well-known need structure suggested by Maslow (1954) assumes that there is an implied hierarchy in the individual's need structure. Physiological needs are defined at the lower level of the hierarchy and psychological motives at the higher end. Maslow suggests that needs at the lower level need to be fulfilled before needs at the higher level are searched after. Maslow distinguishes between physiological needs (e.g. water, sleep and food), safety needs (e.g. security, shelter and protection), social needs (e.g. love, friendship and acceptance by others), esteem needs (e.g. prestige, status and accomplishment), and self actualization needs (self fulfilment and enriching experiences)

Maslow's need hierarchy is just one of the many classifications that have been proposed in the literature (see e.g. MacFadyan, 1986 for an overview). It is non-product choice specific in that the motives guide human behaviour in general. Others have proposed classifications that are consumption or even product choice specific (see e.g. Sheth, 1975; McGuire, 1976; Hanna, 1980). Building on the work of Fennell (1975, 1978), Rossiter and Percy (1987) developed a classification that distinguishes between eight basic motivations for product choice behaviour, which are classified as either positive or negative. Negative motivations are those that have an aversive origin and primarily reflect utilitarian aspects of consumption. Positive motivations on the other hand have an appetitive origin and reflect more hedonic aspects of consumption such as sensory pleasure, intellectual stimulation and conspicuous consumption for the sake of social approval. A third category of choice motivations reflects those of a mildly negative origin, where the consumer runs out of a product and routinely repurchases the previous brand. The distinction between utilitarian and hedonic motivations in consumption is receiving increasing attention in the consumer behaviour literature, based on the recognition that (food) consumption may be a goal in and of itself (hedonic consumption) as well as an instrumental means to achieve some further goal (utilitarian consumption).

Consumer specific classifications, although primarily taxonomic tools too, have an advantage over general classifications in that they may provide more concrete guidance for marketing efforts. These motivations manifest themselves in specific product requirements, which can be fulfilled by specific marketing policies. In this respect it is important to realize that specific consumption acts may be guided by combination of underlying motivations. Also, different consumers may satisfy the same basic motivations in different ways. The particular form of consumption used to satisfy a need is termed a 'want' and will be discussed in the evaluation of alternatives stage of the model. How consumers translate needs into wants largely depends on the consumer's history, learning experience and value structure.

7.5.2 Search behaviour

Having recognized a consumption problem, the consumer is supposed to search for consumption alternatives or consumption behaviours that may solve the consumption problem. Internal search relates to the information actively or passively stored in memory. In case this internal information is not deemed sufficient for solving the problem, the consumer is hypothesized to engage in additional external search behaviour about the alternatives and the extent to which they may contribute to solving the consumption problem. Examples of external search would include consultation of consumer reports, and other sources (e.g. labelling, etc.).

Based on internal and external search behaviour, the consumer forms a perception about marketing stimuli. Perception is the process of selection, organization and interpretation of stimuli to a meaningful picture of the world around us. It includes both cognitive (the knowledge consumers have about foods) and affective factors (feelings and emotions). Perception is a complex process, insight into which provides the marketing manager with a lot of relevant information. One of the key characteristics is that information has to go through a perceptual filter before it is stored in the consumer's knowledge structures. Perception is a selective process both in terms of exposure, attention, comprehension, acceptance and retention of available marketing information. More specifically this implies that the consumer may perceive and interpret information in a different way than it was intended by the source. Insight into the beliefs consumers hold about the product (or brand) in relation to competing brands in the product category is of particular importance to the firm in relation to (re-)positioning strategies. These insights are obtained through what is commonly referred to as perceptual mapping or image research (e.g. Hauser and Koppelman, 1979).

7.5.3 Evaluation of alternatives

After internal and/or external search for information about the alternatives that may potentially solve the consumption problem, the consumer is hypothesized to engage in an evaluation process. The purpose of this stage is to determine the most suitable alternative that satisfies the need in line with the consumer's want structure. Alternatives are evaluated on a limited number of choice criteria (e.g. taste, health, price, etc.). Each of these criteria has a weight attached to it reflecting its relative importance in the overall evaluative judgements (attitude, preference, image, etc.). Which aspects are considered by the consumer and the relative importance attached to them, depends on the underlying motivations, distinctiveness of the available alternatives, personality and lifestyle, reference groups, culture and the information about and prior experiences with the choice alternatives. Several decision rules (both compensatory and non-compensatory) have been proposed in the consumer behaviour literature about how consumers form overall evaluative judgments (e.g. Bettman, 1979). A popular compensatory model for the integration of beliefs into an overall evaluative judgment is the Fishbein attitude model, which models the overall judgment about the product (attitude toward the object) as a weighted linear combination of a selected number of salient beliefs. Whether or not this attitude toward the object translates into a behavioural intention to buy the product depends on the influence of relevant others, i.e. the social context in which the decision is made. This social norm element in the Fishbein model represents the individual's perception of what other relevant individuals expect from the consumer's behaviour. The extended Fishbein-model has the attitude toward the object and the social norm as its two components. Social norm is modelled as a function of the normative beliefs and the individual's motivation to comply. The Fishbein model is more extensively discussed in the chapter by Shepherd.

7.5.4 Choice and outcomes of choice

Choice is modelled as the resultant of two forces: purchase intention and unanticipated circumstances. These unexpected circumstances can be of many different forms such as changes in the available resources and out-of-stock conditions. After choice, the product will actually be consumed and the outcomes of choice can be assessed. Satisfaction with the choice will act as a reinforcer of the choice process, whereas dissatisfaction will lead to changes in the decision process.

7.5.5 Consumer behaviour under low involvement

The structure of consumer choice behaviour discussed so far assumes that consumers go through an extensive and elaborative process of decision making. The primary strength of such models is that they structure the influential factors in consumer decision making. However, in practice it would be unrealistic to assume that for every single purchase the consumer would go through such a detailed process. Consumers are 'cognitive misers' that attempt to limit the time and effort devoted to consumption behaviour. Involvement and prior experience with the product category are two primary determinants of the extensiveness of the decision making process. Involvement is the degree of personal relevance a product has to the consumer and to a large extent depends on the perceived financial, functional, social and physical risk associated with consumption of the product.

Several authors have claimed that many consumption decisions are of a low involvement nature and that consumers devote very limited cognitive effort to their purchases (e.g. Olshavsky and Granbois, 1979; Hoyer, 1984). The distinction between decision making under low and high involvement is often represented in terms of the beliefs → choice → evaluation hierarchy of effects scheme, rather than the beliefs → evaluation → choice scheme for high involvement decision making. The consumer learns about the product passively, makes a choice and on the basis of experience with the product evaluates whether or not the product will be repurchased. Petty and Cacioppo (1986) use this distinction in their Elaboration Likelihood Model (ELM) for persuasive communication. In high involvement situations, information is processed through the central route in which the content of the message is evaluated extensively in terms of pros and cons of a product or a particular consumption behaviour. Under low involvement situations, the message is processed through the peripheral route, where cognitive elaboration is limited and consumers form attitudes on the basis of simple cues in the message or the context of the message, such as the source authority or the number of time the message is repeated. This illustrates the importance of insight into consumer involvement for marketing communication strategies.

Extensiveness of the decision making process is further influenced by the degree of experience with the product or the product category (Howard, 1989), reflected in the product's position in the product life cycle. During the introduction phase consumers are relatively unfamiliar with the products in the product category. They have insufficient knowledge about which choice criteria to use and how the available choice alternatives would rate on these choice criteria. The consumer decision making strategy at this stage is characterized by an extensive use of available product information and by slow speed. In the product growth phase of the life cycle, most consumers have developed choice criteria on which they evaluate the product, but they are not yet familiar with all available alternatives and how they rate on these choice criteria (e.g. because of new brands introduced into the market place). Decision making at this stage will be of the Limited Problem Solving type. Consumers engage in less detailed information processing (only to get insight into the (new) alternative's ratings on the choice criteria) and speed in the decision making is higher than in Extended Problem Solving. At the maturity stage of the life cycle most consumers have considerable experience with the product category and the available choice items, so that both the evaluative criteria and the product ratings are well established. In such situations, consumers hardly process any information and decision making occurs at a high speed. Consumers have developed relatively stable preferences as well as a so-called evoked set of a limited number of alternatives that they will actively consider in the choice situation. This type of decision making is referred to as Routinized Response Behaviour (Howard, 1989).

7.6 Food choice behaviour in affluent societies

Western food markets have developed into true buyers' markets in which consumers are critical and demanding. High purchasing power and expenditure on foods, in combination with abundant product supplies in terms of quantity and variety, has put the food consumer in a relatively comfortable position. Food consumption has reached high levels, up to a certain point of satiation where physiological needs are concerned. If we assume that consumers strive to maximize utility derived from food consumption, it may be concluded that increases in quantity of consumption hardly contribute to this objective. For many products, utility can more effectively be increased by 'qualitative' rather than 'quantitative' changes in consumption. These qualitative improvements can be achieved through the consumption of food products that better match the idiosyncratic product demands. As a result, psychological motives have come to the forefront as primary motivators of food consumption behaviour.

At the same time, basic quality standards of food supply have increased substantially during the last decade, due to the increased attention for product quality (e.g. Steenkamp, 1989). In modern food markets it is hard to find products that do not meet the basic standard in terms of hygienic and sensory quality. This decreasing horizontal quality differentiation (Abbott, 1955) has considerably reduced the overall quality risk associated with choices from a particular product category. Despite the fact that hygienic and sensory quality are still the most *important* choice criteria that need to be fulfilled, we believe that the availability of a broad product assortment with comparable performance on these criteria is reducing their role as a *decisive* criterion for consumer food choice behaviour. Sensory and hygienic quality are gradually becoming necessary rather than sufficient conditions for market success.

We believe that current market conditions have allowed 'new' choice motives to come to the forefront as determinant food choice criteria. Examples would include 'product-pluses' such as convenience, environmental characteristics, and conspicuous consumption among other factors. We refer to these choice criteria as 'secondary' (*cf.* Van Trijp, 1995) to indicate that in an absolute sense they are probably not as important to the consumer as the attributes at the core of overall product quality (e.g. taste, absence of dangerous substances etc.). However, they are hypothesized to be *determinant* in shaping choice behaviour as long as adequate basic product quality is ensured. Therewith they will become the potential competitive advantages that should guide food marketing strategies of the future. We further believe that contrary to the basic quality criteria, the secondary choice motives are more idiosyncratic in nature. This will result in increased fragmentation of consumer demand in food markets.

Several macro-level developments also exert an influence on food demand, resulting in even more specific food choice criteria. For example, in terms of demographic changes many Western countries witness stagnation in population growth, increasing shares of elderly in the population and smaller household sizes, stimulating the demand for smaller portion sizes among other factors. Intensifying international competition and increasing migration stimulate among other factors the demand for exotic food products. Technological developments have an impact on food marketing through the availability of new technology in preparation (e.g. microwaves), production (e.g. biotechnology), packaging and conservation.

7.7 Implications for food marketing

The fragmentation in consumer demand, combined with the intensifying competition and the high quality standards in product supply has important implications for food marketing. We will discuss several of these implications with reference to relevant stages of the consumer decision process (section 7.5).

7.7.1 *Market segmentation*

In light of the increasing fragmentation in consumer demand, effective market segmentation will become increasingly important, and increasingly difficult. Because of the individualization in life-styles and consumption motivations, traditional market segmentation techniques based on socio-demographic and economic factors fall short. There will be an increased need for market segmentation based on food consumption-specific values and life styles. Due to fragmentation, conformance to the criteria for effective market segmentation, in terms of segment size, measurability, accessibility and actionability will become increasingly important.

7.7.2 *Product differentiation*

Product differentiation is another strategy that will gain in importance in light of the abundant product supplies and freedom of choice to the consumer. High creativity, based on insight into the idiosyncratic product demands in the target group, will be necessary to distinguish the product from the competition and to build a sustainable competitive advantage in the market place. It will become increasingly important to differentiate the product in terms of consumer benefits, rather than product characteristics. More explicitly than before, food products will have to be positioned as value-satisfiers rather than physical entities. Marketing should emphasize the communication of 'what the product does to the customer' in terms of its contribution to the achievement of the customers' values deemed important in life, rather than 'what the product is'. A useful way of thinking about this problem is in terms of means–end chain theory (e.g. Gutman, 1982). Means–end theory formalizes the link between the consumer's concrete demand for product attributes and its contribution to value achievement, in terms of a simple knowledge structure connecting interconnected meanings about attributes, consequences and values. It assumes that product attributes are means to further ends either in terms of the immediate consequences they are believed to lead to (e.g. 'what is it good for in terms of benefits and risks?') and their contribution to value achievement (i.e. 'to what extent does it help me achieve the things I deem important in life?'). Figure 7.3 gives a graphical representation of the basic elements of the means–end chain.

One of the key elements of effective product differentiation will be the identification of those 'secondary' choice motives that are decisive in choice behaviour of the target market and may provide the firm with a sustainable competitive advantage. Insight into attribute importance alone, may be

Product knowledge			Self-knowledge		
concrete attribute >	abstract attributes >	functional consequences >	psychosocial consequences >	instrumental values >	terminal values

Figure 7.3 Basic elements of means–end chains.

insufficient as fulfilment of many of the most important choice criteria have become necessary rather than sufficient conditions for market success. The reader is referred to Myers and Alpert (1977) for a more elaborate discussion of the identification of important and determinant attributes.

The development of a strong brand is one way of establishing product differentiation (e.g. Aaker, 1991). While this strategy is very common in the marketing of processed foods, it is not yet very common at the level of agricultural fresh produce. Brand development as a product differentiation strategy for agricultural fresh produce may be expected to become more important in the future as a means to differentiate product supply and to provide the customer with a basic guarantee for product quality.

Marketing communication will become important as a tool in product differentiation. It should communicate the Unique Selling Points to the customer, taking into account the low levels of involvement. Unless there is a strong recognizable competitive advantage *vis-à-vis* competitive product supply, communication should emphasize the peripheral route of communication. Note that this not only applies to food marketers but equally to nutritional educators who try to influence food consumption into a more desirable direction.

7.7.3 Product development and innovation

Product development and innovation will become increasingly important in the light of the dynamics in the marketing environment. These dynamics include competitive innovation activity, changing consumer demands, availability of new technology, and changing legislation in the food sector. As these dynamics are likely to intensify, product life cycles are likely to shorten even further, making product development and innovation a must to the food company. Also, it is becoming more important that product development and innovation strategies are adjusted to identified needs in the market. More than before, this will require the 'translation' of consumers' abstract choice criteria into tangible specifications which provide the Research and Development department with sufficiently detailed guidance for product development. The ultimate aim of this process is that the engineers develop a new product which carries the tangible characteristics that will deliver the abstract need satisfaction consumers demand. This is a very complex and difficult process (see e.g. Gupta *et al.*, 1985, 1986; McBride, 1990; Van Trijp and Schifferstein, 1995).

The organization of product development activities, in particular the cooperation between marketing and R&D will require more attention in the near future. Quite recently Quality Function Deployment as a technique for formalizing the integration and communication between marketing, R&D, engineering and manufacturing has been introduced in the marketing literature (e.g. Hauser and Clausing, 1988). This technique first introduced by the Japanese in 1972 carefully structures the communication among functional groups about how the abstract consumer demands in the market place can be translated into concrete and tangible characteristics of the physical product and its production process. These structured procedures may be expected to become increasingly important in the near future.

7.7.4 *Marketing organization*

Fierce competition and highly specific customer demands with respect to food, have implications for the organization of marketing activities. In addition to the need for a smooth marketing organization within the food company, there will be an increasing need for coordination of the marketing activities with suppliers of input and clients (wholesalers and/or retailers). This evolution of marketing organization has stimulated *chain marketing*, a joint operation of two or more subsequent companies in the marketing channel, e.g. farmer—food industry—retailer. Also the need for efficient logistical operations and for integrated quality control favour vertical coordination in the food marketing channel. This may either be achieved through *administered* vertical marketing systems, where marketing is coordinated between companies by deliberation and persuasion, by *contractual* vertical marketing systems, where marketing between companies is coordinated by formal contracts, or by *corporate* vertical marketing systems where co-ordination is brought about by vertical integration of companies (Stern and El-Ansary, 1992). At present many food companies already put much effort in the development of relationships with suppliers and clients. It is to be expected that this development will continue and may result in wider adoption of vertical food marketing systems, probably with the food retail companies in a dominant position because of their bargaining power.

7.8 Illustration: consumer oriented product development

In this section we will illustrate some of the basic marketing issues outlined above, in the context of consumer oriented product development. Let us assume that the company wants to achieve growth, and that a particular product category has been chosen to achieve this goal. The marketing manager will now make an in-depth analysis of growth opportunities (Ansoff, 1957). Growth may be achieved through intensifying marketing efforts for the present products in current markets (market penetration), but also by approaching new markets with the present product (market development). An example of the latter case would be the introduction of a high-energy-drink, previously positioned in the professional sports market into the non-professional market. The company may also decide to realize the desired growth through the introduction of a new product, either in an existing market (product development) or a new market (diversification).

Let us assume that the firm opts for new product development and takes a pro-active consumer oriented approach. Then the decision process may be described in five steps (Urban and Hauser, 1993): (i) opportunity identification, (ii) design of the new product, (iii) market testing of the new product, (iv) introduction, and (v) life-cycle management. The opportunity identification stage concerns the definition of the best market (segment) to enter and the generation of new product ideas that could be the basis for entry. The market will be analysed in terms of its structures and component segments to identify viable market opportunities (growing, profitable markets that can be entered) that match the strengths and weaknesses of the innovating company and its initial new product ideas. A large number of market structure models is available (see e.g. DeSarbo *et al.*, 1993) including macro-type analyses on actual purchase behaviour (e.g. scanner and panel data) and perceptual approaches. Based on an evaluation of characteristics (such as growth potential, competitive attractiveness, required investments) of potential markets, the firm will conduct a market profile analysis to identify the most promising product-market combinations. For these markets the firm will search for creative new product ideas that might generate value for customers in these market segments. Sources for new product ideas are diverse and may arise from new technologies, competitive actions, and consumer needs among others. Several techniques are available for the generation and evaluation of new product ideas (see Urban and Hauser, 1993). These initial product ideas for selected target markets are brought into the design phase. At the design phase the purpose is: (i) to identify the key benefits the product is to provide to customers, (ii) the positioning of these benefits versus competitive products, and (iii) the development of the physical product, marketing strategy, and service policy to fulfil the key benefits.

Marketing research plays an important role in the design phase. Qualitative research techniques (e.g. focus groups) are utilized in early stages of the design process to obtain a qualitative signal of how consumers go about purchasing from the product category, what they consider important in their choice behaviour, how they gain information about the product category, what they consider shortcomings in present product supply and so forth. Results from this qualitative research provide insight into some basic issues in the design process, and also serve as input for quantitative consumer research. The purpose of the quantitative analysis is to gather input for more formal quantitative modelling of consumer choice behaviour. Research into consumer perception plays a central role at this stage. Typically consumers will be asked to rate a number of products (potentially including one or more new product ideas) on a number of criteria deemed relevant by the consumer. Basic cognitive dimensions underlying consumer perception may be identified through multivariate techniques such as factor-analysis, discriminant analysis, multi-dimensional scaling and free choice profiling. Relationships between consumer perceptions and physical product features will be identified as these provide guidance for production (Steenkamp and Van Trijp, 1989). Relationships between consumer perceptions and preference will be assessed as a means for setting priorities in terms of product positioning. Market segmentation will be conducted to identify relevant subgroups of consumers that are homogenous in terms of preference formation and choice behaviour, probably justifying separate marketing strategies. Based on these consumer models, 'what-if' forecasts may be made in terms of expected market share for new product options, taking into account expected consumer awareness for the new product as well as expected availability of the new product at the point of sale. These analyses allow the firm an evaluation of the expected profitability of the new product. Depending on the go on/no go decision, the product may be carried into the next phase of product development, be refined or eliminated.

If the prospects for the new product are promising it may be considered for introduction. Marketing efforts for the product will be fine-tuned and specific marketing programs will be developed in terms of product packaging, product name, price setting, choice of outlets through which the product will be distributed, and communication efforts (advertising, promotional activity). In certain situations it may be desirable to carry the fully developed new product into a test-market to make a more formal assessment of expected market success. If all these evaluations yield satisfactory results, the product may actually be introduced into the market. After introduction, the marketing efforts will be adjusted to the stages of the Product Life Cycle the new product is going through.

7.9 Conclusions

This chapter has discussed basics of food marketing from a marketing management perspective. The consumer orientation of this approach implies that the needs, wants and specific demands of potential customers should guide the firm's marketing efforts. We discussed several developments in consumer behaviour in Western food markets as well as their implications to food marketing. To a large extent these developments can be attributed to the fact that most food markets have developed into true buyer's markets. They are characterized by an abundant supply of generally high quality products. As a result consumers have become very demanding in an attempt to satisfy their higher order consumption needs in a very idiosyncratic way. There is little reason to believe that this situation will change in the future. It is more likely that product competition and product assortments will further increase. This situation sets high requirements on the food marketer to whom it will become even more important to understand the consumer and his/her behaviour and to satisfy the idiosyncratic product demands.

References

Aaker, D.A. and Myers, J.G. (1987) *Advertising Management*, 3rd edn, Prentice Hall, Englewood Cliffs, NJ.

Aaker, D.A. (1991) *Managing Brand Equity*, Free Press, New York.

Abbott, L. (1955) *Quality and Competition*, Greenwood Press, Westwood.

Abell, D. (1980) *Defining the Business: The Starting Point of Strategic Planning*, Prentice Hall, Englewood Cliffs, NJ.

Ansoff, H.I. (1957) Strategies for diversification. *Harvard Business Review*, **35** (Sept/Oct), 113–24.

Bartels, R. (1970) *Marketing Theory and Metatheory*, RD Irwin, Homewood, Ill.

Bettman, J.R. (1979) *An Information Processing Theory of Consumer Choice*, Addison-Wesley, Reading, MA.

Blattberg, R.C. and Neslin, S.A. (1993) Sales Promotion Models, in *Marketing, Handbooks in Operations Research and Management Science*, Vol. 5 (eds J. Eliashberg and G.L. Lilien), North Holland, Amsterdam, pp. 553–610.

Booz, Allen and Hamilton (1982) *New Product Management for the 1980's*, Booz, Allen and Hamilton, New York.

Borden, N.H. (1964) The Concept of the Marketing Mix, *Journal of Advertising Research*, **4** (June), 2–7.

Bowersox, D., Closs, D.J. and Helferich, O.K. (1986) *Logistical Management*, 3rd edn, Macmillan Publishing Company, New York.

Chapman, J. and Holtham, C. (1994) *IT in Marketing*, Alfred Waller Ltd. Publishers, Henley on Thames, Oxon.

Churchman, W. (1968) *The Systems Approach*, Dell Publishing Co, New York.

Copeland, M.T. (1923) The relation of consumers' buying habits to marketing methods. *Harvard Business Review* **1** (Apr), 282–89.

DeSarbo, W.S., Manrai, A.K. and Manrai, L.A. (1993) Non-spatial tree models for the assessment of competitive market structure: an integrated review of the marketing and psychometric literature, in *Marketing, Handbooks in Operations Research and Management Science*, Vol. 5 (eds J. Eliashberg and G.L. Lilien), North Holland, Amsterdam, pp. 193–257.

Engel, J.F., Blackwell, R.D. and Miniard, P.W. (1986) *Consumer Behavior*, 5th edn, The Dryden Press, New York.

Fennell, G. (1975) Motivation research revisited, *Journal of Advertising Research*, **15** (3), 23–28.

Fennell, G. (1978) Consumers' perceptions of the product-use situation. *Journal of Marketing*, **42** (2), 38–47.

Gabor, A. and Granger, C.W.J. (1966) Price as an indicator of quality: report on an inquiry. *Economtrica*, **33**, 43–70.

Gupta, A.K., Raj, S.P. and Willemon, D. (1985) R&D and marketing dialogue in high-tech firms. *Industrial Marketing Management*, **14**, 289–300.

Gupta, A.K., Raj, S.P. and Willemon, D. (1986) A model for studying R&D–marketing interface in the product innovation process. *Journal of Marketing*, **50** (2), 7–17.
Gutman, J. (1982) A means-end chain model based on consumer categorization processes. *Journal of Marketing*, **46** (2), 60–72.
Hanna, J.G. (1980) A typology of consumer needs, in *Research in Marketing*, Vol. 3, (ed J.N. Sheth), JAI Press, Greenwich, CT, pp. 83–104.
Hauser, J.R. and Clausing, D. (1988) The house of quality. *Harvard Business Review* **66** (3), 63–73.
Hauser, J.R. and Koppelman, F.S. (1979) Alternative perceptual mapping techniques: relative accuracy and usefulness. *Journal of Marketing Research*, **16** (Nov), 495–506.
Heijbroek, A.M.A., Van Noort, W.M.H. and Van Potten, A.J. (1994), *The retail food market, Structure, trends and strategies*, Rabobank Nederland, Utrecht.
Houston, F.S. and Gassenheimer, J.B. (1987) Marketing and exchange. *Journal of Marketing*, **51** (Oct), 3–18.
Howard, J.A. (1957) *Marketing Management: Analysis and Decision* RD Irwin, Homewood, Ill.
Howard, J.A. (1989) *Consumer Behavior in Marketing Strategy*, Prentice Hall, Englewood Cliffs, NJ.
Howard, J.A. and Sheth, J.N. (1969) *The Theory of Buyer Behavior*, John Wiley, New York.
Hoyer, W.D. (1984) An examination of consumer decision making for a common repeat purchase product. *Journal of Consumer Research* **11**, 822–829.
Jain, S.C. (1993) *Marketing Planning and Strategy*, 4th edn, South-Western Publishing Co, Cincinnati, Ohio.
Kohls, R.L. and Uhl, J.N. (1990) *Marketing of Agricultural Products*, 7th edn, Macmillan Publishing Company, New York.
Kotler, Ph. (1994) *Marketing Management: Analysis, Planning and Control*, 8th edn, Prentice Hall, Englewood Cliffs, NJ.
Lilien, G.L., Kotler, Ph. and Moorthy, K.S. (1992) *Marketing Models*, Prentice Hall, Englewood Cliffs, NJ.
MacFadyan, H.W. (1986) Motivational constructs in psychology, in *Economic Psychology: Intersections in Theory and Application* (eds A.J. MacFadyan and H.W. MacFadyan), Elsevier, Amsterdam, pp. 67–107.
Maslow, A.H. (1954) *Motivation and Personality*, Harper, New York.
McBride, R.L. (1990) Three generations of sensory evaluation, in *Psychological Basis of Sensory Evaluation* (eds R.L. McBride and H.J.H. MacFie), Elsevier Applied Science, London, pp. 195–205.
McCarthy, E.J. (1960) *Basic Marketing: A Managerial Approach*, RD Irwin, Homewood, Ill.
McCarthy, E.J. and Perreault W.D., Jr (1993) *Basic Marketing: A Managerial Approach*, 11th edn. RD. Irwin, Homewood, Ill.
McGuire, W.J. (1976) Some Internal Psychological Factors Influencing Consumer Choice. *Journal of Consumer Research*, **2**, 302–319.
Meulenberg, M.T.G. (ed.) (1993) *Food and Agribusiness Marketing in Europe*, International Business Press, New York.
Myers, J.H. and Alpert, M.I. (1977) Semantic confusion in attitude research: salience vs. importance vs. determinance, in *Advances in Consumer Research*, Vol. 4 (ed W.D. Perrrault), Association for Consumer Research, Altlanta, pp. 106–10.
Olshavsky, R.W. and Granbois, D.H. (1979) Consumer decision making—fact or fiction? *Journal of Consumer Research*, **6**, 93–100.
Petty, R.E. and Cacioppo, J.T. (1986) The elaboration likelihood model of persuasion, in *Advances in Experimental Social Psychology*, Vol. 19 (ed L. Berkowitz), Academic Press, New York, pp. 123–205.
Porter, M.E. (1980) *Competitive Strategy: Techniques for Analyzing Industries and Competitors*, Free Press, New York.
Porter, M.E. (1985) *Competitive Advantage: Creating and Sustaining Superior Performance*. Free Press, New York.
Rossiter, J.R. and Percy, L. (1987) *Advertising and Promotion Management*, McGraw-Hill, New York.
Schiffman, L.G. and Kanuk, L.L. (1991) *Consumer Behavior*, 4th edn, Prentice Hall, Englewood Cliffs, NJ.

Shaw, A. (1912) Some problems in market distribution. *Quarterly Journal of Economics*, **26** (Aug), 706–765.
Sheth, J.N. (1975) *A Psychological Model of Travel Mode Selection*. Bureau of Economic and Business Research of the University of Illinois, working paper #291, Illinois.
Solomon, M.R. (1992) *Consumer Behavior: Buying, Having and Being*, Allyn and Bacon, Boston.
Steenkamp, J.E.B.M. (1989) *Product Quality*, Van Gorcum, Assen, The Netherlands.
Steenkamp, J.E.B.M. and Van Trijp, H.C.M. (1989) Quality guidance: a consumer-based approach for product quality improvement, in *Marketing Thought and Practice in the 1990s*, Vol. 1. (eds G.J. Avlonites, N.K. Papavasiliou and A.G. Kouremenos), European Marketing Academy, Athens, pp. 717–736.
Stern, L.W. and El-Ansary, A.I. (1992) *Marketing Channels*, 4th edn, Prentice Hall, Englewood Cliffs, NJ.
Urban, G.L. and Hauser, J.R. (1993) *Design and Marketing of New Products*, 2nd edn, Prentice Hall, Englewood Cliffs, NJ.
Van Trijp, H.C.M. (1995) *Variety Seeking in Product Choice Behavior: Theory with Applications in the Food Domain*. Mansholt Series 1, Wageningen Agricultural University, Wageningen, The Netherlands.
Van Trijp, H.C.M. and Schifferstein, H.N.J. (1995) Sensory analysis in marketing practice: comparison and integration. *Journal of Sensory Studies*, **10** (2), 127–147.
Walker, O.C., Boyd, H.W. and Larréché, J.C. (1992) *Marketing Strategy, Planning and Implementation*, RD Irwin, Homewood, Ill.
Weld, L.H.D. (1917) Marketing functions and mercantile organization. *American Economic Review*, **7** (June), 306–318.
Wierenga, B. (1974) *An Investigation of Brand Choice Processes*, University Press, Rotterdam.
Zeithaml, V.A., Parasuraman, A. and Betty, L.L. (1990) *Delivering Quality Services*, The Free Press, New York.

8 Economic influences on food choice—non-convenience *versus* convenience food consumption

JENS BONKE

8.1 Introduction

In recent decades women's participation in the labour market has increased significantly, the implication being more pronounced time pressure in most families all over the developed world (Gershuny, 1994; Bonke, 1995a). In the New Household Economics, the value of time affects consumption (Becker, 1965) for which reason households with working wives consume more time-saving goods and services than do households with full-time homemakers, i.e. the households substitute time-saving services for their own time. Therefore, the question is how households have sustained or enlarged their level of economic welfare, when gaining money and losing disposable time.

With reference to the comprehensive literature in this topic, appropriate time-saving and time-buying strategies are discussed, although the focus in the empirical analyses here is on the substitution of non-convenience goods by convenience goods, and the reduction of household production by the buying of full substitutes. In most literature the focus is either on household expenditure on food away from home or on expenditure on fast food, whereas we analyse the preferences for both categories of foods using the same data but attempting to draw conclusions on the various time-saving strategies for households with different characteristics.

The methodology used is that of analysing the correlation between various food regimes, consisting of foods classified by their preparation time-intensity and goods-intensity respectively, and some socio-economic characteristics of the household. Thus, food regimes are defined by classifying the households according to their expenditure on these foodstuffs. Although most studies find that disposable time and the employment of wives explain expenditure on food-away-from home (McCracken and Brandt, 1987; Soberon-Ferrer and Dardis, 1991), the results are influenced by the methods employed. For instance, multiple regression analysis based on Consumer Expenditure Survey Data with many households with no expenditures on the food categories eating out and fast food causes a sample selection bias to appear, which calls for correction. For this reason we perform estimations by probit procedures to find the probability of being an 'eating outer' and 'fast fooder' respectively—i.e. these terms will be used to refer to people spending money on these goods and services—followed by multiple regressions on expenditures on different food categories by 'eating outers' and 'non-eating outers', and by 'fast fooders' and 'non-fast fooders'.

Finally, the question is raised of whether, contrary to expectations, rich and busy households use more time in meal preparation, because this activity is perceived as a 'luxury good' on which money and time are spent simultaneously.

8.2 Time-saving and time-buying strategies

There are several strategies and combinations of strategies to intensify time use in household production, maintaining constant total quantity and quality of this production. Here the focus is on meal preparation, which is the single most time-consuming household activity (Bonke, 1995b) and the best covered in the literature on household behaviour.

The first strategy is the possible *substitution of women's non-market labour by capital equipment*. According to Mincer (1960) two-income families spend more money on these appliances than do one-income families, because durable goods are investments, and the income earned by the wife is considered transitory as opposed to that of the man, which is 'permanent' income. If, however, the marginal propensity to save out of transitory income is higher than the propensity to save out of permanent income, households spend more on durable goods, Mincer assumes. Another explanation of the relationship between the labour supplied by the wife and the consumption in the family is to perceive women's income as the one that fills the gap between the need for consumption and the income in the life-cycle. In this way the income earned by the wife is dependent on the husband's income (Strober, 1977). However, by multiple regressions Strober shows that for different life-cycle groups, one-income families have the same investment in durable goods relative to income as two-income families; the same is shown by Anderson (1971), Foster *et al.* (1981), Strober and Weinberg (1977, 1980), and Weinberg and Winer (1983) by adjusting for various factors.

Another hypothesis has appeared, namely that of durables and wives' home-time being complements in household production, which means that expenditure on durables decreases as wives' labour supply rises. Thus Galbraith (1973) claims that durables and women's time at home are complements rather than substitutes; Bryant (1988) gives evidence of this by estimating a negative effect of wives' employment on durables expenditure for which reason he concludes that the time of the housekeeper and the household appliances tend to constitute complementary input in household production. The findings by Redman (1980) might also confirm this by showing a negative correlation between the buying of convenience goods and durables, the former time-saving strategy being chosen by double-career families.

The last hypothesis seems more likely now because of the increased labour market participation rates for women and their more permanent status as workers, while the first hypothesis might have been appropriate in earlier times.

Nonetheless, Reilly (1982) argues that there might be some correlation between labour supply and expenditure on durables, because women doing a lot of overtime—termed role-overload—possess more time-saving durables than women doing less overtime. Apparently, the time constraints of dual-earner households with full-time employed wives provided the need and their higher incomes provided the necessary means to purchase technology, according to Oropesa (1993).

Another strategy is that of *substituting convenience goods and services for non-convenience goods* reducing the household's own non-market labour. For many double-career families the opportunity cost of time is too high to prepare everyday meals themselves, which is why the strategies of buying durables and/or convenience goods become insufficient. They buy services in the shape of meals-away-from home to save time in meal preparation. Nonetheless, Douglas (1976) finds no correlation between the employment rate of the wife and the buying of convenience goods. One-income families buy as many of these goods as do two-income families. What is significant is the wife's level of education (see also Redman (1980)), the family income and the husband's occupation, which increase the expenditures for convenience services, i.e. meals-away-from home. However, Waldman and Jacobs (1978), Vickery (1979) and Bonke (1988) find positive correlations between the demand for time-saving services, meals-away-from home, and the employment rate of women—adjusted for different incomes, i.e. double-career families may earn more than single-career families—and so does Yen (1993) applying a statistical model—Double Hurdle—which among others takes into account the likelihood of buying meals-away-from home, and the expenditures on these foods. Thus households with working wives and those with higher incomes are more likely to consume food-away-from home, and of those spending money on these foodstuffs employment and higher incomes mean more expenditures, too.

The relationship between income and the purchase of time-saving services is confirmed by Bellante and Foster (1984), and Prochasha and Schrimper (1973), who demonstrate a positive correlation between the opportunity cost of time and the expenditure for food-away-from home. In Bonke (1992a) the income-elasticities concerning unit expenditure on fast food and meals out are found to be positive, although smaller in Sweden than in Denmark, which means that spending on the two categories of foodstuffs are affected much more by income in Denmark than in Sweden. Another finding for Denmark is that the elasticity of income for convenience food is the largest in households with full-time working wives, i.e. larger than one, which means that expenditure for food-away-from home constitutes an increasing portion of the budget when the income increases (Bonke, 1988). However, these studies apply only multiple regressions.

Finally, Bellante and Foster (1984) found that in families with small children there seems to be a lower demand for meals-away-from home, but, at the same time, they demand other time-saving strategies such as professional child care and hired help.

Formerly, *substituting the labour of paid help for one's own non-market labour* was a widespread time-buying strategy. The increasing employment rate for women partly created by—and in itself creating—a rising income level means, however, that the 'wage-exchange rate' is not favourable enough to substitute the wife's household work by paid household help. The higher the financial equality, the less hired help, it appears. Gershuny (1983) illustrates this by showing that for decades, only wives in working-class families reduced their household working hours, while wives in middle-class families increased the time they spent on housekeeping. The reason is that the latter category could no longer afford to hire help, as their opportunity cost of time increased.

The last strategy which has to be mentioned entails that the *employed wife decreases the time spent in household work, and, at the same time, becomes more efficient.* The first phenomenon is confirmed in various studies of time use in household production (see Robinson, 1977; Walker and Woods, 1976; Bonke, 1989; and Körmendi, 1990) showing an immense decrease in women's household work which is not counterbalanced by an equivalent increase in men's contribution, although there has been some trade off. On the other hand, the question of efficiency is much more difficult to evaluate, and consequently there seems to be no empirical research analysing the correlation between quality and quantity of household production, and the use of time. Nonetheless, Vanek (1974), in her much quoted article 'Time spent in housework', claims that Parkinson's law is in operation, which means that time pressure increases productivity, as does the wife's role as a good mother and housekeeper.

Many of the strategies mentioned here may be combined in the individual household, and therefore an attempt will be made to analyse the relationship between them—something which is rarely done in the literature.

Furthermore, we question whether a household's choice of foods depends on its economic resources, time and money, or whether cultural background characteristics are more likely to explain the behaviour.

8.3 Methodology—data and variables

In economic terms preparation of food is a combination of use of time and food expenditure. Assuming that the correlation between the rate of processing and the price is positive, and the processing in the industry is negatively correlated to the level of preparation in the household, the price and preparation time will become substitutes. Although there may be examples of convenience foodstuffs which are cheaper than provisions bought for preparation in the households—also for foods of equal quality—the empirical analysis here assumes a negative correlation between price and preparation time. The consequence is that either the household is time-intensive or goods-intensive, or something in between in order to attain a certain quantity and/or quality of the meal. The household might also be intensive in the use of both resources, or extensive, thereby gaining another level of satisfaction. Or the household runs several food regimes—i.e. combinations of time use and expenditures—simultaneously, which seems to be the most likely.

In Table 8.1 the households in food regimes A and P are similar in their level of consumption but vary in the allocation of time and money. In pattern A the household chooses a time-intensive household production, as opposed to that in pattern P, where the household chooses a goods-intensive home production, in agreement with Becker's (1965) terminology. This means that the household in A substitutes expenditure for time, whereas the household saves time by paying more money on foods in pattern P. The households in the other groups lie between these extremes.

For the empirical analysis, data from the Expenditure Survey of the year 1987 (Danmarks Statistik, 1992) as well as data from the Time Budget Survey of the same year (Danish Institute of Social Research, see Andersen, 1987) were used.

The study sample included couples in which the husband worked full-time, 35 hours or more per week, for at least 48 weeks of the survey year. Single men also worked full-time, while single women worked full-time or part-time, which means 15 hours or more per week, for at least 48 weeks of

Table 8.1 Food regimes

	Expenditures			
Time	Very small	Small	Large	Very large
Very much	A	B	C	D
Much	E	F	G	H
Little	I	J	K	L
Very little	M	N	O	P

the survey year. Families with more than two adults were excluded, because of the potential for more than one food regime in the same family. These adjustments resulted in a study sample of 1395 families.

The diary statistics include various activities and the Expenditure Survey by Danmarks Statistik include data on expenditure for different foods and services, also expenditure for eating out. Consequently the same statistical survey does not allow a direct study of the connection between time use and expenditure on foods.

However, the expenditure survey renders possible a normative categorization of different foods, where the level of preparation is one of the issues. This means that one has to establish whether various kinds of foods are manufactured in a way which makes them ready for final consumption, and thereby more or less time-consuming for the household. Another issue is the price of the foods, which implies a separation between more or less expensive food. Consequently, every kind of food has to be defined by the level of expenditure-intensiveness and the level of time-intensiveness respectively (see Table 8.2).

By convenience foods we understand items which embody multiple ingredients, provide high levels of time saving and/or energy inputs, and have culinary expertise built in, according to the Capps *et al.* (1983) definition of complex convenience foods. Together with some manufactured convenience foods with no home-prepared counterparts, such items are what is ordinarily thought of as fast- or convenience foods. Lindvall (1989) separates foods into essential and non-essential foods, the argument being that foods which are part of ordinary and everyday activities are more sensitive to economic factors than other foods. The lack of time-intensive substitutes for other items is also a reason Lindvall excluded them from the analyses.

Table 8.2 Classification of foods

Non-convenience	Semi-convenience	Convenience
Preparation meals	*Semi-preparation meals*	*Fast meals*
Non preserved meat and fish	Panned fillet of fish, etc.	Complete meals, soups
Preserved meat and fish		
Potatoes, rice, pasta	Processed potatoes	
Vegetables	Preserved vegetables, etc.	
Dessert ingredients	Preserved fruit	Fast dessert
Cooking ingredients	*Cooking ingredients*	
Fats	Ready made sauce, etc.	
Spices		
Others		
		Meals out
		Bar
		Restaurant

This means that the main meal is separated by the level of convenience, i.e. preparation foods, semi-preparation foods or fast foods, and at the same time every meal is assumed to consist of two parts, a meal-base and a meal-complement. This distinction, made by Lindvall, ensures that the household can choose a dinner where part of the food is time-intensive, while other parts are goods-intensive. The categorization of the foodstuffs here is slightly different to Lindvall's, the current one being carried out according to Danish eating traditions (Bonke, 1992a).

Food-away-from home is considered the most convenient way of eating, because the meal is prepared by others, and transportation and the time spent on eating are not perceived as productive activities (Bonke, 1995b).

The relative size of expenditures for a certain group will be taken as a measure of the preference for that particular category. We also measure the significance of each group in absolute terms as the expenditure in the relevant currency. By employing an equivalence scale we have a *unit* expenditure measure—the cost per consumer—which renders food expenditures for different households comparable.

The expenditures for different food-categories are shown in Table 8.3.

The *income* concept has to include public transfers and exclude taxes, i.e. disposable income, in order to measure the opportunity to buy goods, services etc. However, the disposable income is not always equal to the actual consumption. Therefore, the total expenditure is preferred as a

Table 8.3 Expenditure distribution on different food categories ($N = 1395$)

	Percentage		Unit expenditure			
	Expenditure all households				Per expenditure household[a]	
Category	Mean DKK/ household	RSD[b]	Mean DKK/ household	RSD[b]	Mean	RSD[b]
Convenience foods						
meals out	7	2	889	2	1581	1
fast meals	3	2	382	3	643	2
Semi-convenience foods						
semi-prepared meals	3	1	367	1	427	1
Non-convenience foods						
prepared meals	24	<1	3029	1	3120	1
cooking ingredients	8	1	851	1	897	1
All food[c]	100		11988	<1	11988	<1

[a] Households having non-zero expenditures in the respective category. [b] Relative Standard Deviations (SD/Mean). [c] Including foodstuffs for breakfast and lunch, and snacks.

measure of comprehensive consumption, i.e. permanent income (Wagner and Hanna, 1983), and here used in the statistical analysis under the designation disposable income. Referring to variable income the expenditure for housing is subtracted from the total expenditure (Homan, 1988), which because of the privilege of deducting interest on loans on income tax returns make comparisons between owners and tenants possible.

Similar to the correction of food expenditure for different household compositions, an adjustment is also made to the disposable income by an estimated equivalent scale.

In order to find the *time* disposable for household work and leisure activities, daily personal care including sleep has to be subtracted from the 24 hours. The time for child-minding is not disposable in this context. Some tasks have to be done, which means that children require committed time (Ås, 1982). Working hours—i.e. contracted time—and commuting time are also excluded, leaving the disposable time required here. Nonetheless, some economists argue that the decisions of supplying time to the labour market and household work are taken jointly, and even that of child care (Gustafsson, 1991) and sleep (Biddle and Hamermesh, 1989), the implications of which are simultaneous equation bias for which reason predicted labour hours, time for child care and sleep should be employed. The main argument for not doing this here is that Danish women encouraged by the highest coverage of public-financed child care provisions in Western Europe (Borchorst, 1993) seem to be as labour market oriented as Danish men, and that the trade-off found by Biddle and Hamermesh, though significant, is very small; thus the so-called time-puzzle in the families concentrates mostly on the daily household work.

In the Expenditure Survey we find no information on time-use for which reason the coefficients found in the regression analyses of the Time Budget Survey—see Table 8.4—are applied to the household characteristics in the Expenditure Survey to simulate time budgets, i.e. disposable time. Because the two surveys are different—the first one is at the household level and the second one is an individual test sample—'synthetic households' were generated by the Time Budget Survey's using information on the spouse. A comparison of mean disposable time generated by the two surveys shows only small differences, however, mainly due to a higher relative number of preschool children in the Time Budget Survey compared to the findings in the Expenditure Survey. Nonetheless, creating the disposable time in this way causes a variable with measurements errors, which may explain the low levels of significance in the regressions including this variable.

The food consumption level is not merely determined by the amount of time and money spent, but also by the *productivity* in the food-processing, which means that some measures of productivity must be incorporated into our analyses.

Of course, a modern household possesses a minimum of household appli-

Table 8.4 Disposable time coefficients in regression analyses[a]

Variable	Men ($N = 682$)		Women ($N = 772$)	
	Mean		Mean	
Dependent variable[d]				
DISPTIME	406.7		568.8	
	B	RSD[c]	B	RSD[c]
Independent variables[d]				
Constant	442.9[e]	0.04	794.3[e]	0.02
COUPLES	−21.9	0.90	29.3	0.58
CHILD	16.7	0.77	25.0[g]	0.47
CHILDPR	−56.0	0.52	−99.8[e]	0.26
CHILDSC	−23.8	1.16	−33.2	0.77
PARTTIME[b]	−10.9	1.63	−229.4[e]	0.09
FULLTIME[b]			−312.6[e]	0.06
EDUM/EDUW	−7.1	0.61	−13.1[g]	0.39
R^2	0.02		0.34	
$adj. R^2$	0.01		0.33	

[a] Time budget survey, see Andersen (1987). [b] Men work full-time in this sample, therefore part-time and full-time refer to women. [c] Relative Standard Deviations (SD/Mean). [d] Full definitions of the variables are given in Table 8.5. [e,f,g] Pearson's correlation coefficients significant on 0.001, 0.01 and 0.05 levels respectively.

ances for food-processing, with some appliances which are not necessary, but function as time-saving investments. Freezers, dishwashers, and microwave ovens may increase productivity in household production, i.e. allowing a meal of an equivalent quality to be cooked in a more convenient way, which means that households either work fewer hours to produce the same amount of goods, or they work as many hours as before and therefore produce more goods. Which of the strategies is chosen is of course an open-ended question.

A positive correlation between productivity and the home maker's educational level is anticipated. For couples this means that highly educated women use the modern technology more effectively, and choose more time-saving strategies, compared to women with a lower education. Nevertheless, education might give preference to meal preparation too, according to a life-style, of which home made foods—i.e. meal preparation—is one of the characteristics, and child caring another (Bonke, 1988).

Furthermore, it is assumed that the experience obtained by learning by doing increases with the age of the housekeeper, which gives a higher productivity. The age, however, also measures the effect of birth cohorts in cross-sectional samples like the one applied here for which reason the interpretations become more complex.

Finally, the number of household members affects productivity in food-processing. At the same time there are economies of size and joint produc-

Table 8.5 Resources in the household—means and standard deviations of independent variables[a]

Variable	Definition	Means	RSD[b]
DISPINC	Total expenditure	175 184	0.5
VARINC	*DISPINC* – housing costs	118 396	0.5
UNITINC	*DISPINC/UNITC*	95 114	0.4
DISPTIME	1440(2880) minutes – (personal care + working time incl. transportation + childcare time)	825	0.26
EDUTIME	Years of formal education	2.5	0.47
AGE	Age of adult, the wife in couples	38	0.33
UNITC	Number of members in the household	1.83	0.30
COUPLES	Couples/all households	0.69	0.68
CHILD	Number of children (0, 1, 2, 3, 4+)	0.84	1.20
CHILDPR	Youngest child of preschool age <7 years old/all households	0.23	1.83
CHILDSC	Youngest child of school age 7–17 years old/all households	0.25	1.78
APPL	Households with dishwashers and/or microwave oven/all households	0.38	1.28
DOMSERV	Expenditures for domestic servant (>0/0)	0.06	3.86
URBAN	Size of the biggest town in the district; 1–4, growing	2.56	0.48
OWNREN	Owners of houses/all households	0.61	0.80

[a] For more explicit explanations of the variables, see Bonke (1992a). [b] Relative Standard Deviations (SD/Mean).

tion, i.e. child-minding and meal-preparation, which means that fewer resources are required to produce one portion of a meal.

In Table 8.5 we find the means and standard deviations of the resource variables in this chapter.

8.4 Food regimes

By ranking the households according to increasing disposable unit income, we designate the first quartile as very poor households, and the last quartile as very rich households. The disposable time is categorized correspondingly, where very busy households belong to the lowest time-quartile and very idle households the last quartile. In this way we create 16 categories of household, which seem to be distributed very evenly, although the time resource is not in general distributed as evenly as the income resource (Bonke, 1992a).

Considering only main meal foodstuffs, which cover nearly half the total expenditures (Bonke, 1992a), the unit expenditures on non-convenience foods are not lower the more money and less time the households have, while expenditures on semi-convenience foods and convenience foods—fast food and meals out—are distributed between the household categories in an economically rational way. The same applies for the percentage distribution of expenditures, where no statistically significant differences between the expenditures by household categories compared par-wise are found either.

Poor and idle household do not completely match the above pattern, because they spend the largest amount of money on non-convenience foods and also relatively more money on meals out, see Table 8.6. The reason might be socio-economic as the age of adults (the wife in couples) is low, and the ownership of appliances and houses is high in this category of households.

Nevertheless, the finding that very rich and very busy households spend four to five times as much money on meals out than do very poor and very idle households (Table 8.6)—grouping the former under food regime P and the latter under food regime A—seems to confirm the hypothesis that households do to some extent allocate their time and money resources rationally when buying these kinds of convenience foodstuffs.

8.5 Statistical model

In the data analysed there are many households without expenditures on fast food and particularly without expenditures on meals-away-from home. One of the reasons is the way of reporting the expenditures in family budget surveys (Eurostat, 1989, 1991). During a rather short period of registration (two-weeks in the Danish survey) some households may not purchase any meals-away-from home, and thus are not defined as eating-outers, although they sometimes spend money on these kinds of meals, and the same for expenditure on fast food, which might not be spent in every fortnight's period of time. Nevertheless, we believe that it is appropriate to distinguish between different groups, and the most simple and practical way to do this is by separating eating outers from non-eating outers, and fast fooders from non-fast fooders, by their registered expenditures (>0 versus =0) on these food categories, and subsequently analysing the behaviour of the households in these different food-categories.

A two-step model is specified to estimate the expenditure on eating out and fast food respectively; see Lee and Brown (1986) and Bonke (1993). In the first step the probabilities of eating out and of buying fast food are estimated, and in the second step we estimate the expenditures on these food categories for those households reporting such expenditures. This procedure is due to the fact that the classified groups may be distributed

Table 8.6 Unit expenditures on foodstuffs in different households

Household categories[a]	N	Non-convenient	Semi-convenient	Convenience		
				All	Fast meals	Meals out
			Unit expenditure relative to all households' unit expenditure/(RSD[b])			
Very rich and very busy	81	100 (0.89)	124 (1.24)	196 (1.57)	143 (1.89)	219 (1.96)
Rich and busy	90	97 (0.59)	101 (0.86)	78 (1.24)	97 (1.49)	70 (1.77)
Poor and idle	82	104 (0.71)	86 (0.92)	101 (1.37)	95 (1.51)	104 (1.82)
Very poor and very idle	96	73 (0.77)	80 (1.04)	41 (2.05)	44 (1.79)	40 (2.84)
All households	1395	100 (0.80)	100 (1.23)	100 (1.72)	100 (2.86)	100 (2.10)
			Percentage			
Very rich and very busy	81	19.4 (0.63)	3.2 (1.25)	16.7 (1.04)	3.9 (2.03)	12.8 (1.29)
Rich and busy	90	22.9 (0.41)	3.1 (0.78)	7.7 (1.11)	2.6 (1.20)	5.1 (1.65)
Poor and idle	82	25.7 (0.49)	2.7 (0.86)	10.6 (1.30)	2.8 (1.35)	7.8 (1.73)
Very poor and very idle	96	23.7 (0.51)	3.4 (0.94)	5.0 (1.65)	1.8 (1.69)	3.3 (2.34)

[a] The categories correspond to the food-regimes P, K, F, A in Table 8.1. [b] Relative Standard Deviations (SD/Mean).

differently on the relevant variables giving rise to a sample selection bias. Therefore the first estimation yields a correction term, λ, which is used subsequently as an independent variable in the regressions of expenditure on meals out and on fast food.

Expenditure is divided into away-from-home expenditure (*AFH*) and home-food expenditure (*HF*)—and the same is done for fast food expenditures—with the vectors X_1 and X_2 as explanatory variables, and a Z-vector of variables explaining the decision to eat out (*EATOUT**). The sets of explanatory variables are generally assumed to be different between most of these equations, which constitute the models

$$AFH = \beta_1' X_1 + \upsilon_1 \tag{8.1}$$

$$AH = \beta_2' X_2 + \upsilon_2 \tag{8.2}$$

$$\text{if } EATOUT^* = \gamma' Z - \varepsilon = 0,$$

and likewise

$$AFH = 0 \tag{8.3}$$

$$AH = \beta_3' X_2 + \upsilon_3, \tag{8.4}$$

where γ', β_1', β_2' and β_3', are parameter vectors, and ε, υ_1, υ_2 and υ_3 are error terms.

However, *EATOUT** is an unobserved variable, which might be approximated by the dummy variable *EATOUT*, i.e. $EATOUT = 1$ if $EATOUT^* \geq 0$ and $EATOUT = 0$ if $EATOUT^* < 0$.

If the residuals υ_1, υ_2, υ_3 and ε have multivariate normal distributions, with mean vector 0 and a specified co-variance matrix, a two-stage estimation method, i.e. a switching simultaneous equation model (Lee *et al.*, 1980), can be used to estimate equations (8.1), (8.2), and (8.4).

In the first stage the γ-values are estimated by probit maximum likelihood procedures, and the marginal effects are displayed as the coefficients multiplied by the mean value of the density function, which may be interpreted as the partial effect on the dependent variable of a unit change in the average value of the independent variable, as is the case of OLS-coefficients. The marginal effects of the variables *LOGVINC* and *LOGTIME* may be written as $(\delta\text{probit}/\delta(VARINC))^*VARINC$ and $(\delta\text{probit}/\delta(DISPTIME))^*DISPTIME$, where probit is the probability of eating out and buying fast food respectively.

Secondly, equations (8.1), (8.2) and (8.4) are estimated by ordinary least squares (OLS) with the addition of a variable $\hat{}\phi_i/\hat{}\Phi_i$ in equation (8.1) and (8.2), and in (8.4) as $\hat{}\phi_i/(1-\hat{}\Phi_i)$, i.e. $\hat{}\phi$ is the standard normal density function and $\hat{}\Phi$ the cumulative normal distribution function, where both are evaluated at $\hat{}\gamma' Z_i$. This means that equation (8.1) may be reformulated as

$$AFH_i = \beta_i' X_i - \sigma_{1\varepsilon} \hat{\ } \phi_i / \hat{\ } \Phi_i + \eta_{1i} \tag{8.5}$$

and equation (8.2) as

$$AH_i = \beta_3' X_i - \sigma_3 \hat{\ } \phi_i / \left(1 - \hat{\ } \Phi_i\right) + \eta_{3i} \tag{8.6}$$

and estimated by OLS, where η_{1i} and η_{3i} are error terms with zero conditional means; see Lee and Brown (1986) and Lee *et al.* (1980).

The reason for dividing the data into two sets—eating outers/non-eating outers and fast fooders/non-fast fooders—and for estimating the probability of eating out and fast food separately, is the relatively small correlations between these expenditures.

8.6 Estimation results

The probability of eating out is estimated by different models (Table 8.7), and the coefficients of the probability of eating out and the calculated marginal effects are shown. Surprisingly, increasing unit income (*LOGVINC*) gives rise to fewer eating outers, i.e. an income increase of DKK 10000[a] decreases the probability of eating out by 6–7%. And even more surprisingly, eating outers' expenditure declines with increasing income (see Table 8.9). These findings are the opposite to those of McCracken and Brandt (1987); the explanation might be found in the frequency of eating outers buying food away from home, which is possibly lower for high income groups than for low income groups. Thus, high income Danish households may choose expensive eating places when they do go out—i.e. confirming McCracken and Brandt's relationship between income and type of food facility chosen—and buy more expensive meals.

Table 8.7 also shows that more disposable time increases the probability of eating out by nearly 1% of every marginal hour available per day (Marginal effect divided by average disposable time). Looking at the variables determining disposable time, see model B in Table 8.7, we find women's part-time jobs (*PARTTIME*) increase the probability of eating out. Calculated as frequencies, we find that 39% of families with full time homemakers are eating outers, whereas the corresponding figures for families with part-time working wives and double-career families (*FULLTIME*) are 49% and 43%, respectively.

For singles without children, corresponding couples and couples with pre-school children the eating outer frequencies are calculated to about 0.45, while it is only 0.36 in families with school children. However, the life-cycle effect is more marked in reality, because of the positive correlation

[a] DKK 10000 = UK £1165 = US$1800 (1995 exchange rates).

Table 8.7 The decision of eating out—probit estimations[a] (N = 1247)

Independent variables[b]	Model A	Marginal effect	Model B	Marginal effect	Model C	Marginal effect
CONSTANT	9.954[c] (0.214)		11.949[c] (0.139)		7.884[c] (0.227)	
LOGVINC	–1.107[c] (0.134)	–0.476	–1.127[c] (0.134)	–0.478	–0.916[c] (0.170)	–0.382
LOGTIME	0.289 (0.668)	(0.124)	"	"	"	"
AGE	"	"	"	"	0.031[c] (0.167)	0.013
UNITC	"	"	"	"	(0.162) (0.952)	(0.068)
COUPLES	"	"	0.072 (3.009)	(0.031)	"	"
CHILD	"	"	0.080 (1.441)	(0.034)	"	"
CHILDPR	"	"	–0.067 (3.797)	(–0.028)	"	"
CHILDSC	"	"	–0.445 (0.540)	(–0.189)	"	"
PARTTIME	"	"	0.422 (0.523)	(0.179)	"	"
FULLTIME	"	"	0.205 (0.958)	(0.086)	"	"
EDUTIME	"	"	–0.010 (4.836)	(–0.004)	–0.021 (2.411)	(–0.009)
APPL	"	"	"	"	–0.175 (0.781)	(–0.073)
DOMSERV	"	"	"	"	–0.362 (0.698)	(–0.151)
URBAN	"	"	"	"	0.166[d] (0.318)	0.069
OWNREN	"	"	"	"	0.129 (1.264)	(0.054)
$-2\log L$	1640.729[c]		1628.735[c]		1587.076[c]	
Concordant pairs	61.9%		63.5%		66.3%	

[a] Asymptotic relative standard deviation (SD/Mean) in parentheses. Marginal effects are calculated by $\delta/\delta x_j \text{prob}(eatout = 1/X)$, and numbers in parentheses indicate insignificant coefficients. [b] Full definitions of the variables are given in Table 8.5. [c,d,e] Pearson's correlation coefficients significant on 0.001, 0.01 and 0.05 levels respectively.

Table 8.8 Expenditures on foodstuffs in eating out, non-eating out, fast food and non-fast food households—means and standard deviations

	Means DKK/household	RSD[a]	Means DKK/household	RSD[a]
	Eating outers		*Non-eating outers*	
Non-convenience	2924	1	3163	1
Semi-convenience	374	1	358	1
Convenience				
fast meals	364	2	405	4
food away from home	1581	1		
	Fast fooders		*Non-fast fooders*	
Non-convenience	3389	1	2500	1
Semi-convenience	414	1	299	1
Convenience				
fast meals	643	2		
food away from home	906	2	865	2

[a] Relative Standard Deviations (SD/Mean).

between the independent variables pre-school children and part-time working mothers, the implication of which is an eating outer frequency calculated to 0.51.

Not only does the probability of eating out decrease when the family includes school children (Table 8.7), but expenditure on this type of meal is also lower for eating outers. There is a difference of nearly DKK 500[b] in the spending on meals out between families with school children and families without school children (Table 8.9). The number of children also counts negatively in this respect, the figures show.

Although a negative correlation between the housekeeper's age (the wife in couples) being a proxy of experience in household production—and cohorts—and expenditure on convenience food (fast food and eating out) was to be expected, we found a positive coefficient between the probability of eating out and the housekeeper's age (Table 8.7), and among eating outers the expenditure on meals away from home increases with the age of the housekeeper (Table 8.10). However, as the sample is cross-sectional the results refer to generational differences not to the effect of age/experience inside a sudden generation.

The productivity variables technology (*APPL*) and domestic servants (*DOMSERV*) both reduce the probability of eating out, as well as expenditures on this kind of meal. Thus, households behave rationally in performing their potential capacities on eating out (Tables 8.7 and 8.10).

Finally, we find that there are co-variances between the decision to eat

[b] DKK 500 = UK £58 = US$90 (1995 exchange rates).

Table 8.9 The expenditures on different kind of foodstuffs for eating outers and non-eating outers[a]—socio-economic variables, 1987

	Eating outers ($N = 707$)				Non-eating outers ($N = 538$)		
Independent variables[b,c]	Non-convenience	Semi-convenience	Convenience (fast food)	Convenience (eat out)	Non-convenience	Semi-convenience	Convenience (fast food)
CONSTANT	−53 366.0[e] (0.4)	−3 266.1 (1.0)	−78.3 (73.0)	27 126.0 (0.6)	−50 690.0 (0.6)	−1 455.5 (3.4)	16 233.2[d] (0.6)
LOGVINC	5 985.5[e] (0.4)	351.0 (1.0)	38.8 (16.5)	−3 122.0 (0.6)	4 581.9[f] (0.5)	168.8 (2.3)	−1 210.1 (0.6)
COUPLES	−784.7 (0.73)	54.7 (1.8)	−140.5 (1.2)	731.9 (0.7)	−86.9 (7.1)	−60.7 (1.7)	72.2 (2.9)
CHILD	−123.7 (1.4)	−11.0 (2.7)	−87.4 (0.6)	−299.1 (0.5)	−227.5 (0.9)	−50.5 (0.7)	−1.9 (34.0)
CHILDPR	221.1 (1.7)	18.4 (3.6)	131.4 (0.9)	−270.1 (1.3)	56.7 (7.8)	154.3[f] (0.5)	−146.0 (1.0)
CHILDSC	238.5 (1.5)	105.4 (0.6)	82.7 (1.3)	−472.3 (0.7)	6.0 (68.7)	69.6 (1.0)	−169.4 (0.8)
PARTTIME	551.3 (0.7)	−9.9 (6.4)	−30.6 (3.6)	−311.1 (1.1)	−282.9 (1.4)	40.9 (1.7)	−144.9 (0.9)
FULLTIME	473.8 (0.7)	−13.7 (4.3)	25.6 (4.1)	−396.1 (0.8)	−42.9 (8.7)	60.3 (1.1)	−147.8 (0.9)
λ	−1 012.2[f] (0.5)	−38.4 (2.0)	15.0 (9.1)	956.9[f] (0.4)	608.7 (0.9)	−13.9 (6.6)	−344.2 (0.5)
$adj.R^2$	0.05	0.04	0.01	0.12	0.11	0.07	0.04

[a] The probit estimation determining the λ-coefficient is estimated by model A, see Table 8.7. [b] The variables also inclue EDUTIME. [c] Full definitions of the variables are given in Table 8.5. Parentheses denote Relative Standard Deviations (SD/Mean). [d,e,f] Pearson's correlation coefficients significant on 0.001, 0.01 and 0.05 levels respectively.

Table 8.10 The expenditures on different kind of foodstuffs for eating outers and non-eating outers[a]—productivity variables, 1987

	Eating outers ($N = 707$)				Non-eating outers ($N = 538$)		
Independent variables[b,c]	Non-convenience	Semi-convenience	Convenience (fast food)	Convenience (eat out)	Non-convenience	Semi-convenience	Convenience (fast food)
CONSTANT	32191.0[f]	−4163.0	9294.7[f]	9553.3	1501.3	80.7	3162.1
	(0.4)	(0.6)	(0.5)	(1.4)	(11.7)	(37.7)	(1.9)
LOGVINC	4299.4[f]	536.0	−1388.3[f]	−1889.2	191.0	31.2	−210.4
	(0.5)	(0.7)	(0.4)	(1.0)	(8.3)	(8.8)	(2.6)
AGE	−68.5	−10.3	50.9[f]	102.6	76.4	5.8	13.9
	(1.1)	(1.2)	(0.4)	(0.6)	(0.7)	(1.7)	(1.4)
EDUTIME	48.1	4.2	−32.6	−70.7	−122.6	29.4	−5.1
	(1.8)	(3.6)	(0.8)	(1.2)	(0.7)	(0.5)	(5.9)
APPL	443.6	18.0	−281.6[f]	−700.6	57.6	41.8	38.8
	(1.0)	(4.2)	(0.5)	(0.6)	(6.7)	(1.6)	(3.4)
DOMSERV	1115.1	68.2	−695.7[f]	−877.6	−1633.9[f]	−144.7	−241.9
	(0.8)	(2.3)	(0.4)	(1.0)	(0.4)	(0.9)	(1.0)
URBAN	−802.5[f]	−61.9	247.9[f]	511.2	137.5	3.4	66.4
	(0.5)	(1.1)	(0.5)	(0.7)	(2.1)	(15.4)	(1.5)
OWNREN	−332.0	−51.6	259.2[f]	315.7	−325.8	57.9	−144.1
	(1.1)	(1.3)	(0.4)	(1.1)	(1.1)	(1.1)	(0.9)
λ	−800.4	−89.6	394.5[f]	832.2	−485.3	−51.7	−123.1
	(0.7)	(1.0)	(0.4)	(0.6)	(0.9)	(1.5)	(1.2)
$adj.R^2$	0.08	0.04	0.02	0.12	0.13	0.08	0.05

[a]The probit estimation determining the λ-coeffficient is estimated by model C, see Table 8.7. [b]The variables also includes *COUPLES*, *CHILD*, *CHILDPR*, *CHILDSC*. [c]Full definitions of the variables are given in Table 8.5. Parentheses denote Relative Standard Deviations (SD/Mean). [d,e,f]Pearson's correlation coefficients significant on 0.001, 0.01 and 0.05 levels respectively.

out and expenditure on meals out, which indicates the presence of selectivity bias, justifying the inclusion of the correction term, λ.

By comparing eating-outers and non-eating-outers (Tables 8.9 and 8.10) it appears that higher income in both groups increases the buying of preparation foodstuffs significantly, and that the non-eating outers simultaneously decrease expenditure on fast meals, while eating outers decrease expenditure on eating out. Thus, more consumption power seems to imply that both groups of households become more home-oriented the higher the income.

For eating-outers, meals prepared at home are partly substituted by meals away from home when singles become couples. It also appears that having more children decreases expenditure on most of the categories of home foodstuffs for eating outers as well as for non-eating outers. The explanation may be that of economies of size in household production. However, some of the cost reductions are counterbalanced by more semi-preparation meals by non-eating-outers, when they have children of pre-school age, and by eating-outers, when there are school children in the household. In this way the buying of semi-convenience food functions as a time-saving strategy.

For eating outers, household appliances mean increased expenditure on preparation foodstuffs—which might agree with the US findings of Bryant (1988) saying that appliances and women's time are complements—and the same is true for domestic servants. In both cases fast meals seem to be substituted, the coefficients are highly negative, and the relatively small expenditure on fast food (Table 8.8) is taken into account.

The estimated probability of buying fast food is shown in Table 8.11, and we find *LOGVINC*, *LOGTIME*, *UNITC*, *URBAN* and *OWNREN* being the main explanatory variables. With every DKK 10000 increase in disposable unit income, the probability decreases marginally by 3%, which is opposite to expectations. However, whether this is counterbalanced by increasing expenditure on fast food, when being a fast fooder, is uncertain, because positive as well as negative regression coefficients were found between *LOGVINC* and fast food in the models in Tables 8.12 and 8.13, which indicates co-linearity of the regressors.

The marginal effect of more disposable time is a decreased probability of buying fast food, i.e. every hour of extra free time means nearly 1% fewer fast fooders, which indicates rational time allocation of the consumers, although the explanatory variables of disposable time taken separately, are all highly insignificant in explaining expenditures on fast food, see model B (Table 8.11). However, the UNITC-variable has the same marginal effect on expenditure as disposable time, indicating economies of size, or less satisfaction achieved by this food category, the bigger the families are.

The educational level seems to increase the probability of buying fast food slightly, while the home-owner effect is significant, as these living conditions increase the likelihood of being a fast fooder by 13% compared

Table 8.11 The decision of buying fast food—probit estimations[a] (N = 1247)

Independent variables[b]	Model A	Marginal effect	Model B	Marginal effect	Model C	Marginal effect
CONSTANT	6.781[d] (0.304)		4.413[d] (0.349)		4.551[d] (0.372)	
LOGVINC	–0.469[c] (0.297)	–0.181	–0.448[d] (0.312)	–0.174	–0.449[d] (0.323)	–0.174
LOGTIME	–0.313 (0.601)	(–0.121)	"	"	"	"
AGE	"	"	"	"	–0.002 (2.367)	(–0.001)
UNITC	"	"	"	"	–0.311[e] (0.484)	–0.121
COUPLES	"	"	–0.022 (9.402)	(–0.008)	"	"
CHILD	"	"	–0.008 (14.448)	(–0.003)	"	"
CHILDPR	"	"	0.056 (4.503)	(0.022)	"	"
CHILDSC	"	"	–0.174 (1.370)	(–0.067)	"	"
PARTTIME	"	"	0.010 (21.653)	(0.004)	"	"
FULLTIME	"	"	–0.116 (1.664)	(–0.045)	"	"
EDUTIME	"	"	0.062 (0.780)	(0.024)	0.066 (0.744)	(0.025)
APPL	"	"	"	"	–0.081 (1.660)	(–0.031)
DOMSERV	"	"	"	"	0.250 (0.944)	(0.097)
URBAN	"	"	"	"	0.089 (0.580)	(0.035)
OWNREN	"	"	"	"	0.340[e] (0.477)	0.132
–2log$L$	1651.439[d]		1648.778[e]		1639.971[d]	
Concordant pairs	55.4%		55.7%		56.7%	

[a] Asymptotic relative standard deviation (SD/Mean) in parentheses. Marginal effects are calculated by $\delta/\delta x_j$prob(*eatout* = $1/X$), and numbers in parentheses indicate insignificant coefficients. [b] Full definitions of the variables are given in Table 8.5. [c,d,e] Pearson's correlation coefficients significant on 0.001, 0.01 and 0.05 levels respectively.

Table 8.12 The expenditures on different kind of foodstuffs for fast fooders and non-fast fooders—socio-economic variables, 1987

	Fast fooders ($N = 763$)				Non-fast fooders ($N = 482$)		
Independent variables[b,c]	Non-convenience	Semi-convenience	Convenience (eat out)	Convenience (fast food)	Non-convenience	Semi-convenience	Convenience (eat out)
CONSTANT	7995.8 (0.9)	–941.9 (1.4)	–21 488.0[d] (0.3)	–9369.9[d] (0.3)	16848.0 (1.2)	–681.3 (5.2)	–17952.0 (1.0)
LOGVINC	–1890.9 (0.6)	25.4 (7.7)	2655.0[e] (0.3)	1463.3[d] (0.3)	–658.5 (2.0)	97.5 (2.4)	1527.5 (0.7)
COUPLES	–1572.9[f] (0.4)	–55.8 (2.1)	670.3 (0.8)	335.3 (0.7)	–626.2 (1.4)	–54.9 (2.8)	–104.1 (7.2)
CHILD	–305.4 (0.6)	–72.9[f] (0.4)	–212.1 (0.6)	–67.9 (0.9)	–46.2 (4.3)	37.1 (0.9)	–73.8 (2.3)
CHILDPR	422.6 (0.9)	169.3[e] (0.4)	–89.3 (3.2)	20.3 (6.7)	15.0 (28.6)	–28.4 (2.7)	–565.8 (0.7)
CHILDSC	190.9 (1.8)	157.8[e] (0.4)	–92.8 (2.8)	–26.4 (4.8)	–135.4 (3.1)	–22.5 (3.3)	–514.0 (0.7)
PARTTIME	883.6[f] (0.4)	52.8 (1.2)	–526.8 (0.5)	–259.8[f] (0.5)	–307.7 (1.4)	2.0 (39.1)	–27.3 (13.7)
FULLTIME	813.8[f] (0.5)	34.9 (1.8)	–564.6[f] (0.5)	–273.1[f] (0.5)	34.8 (12.6)	52.2 (1.5)	110.5 (3.4)
λ	1713.1[e] (0.3)	103.8 (0.9)	–672.1 (0.6)	–612.0[e] (0.3)	–1085.2 (0.7)	–21.1 (6.7)	390.2 (1.7)
adj.R^2	0.07	0.05	0.10	0.06	0.05	0.02	0.08

[a] The probit estimation determining the λ-coefficient is estimated by model A, see Table 8.11. [b] The variables also include *EDUTIME*. [c] Full definitions of the variables are given in Table 8.5. Parentheses denote Relative Standard Deviations (SD/Mean). [d,e,f] Pearson's correlation coefficients significant on 0.001, 0.01 and 0.05 levels respectively.

Table 8.13 The expenditures on different kind of foodstuffs for fast fooders and non-fast fooders[a]—productivity variables, 1987

Independent variables[b,c]	Fast fooders ($N = 763$)				Non-fast fooders ($N = 482$)		
	Non-convenience	Semi-convenience	Convenience (eat out)	Convenience (fast food)	Non-convenience	Semi-convenience	Convenience (eat out)
CONSTANT	−49 955.0 (0.6)	−2 269.5 (2.3)	35 348.0 (0.6)	6 383.3 (1.7)	31 055.0 (1.1)	1 624.8 (4.0)	−10 972.0 (2.8)
LOGVINC	8 210.0 (0.7)	220.5 (4.5)	−7 406.2 (0.6)	−1 264.6 (1.6)	−1 758.2 (1.4)	−64.0 (7.0)	1 111.3 (2.0)
AGE	59.0[f] (0.5)	0.4 (11.3)	−61.0[e] (0.4)	−8.3 (1.3)	24.3 (0.6)	−0.3 (7.5)	−15.8 (0.8)
EDUTIME	−1 126.5 (0.8)	13.7 (10.6)	1 264.9[f] (0.5)	232.7 (1.3)	538.2 (0.7)	48.8 (1.4)	−29.0 (11.3)
APPL	1 477.5 (0.7)	−14.6 (12.4)	−1 483.8 (0.5)	−222.9 (1.7)	−753.7 (0.7)	1.9 (48.7)	−2.5 (179.3)
DOMSERV	−4 108.6 (0.8)	−102.2 (5.4)	5 534.3[f] (0.4)	723.5 (1.6)	975.0 (1.5)	67.6 (3.9)	−140.2 (9.0)
URBAN	−1 398.7 (0.8)	1.2 (170.9)	1 572.2 (0.5)	290.5 (1.4)	261.2 (1.8)	20.6 (4.3)	−114.8 (3.7)
OWNREN	−5 373.4 (0.8)	26.3 (28.6)	6 409.6[f] (0.5)	1 123.7 (1.4)	2 336.9 (0.8)	188.7 (1.8)	−305.2 (5.3)
λ	−3 431.7 (0.9)	13.2 (38.5)	4 408.3[f] (0.5)	775.7 (1.4)	−1 852.2 (0.8)	−121.1 (2.2)	191.1 (6.9)
adj.R^2	0.08	0.05	0.12	0.05	0.11	0.02	0.09

[a]The probit estimation determining the λ-coefficient is estimated by model C, see Table 8.11. [b]The variables also includes *COUPLES*, *CHILD*, *CHILDPR*, *CHLDSC*. [c]Full definitions of the variables are given in Table 8.5. Parentheses denote Relative Standard Deviations (SD/Mean). [d,e,f]Pearson's correlation coefficients significant on 0.001, 0.01 and 0.05 levels respectively.

to the situation of tenants. And further, for a fast fooder, expenditure increases correspondingly to home-ownership, as does expenditure on eating meals-away-from home even significantly (Table 8.13).

Spending money on fast food by more highly educated women entails more expenditure on meals out and less on preparation meals, whereas more household appliances as well as two full-time jobs mean less expenditure on meals out and more on meals prepared at home (Table 8.12). A double-career family, highly educated, with time-saving technology, sometimes eating fast food, therefore seems to exhibit traditional food choice behaviour, because the underlying effects outweigh each other. Therefore, it might be argued that busy and rich families prefer non-convenience foods to convenience foods, because they belong to life-styles in which home-made food is a luxury good requiring both time and money. This leads us to question the hypothesis of rational behaviour concerning food choice.

Moreover, like the other regressions in this article, the models exercised here explain only a fraction of the whole variation of expenditure on different food-categories. However, the models of the meals-away-from-home foodstuffs are the most appropriate, having the highest *adj.R2*.

8.7 Conclusion

The overall finding is that very rich and very busy households spend more on convenience foods than poor and time idle households—both relatively and absolutely—while the latter spend relatively more on non-convenience food. This at first seems to confirm the hypothesis of rational behaviour among households.

The household production is more or less efficient depending on the size of this production, the experience and education of the producers, the technical equipment available, and the conditioning surroundings, all of them affecting the choice of foodstuffs with varying degrees of convenience. By including these variables as proxies of productivity and making explicit specifications of the models we improve the explanations of the choice of foods. Mentioning some of the findings, the age of the housekeeper (the wife in couples) means more life experience and thereby higher productivity in household production; nonetheless, we find that not only is older individuals' probability of eating out bigger than younger individuals', as are eating outers' expenditure on this food category increasing by the age of the housekeeper.

Expenditure on non-prepared food declines with more education in both Denmark and Sweden, while higher-educated Swedes buy more fast food than lower-educated Swedes (Lindvall, 1989), and Danes more semi-prepared food the higher the educational level. The interpretation may be that the modernization of the Danish food pattern is slower than the displayed internationalization of the Swedish way of eating.

In analysing food behaviour the population is divided into eating-outers and non-eating outers, and fast fooders and non-fast fooders, according to their registered expenditures on these food categories. Subsequently the behaviour of the households in these different food-regimes are compared.

Surprisingly, increasing unit income gives rise to fewer eating outers, and even more surprisingly, eating outers' expenditure declines with increasing income. The non-eating outers simultaneously decrease expenditure on fast meals. In this way more consumption power seems to imply that both groups become home-oriented concerning meal preparation.

For eating outers, household appliances and domestic servants mean increased expenditure on non-convenience foodstuffs, and the substitution of fast meals, finding highly negative coefficients for this food category.

Women's labour market participation, and especially if they have a part-time job, increases the probability of eating out, which is also the case concerning double-career families with two full-time jobs; although the latter effect is smaller than the former effect on the probability of being an eating outer. The marginal effect of more disposable time is a decreased probability of buying fast food, although the variables explaining disposable time are all highly insignificant in explaining expenditures on fast food.

Spending money on fast food education entails higher expenditure on meals out and less on preparation meals, whereas more household appliances as well as two full-time jobs mean less expenditure on meals out and more on preparation meals. A double-career family, highly educated, with modern time-saving technology, sometimes eating fast food, therefore has an ordinary food choice behaviour. Thus busy and rich families prefer non-convenience foods to convenience foods, probably because of belonging to life-styles in which home-made food is a luxury good requiring both time and money; this might therefore explain why these households do not adopt time-saving strategies.

All the probit-estimations and regressions applied, however, explain only a minor part of the variations in the models which have to be taken into account when interpreting the findings. The quality of the data in the family budget survey is the main problem, but as long as these are the only data available, this kind of empirical research work must be done on that basis.

Acknowledgement

The MAPP-program (Market-based Process and Product Innovation in the Food Sector) is gratefully acknowledged for financing this project.

References

Ås, D. (1982) *Measuring the Use of Time*. Special Study No. 7. Paris: The OECD Indicator Development Program.
Andersen, D. (1987) *Den danske befolknings tidsanvendelse 1987*. Copenhagen: Danish Institute of Social Research, Working paper.
Andersen, W.T. (1971) *The Convenience Oriented Consumer*. Austin, TX: Bureau of Business Research. University of Texas.
Becker, C. (1965) A theory of the allocation of time. *The Ecomic Journal* September LXXV, 493–517.
Bellante, D. and Foster, A.C. (1984) Working wives and expenditure on service. *Journal of Consumer Research* **11**, 700–707.
Biddle, J.E. and Hamermesh, D.S. (1989) *Sleep and the Allocation of Time*. Working Paper No. 2988. National Bureau of Economic Research (NBER). Cambridge, MA.
Bonke, J. (1988) *Husholdninger og husholdningsproduktion—socioøkonomiske forklaringer på efterspørgslen efter varer til husholdningsproduktion og alternativ service* (Households and Household Production—Socioeconomic Explanations of the Demand of Goods to Household Production and Alternative Services). Copenhagen: University of Copenhagen, Department of Economics, Memo 166.
Bonke, J. (1989) *Ligestilling i husholdningsarbejdet 1964–87—en belysning af udviklingen i tidsmæssig arbejdsdeling mellem kvinder og mænd, og nogle forklaringer på forandringerne* (Equality in Household Work 1964–87—An Enlightment of the Development in the Division of Labor Between Women and Men, and Some Explanation of the Changes). Copenhagen: University of Copenhagen, Department of Economics, Memo 176.
Bonke, J. (1992a) *Choice of Foods—Allocation of Time and Money, Household Production and Market Services*. MAPP Working Paper no 3.
Bonke, J. (1993) *Choice of Foods—Allocation of Time and Money, Household Production and Market Services. Part II*. MAPP Working Paper No 9.
Bonke, J. (1995a) *Arbejde, tid og køn—i udvalgte lande* (Work, Time and Gender—in Selected Countries). The Danish National Institute of Social Research. Copenhagen.
Bonke, J. (1995b) *Faktotum—husholdningernes produktion* (Factotum—Household Production). Ph.D. dissertation. Copenhagen: University of Copenhagen. Department of Economics.
Borchorst, A. (1993) Working lives and family lives in Western Europe. In S. Carlsen and J.E. Larsen (eds) *The Equality Dilemma*. Copenhagen: The Danish Equal Status Council.
Bryant, W.K. (1988) Durable and wives' employment yet again. *Journal of Consumer Research* **15** (June) 37–47.
Capps, O., Tedford, J.R. and Havlicek, J. (1983) Impacts of household composition on convenience and nonconvenience food expenditures in the South. *Southern Journal of Agricultural Economics* December, 111–118.
Danmarks Statistik (1992) *Forbrugsundersøgelsen 1987. Formål, metode og hovedresultater*. Statistiske Efterretninger 1992:1.
Douglas, S. (1976) Cross-national comparisons and consumer stereo-types: A case study on working and non-working wives in the U.S. and France. *Journal of Consumer Research* **3** (June).
Eurostat (1989) *Harmonization of Family Budget Surveys*, DOC BF 44/90. November.
Eurostat (1991) *Improving Survey Methods, Content and Utilization: Main Recommendations*. DOC BF 55/91 EN. April.
Foster, A.C., Abdel-Ghany, M. and Ferguson, C.E. (1981) Wife's employment—Its influence on major family expenditures. *Journal of Consumer Studies and Home Economies* **5** (June), 115–124.
Galbraith, J.K. (1973) *Economics and the Public Purpose*. London.
Gershuny, J. (1983) *Social Innovation and the division of labour*. Oxford University Press.
Gershuny, J. (1994) *Economic Activity and Women's Time Use. Time Use of Women in Europe and North America*. UN. Economic Commission of Europe.
Gustafsson, S. (1991) Ekonomisk teori för tvåförsörjarfamiljen (Economic theory of double career families). *Ekonomisk Debatt*. **6/91**.

Homan, M.E. (1988) *The Allocation of Time and Money in One-Earner and Two-Earner Families; an Economic Analysis*. Offsetdrukkerij Kanters Alblasserdam.

Körmendi, E. (1990) Work sharing at home. In G. Viby Mogensen (ed.) *Time and Consumption*. Copenhagen: Danmarks Statistik.

Lee, J-Y. and Brown, M.G. (1986) Food expenditure at home and away from home in the United States—A switching regression analysis. *The Review of Economics and Statistics* **68**, 142–47.

Lee, L-F., Maddala, G.S. and Trost, R.P. (1980) Asymptotic covariance matrices of two-stage probit and two-stage tobit methods for simultaneous equation models with selectivity. *Econometrica* **48** (2) (March), 491–503.

Lindvall, J. (1989) *Expensive Time and Busy Money*. Linkøping: Linkøping University. Sweden.

McCracken, V.A. and Brandt, J.A. (1987) Household consumption of food-away-from-home: Total expenditure and by type of food facility. *American Journal of Agricultural Economics* **169** (May), 274–284.

Mincer, J. (1960) Labour supply, family income and consumption. *American Economic Review* **50** (May), 574–83.

Oropesa, R.S. (1993) Female labor force participation and time-saving household technology: A case study of the microwave from 1978 to 1989. *Journal of Consumer Research* **19** (March) 567–579.

Prochasha, F.J. and Schrimper, R.A. (1973) Opportunity cost of time and other socioeconomic effects on away-from-home food consumption. *American Journal of Agricultural Economics* **55** (November), 595–603.

Redman, B.J. (1980) The impact of women's allocation on expenditure for meals away from home and prepared foods. *American Journal of Agricultural Economics* (May), 234–237.

Reilly, M.D. (1982) Working wives and convenience consumption. *Journal of Consumer Research* **8** (March), 407–418.

Robinson, J.P. (1977) *How Americans Use Time*. New York: Praeger.

Soberon-Ferrer, H. and Dardis, R. (1991) Determinants of household expenditures for services. *Journal of Consumer Research* **17** (March), 385–397.

Strober, M.H. (1977) Wives' labour force behavior and family consumption patterns. *Amer. Economic Association* **67** (1) (February), 410–417.

Strober, M.H. and Weinberg, C.B. (1977) Working wives and major family expenditures. *Journal of Consumer Research* **4** (3), 141–147.

Strober, M.H. and Weinberg, C.B. (1980) Strategies used by working and nonworking wives to reduce time pressures. *Journal of Consumer Research* **6**, 338–48.

Vanek, J. (1974) Time spent in housework. *Scientific American*, 231 (November); The economics of women. In C. Wardle (1977) *Changing Food Habits in the UK*. London: An Earth Resources Research Publication.

Vickery, C. (1979) *Women's economic contribution to the family*. In R.E. Yen, S.T. (1993). Working wives and food away from home: The Box–Cox Double Hurdle Model. *American Journal of Agricultural Economics*, **75** (November), 884–895.

Wagner, J. and Hanna, S. (1983) The effectiveness of family life cycle varables in consumer expenditure research. *Journal of Consumer Research* **10** (December), 281–291.

Waldman, E. and Jacobs, E. (1978) Working wives and family expenditures. In *Proceedings of the American Statistical Association Annual Meeting*. Washington, DC: American Statistical Association, pp. 41–49.

Walker, K.E. and Woods, M.E. (1976) *Time Use: A Measure of Household Production of Family Goods and Services*. Washinton. DC: American Home Economics Association.

Weinberg, C.B. and Winer, R.S. (1983) Working wives and major family expenditures: replication and extension. *Journal of Consumer Research* **10** (September) 259–263.

9 Food choice, mood and mental performance: some examples and some mechanisms

PETER J. ROGERS

9.1 Introduction

It is well established that diet can influence mood and mental performance. Indeed, certain variations in nutritional status can have very marked effects on mental functioning (Table 9.1). This is demonstrated by, for example, the consequences of chronic and severe food restriction (Smart, 1993), thiamin deficiency (Kanarek and Marks-Kaufman, 1991), and iron deficiency (Pollitt, 1987). Acute effects of food and fluid ingestion have also been described, and this area together with an examination of interrelationships between mood and eating is the main concern of the present review. Influences of mood on eating (and drinking) as well as effects of food on mood are considered. The material covered has been included to provide examples of the main findings of this research, and to illustrate certain methodological issues. Possible mechanisms underlying these diet–behaviour relationships are also examined. The relevant literature, however, is very large and cannot be covered fully in a single chapter. Therefore, some subjects such as the effects of alcohol (see Finnigan and Hammersley, 1992 for a review) are not discussed in detail.

9.2 Effects of foods and food constituents on mood and mental performance

9.2.1 *Breakfast and lunch*

The effects of meals on cognitive efficiency have been recently reviewed in detail by Smith and Kendrick (1992). Relatively little work has been carried out in this area, although some of the earliest studies were reported over 40 years ago. A common finding is that performance is impaired in the early afternoon compared with performance measured in the late morning (the so called post-lunch dip). For some tasks this occurs whether or not lunch has been eaten, although consuming a meal at lunchtime, particularly if this is a large meal, does appear to contribute significantly to the post-lunch dip. Tasks which are most typically affected are those requiring sustained attention (Smith and Kendrick, 1992). There is, however, a problem concerning

Table 9.1 Examples of nutritional variables known to affect mental performance

- Food restriction:
  - Early life undernutrition
  - Chronic semi-starvation
  - Dieting to lose weight[a]
  - Short-term fasting (e.g., missing a meal)[a]
- Pre- versus post-prandial state
- Dietary macronutrient composition:
  - Acute effects[a]
  - Chronic effects[a]
- Micronutrient deficiencies:
  - Thiamin (Wernicke–Korsakoff syndrome)
  - Iron
- Phenylketonuria
- Diabetes (hypoglycaemia, and chronic effects[a])
- Alcohol:
  - Acute intoxication
  - Chronic effects, including fetal alcohol syndrome
- Caffeine, caffeine withdrawal[a]

[a] Less well established effects. Adapted from Rogers and Lloyd (1994).

the interpretation of data showing impairments when no lunch is eaten—which might indicate effects related either to fasting or to an underlying circadian variation in, for example, alertness. If the latter were responsible, performance would be expected to recover later in the afternoon. Additionally, it is important to note that post-lunch impairments in performance are by no means always observed. In particular, several results suggest that meals consumed at lunchtime have least effect on performance efficiency under conditions of high arousal; for instance, when subjects are given caffeine (see §9.2.8), or when they are exposed to a high level of background noise (Smith and Miles, 1986). The effects of lunch on performance also appear to be significantly related to personality factors such as extraversion and neuroticism (Craig *et al.*, 1981; Smith and Miles, 1986).

In contrast to the effects of lunch, there is a popular belief that missing breakfast can have an adverse effect on cognitive efficiency. When taken together, however, the published experimental evidence does not substantially support this belief. A series of studies conducted by Tuttle and colleagues (reviewed by Editor, 1957; Smith and Kendrick, 1992) indicated somewhat poorer reaction time performance in subjects who had not been given breakfast, and recently a similar result has been reported for memory performance (Benton and Sargent, 1992). In addition, 9- to 11-year-old children were found to score significantly less well on a problem solving task (matching familiar figures test) when tested in the morning following an overnight fast compared with when they were tested following breakfast (Pollitt *et al.*, 1981, 1983). This latter work has received a considerable amount of attention, but the finding of a detrimental effect of missing breakfast should be viewed in the context of other studies where this was not observed (e.g. Dickie and Bender, 1982; Smith and Kendrick, 1992). Also, given the fairly long period of food restriction imposed in the experiments by Pollitt and colleagues (the children were tested either 3 hours or 18 hours after their last meal), it could be argued reasonably that the lack of more substantial effects indicates that cognitive performance is relatively invulnerable to short-term fasting. In fact, of the various measures of performance taken only one showed a statistically significant deterioration related to missing breakfast, while one other (frequency of recall of the last item on a test of immediate memory) showed a small but significant improvement (Pollitt *et al.*, 1981). This conclusion is further supported by the results of our recent study which showed that even more severe short-term food deprivation failed to impair mental performance (Green *et al.*, 1995). Young adults were tested on a battery of cognitive tasks after eating normally and also after missing one meal, two meals and all three meals in a day. In contrast to previous results showing poorer performance on these same tasks in dieters (e.g. Green *et al.*, 1994), acute food deprivation was without effect.

Rather little information is available on the effects of meals eaten at other times of the day (see Smith and Kendrick, 1992); however perhaps a more important priority for future research is the investigation of the influences of eating pattern. In particular it would be of interest to establish the benefits or otherwise of adopting a snacking or grazing eating style against the more traditional pattern of three meals a day.

9.2.2 Meal composition—serotonin and carbohydrate versus protein

Studies of the effects of meal macronutrient composition on mood and mental performance were stimulated by specific theoretical predictions concerning possible short-term dietary influences on brain functioning. This concerns the hypothesis proposed by R.J. Wurtman and colleagues which links carbohydrate and protein intake, brain serotonergic (5-hydroxytryptamine, 5-HT) function, and mood and behaviour. In outline the hypothesis is as follows (reviewed by Wurtman *et al.*, 1981). Consumption of a high carbohydrate meal increases the ratio of the plasma concentration of the amino acid tryptophan relative to the other 'large neutral amino acids' (e.g. tyrosine, phenylalanine, leucine, isoleucine and valine) (Trp:ΣLNAA). This occurs because insulin released in response to the carbohydrate load facilitates the uptake of most amino acids, but not tryptophan, into peripheral tissues such as muscle. Uniquely for amino acids, tryptophan is bound to albumin in the blood stream, and the affinity of albumin for tryptophan actually increases in response to insulin as free fatty acids (FFA) are stripped off the circulating albumin (due to the effect of insulin on the removal of FFA from the circulation). Tryptophan is the precursor of the neurotransmitter serotonin and, since tryptophan and the other large neutral amino acids compete for entry into the brain and the rate-limiting enzyme for serotonin production (tryptophan hydroxylase) is not fully saturated with substrate under normal conditions, an increase in plasma Trp:ΣLNAA concentration leads to an increase in brain serotonin synthesis and, in turn, to increased serotonergic neurotransmission. In contrast, consumption of a meal high in protein can be expected to have the opposite effect, primarily because most dietary proteins contain relatively very little tryptophan (Wurtman *et al.*, 1981).

The behavioural consequences which have been predicted to follow from these diet-induced changes in neurotransmission include altered food choice and food intake, and changes in pain sensitivity, aggressiveness, mood, alertness and cognitive performance. Specifically, the relative increase in brain serotonergic activity supposedly occurring after consumption of a high carbohydrate versus high protein meal is hypothesised to give rise to, for example, decreased alertness, and consequently a decline in performance efficiency (e.g. Spring, 1986). (Interestingly, this contradicts the popular notion that consumption of carbohydrates, especially sugar, will normally have an energising effect.)

The evidence for such effects, though, is mixed. Several studies found differences in the effects of high carbohydrate and high protein meals in the direction of greater drowsiness, sleepiness and calmness after carbohydrate (see Spring *et al.*, 1987; Young, 1991, for reviews), but these were not consistent across all subject groups, and a majority of mood and performance measures were found not to be sensitive to meal composition. Reaction time and tasks requiring sustained attention were most affected, but even in these studies a large majority of the comparisons between subject groups or performance measures failed to reveal statistically significant effects of meal composition. In addition, administration of tryptophan alone was found to have only a marginal effect on one (simple auditory reaction time) out of four tests of performance, despite the fact that the dose used, approximately 3.5 g, was sufficient to produce a measurable number of subjective changes, including decreased alertness and vigour, and increased fatigue (Lieberman *et al.*, 1983; see also Hrboticky *et al.*, 1985).

More recent studies (Deijen *et al.*, 1989; Christensen and Redig, 1993) have also failed to show any definite carbohydrate versus protein effects on mood, despite confirmation in the latter study of significant effects on Trp:ΣLNAA concentrations. Pivonka and Grunewald (1990) found that a surgar-sweetened drink increased sleepiness and decreased alertness compared with the effects of the same drink sweetened with aspartame or the same volume of water. These differences emerged about 30 minutes after the drinks were consumed. However, other similar studies have not revealed such clear results (e.g. Brody and Wolitzky, 1983), and these experiments did not, of course, test the effects of protein or other nutrient loads. This is also true of a larger number of studies which have examined the effects of oral glucose loads on various aspects of cognitive performance. Nonetheless, contrary to the prediction of the carbohydrate-serotonin hypothesis, performance efficiency was, if anything, improved after consuming glucose compared with a non-nutrient-containing control drink (reviewed by Rogers and Lloyd, 1994).

Other studies examining the effects of meal composition include an investigation by Smith *et al.* (1988) who found that iso-energetic, high-carbohydrate and high-protein meals affected different aspects of attention. However, because the high carbohydrate meals contained 15 percent protein and testing began within an hour after lunch it is very unlikely that this dissociation was mediated by differences in plasma Trp:ΣLNAA concentrations (see §9.2.4).

9.2.3 Carbohydrates and depression

In addition to these effects of meal composition on alertness and mental performance, Wurtman and Wurtman (1989) have developed a further hypothesis which proposes that carbohydrates can relieve depression. This relates specifically to three 'disorders', carbohydrate-craving obesity (CCO), premenstrual syndrome (PMS) and seasonal affective disorder (SAD), the characteristics of which include depressed mood and supposedly a craving for and increased intake of high carbohydrate foods. Together with evidence implicating lowered serotonergic function in the aetiology of depression (e.g. Cowen *et al.*, 1992; Maes and Meltzer, 1995), this has led to the suggestion that the increase in carbohydrate intake constitutes self-medication to relieve the depression (see §9.3.2 for further discussion).

The prediction is, therefore, that consumption of a high-carbohydrate, low protein meal should have different effects in depressed and non-depressed individuals. Relatively few studies have tested this directly, although consistently carbohydrate has been found to be significantly less sedating in CCO, SAD and PMS subjects compared with controls (Lieberman *et al.*, 1986a; Rosenthal *et al.*, 1989; Wurtman *et al.*, 1989). Lieberman *et al.* (1986a) also demonstrated the expected effects on depressed mood in CCO. In PMS subjects a carbohydrate meal was found to markedly improve several aspects of mood, including tension, anger, depression and confusion, all of which were unaffected in control subjects (Wurtman *et al.*, 1989). Both these latter studies, however, failed to establish whether the effects on mood were specific to carbohydrate, because the response to other meals (e.g. high in fat or protein) was not tested (see Rogers and Jas, 1994).

In a study of spontaneous meal patterns de Castro (1987) found a significant negative correlation between self-reported depression and the proportion of carbohydrate (%energy) consumed over a 9-day period. In addition, depression was positively correlated with proportionate protein intake, and feelings of energy were positively correlated with proportionate carbohydrate intake. These relationships, however, were not apparent in an analysis of meal to meal mood and food intake. Nor is the direction of the effects as predicted by Wurtman and colleagues, since the subjects were not depressed to begin with. High carbohydrate, low protein intake is supposed to increase somnolence, but not affect depression, in nondepressed individuals.

9.2.4 Normal dietary variations do not alter serotonergic function in human subjects

A further difficulty for the proposed relationships between the relative protein and carbohydrate content of a meal and mood and performance is that the data relating brain serotonin and meal composition have been obtained from studies on rats, not humans. Studies on humans have shown statistically significant differences in plasma Trp:ΣLNAA concentrations following 'meals' of starch, sugar and protein, but it appears that the magnitude of these effects is probably too small to produce functionally significant changes in brain serotoninergic activity (Leathwood, 1987; Young, 1991). The actual changes observed were small increases in the plasma Trp:ΣLNAA concentrations following starch and sugar meals and a larger decrease in this ratio after high protein meals. Differences in plasma Trp:ΣLNAA concentrations did not reach their maximum until 2 to 3 hours after these meals, with almost no change occurring within the first hour after a carbohydrate meal (e.g. Lieberman *et al.*, 1986b; Christensen and Redig, 1993). Similar effects have been confirmed in SAD and obese subjects (Rosenthal *et al.*, 1989; Pijl *et al.*, 1993). In the study by Christensen and Redig there were no corresponding effects on mood and alertness. Furthermore, meals which are more typical in composition are unlikely to produce any distinct differences in post-prandial Trp:ΣLNAA plasma concentrations, since the presence of a small amount (perhaps as little as 4%) of protein in a high carbohydrate meal is sufficient to block any meal-induced increases in plasma Trp:ΣLNAA concentrations (Teff *et al.*, 1989a; see also Leathwood, 1987; Christensen and Redig, 1993).

Even more striking are the results of a study by Teff *et al.* (1989b) in which samples of cerebrospinal fluid were collected from human subjects 2.5 hours after they had consumed either 100 g of carbohydrate, a high protein load, or water. Compared with the effects of water, the nutrient loads did not alter either the cerebrospinal fluid concentration of tryptophan or cerebrospinal fluid concentration of the serotonin metabolite 5-hydroxyindoleacetic acid. On the other hand, plasma Trp:ΣLNAA concentrations were affected in a similar manner to previous studies. This result would therefore support the view that in human beings differences in meal macronutrient composition do not alter plasma amino acid ratios sufficiently to cause appreciable changes in brain tryptophan levels or serotonin synthesis (but see Fernstrom, 1994).

On the other hand, there is still considerable interest in the possibility of the existence of dietary and metabolic influences on brain serotonergic function. For example, Newsholme and colleagues (Newsholme *et al.*, 1992; Hassmen *et al.*, 1994) have argued that increased brain serotonin is a factor contributing to the development of mental fatigue during sustained exercise. The proposed mechanism involves an increase in plasma FFA concentration occurring during exercise, which raises the free plasma concentration of tryptophan, causing in turn increased entry of tryptophan into the brain and increased brain serotonin activity. This differs from the mechanism suggested by Fernstrom and Wurtman which depends on effects on the total plasma concentration of tryptophan, where a large percentage of tryptophan is normally bound to albumin (Wurtman *et al.*, 1981).

Mechanisms have also been suggested linking cholesterol-lowering diets, brain serotonin and behaviour which, if confirmed, could have very significant epidemiological implications (Muldoon *et al.*, 1991; Kaplan *et al.*, 1994). Furthermore, dieting *per se* (i.e. energy restriction) has been shown to affect both plasma Trp:ΣLNAA concentrations and an indirect measure of serotonergic function in human subjects (Goodwin *et al.*, 1987, 1990), and another study, found significant correlations between carbohydrate intake, Trp:ΣLNAA plasma concentrations and mood during weight-reducing diets (Schweiger *et al.*, 1986). These observations may help to explain, at least in part, some of the psychological consequences of dieting, including the association of weight loss with depression (Cowen *et al.*, 1992) and the finding of an impairment in cognitive performance during dieting (Green *et al.*, 1994; §9.2.7).

9.2.5 Meal composition—carbohydrate versus fat

There have been very few studies reported which have investigated the possible psychoactive effects of dietary fat. This issue is of interest, however, because of Governmental recommendations to reduce fat and increase carbohydrate intake (e.g. Department of Health, 1992), and also because of concern about possible adverse behavioural consequences of consuming low-fat, cholesterol-lowering diets (e.g. Muldoon *et al.*, 1991).

Recently, we have examined the acute effects of manipulating the fat to carbohydrate ratio of individual meals. Subjects' mood and cognitive performance were measured following their consumption of iso-energetic meals, similar in appearance and orosensory properties, but varying in fat and carbohydrate content (e.g. 27 and 62%, 44 and 47%, and 56 and 34% energy as fat and carbohydrate). In the first study, simple reaction time performance and mood were found to be better (e.g. less 'drowsy' and less 'muddled') in the afternoon following a medium fat (medium carbohydrate) lunch compared with either a high or low fat lunch (Lloyd *et al.*, 1994). In contrast, in the second study morning mood was found to be relatively improved following a low fat (high carbohydrate) breakfast compared with either a medium or high fat breakfast (e.g. less 'dejected', less 'drowsy', less 'muddled') (Lloyd and Rogers, 1994). Many of these effects were apparent within 30 minutes of eating and, perhaps most significantly, in both studies optimal mood and performance were associated with the meal which was closest in macronutrient composition to the subjects' typical intake at that particular time of day.

This is likely to have implications for the immediate effects making dietary changes. For example, preliminary evidence from a longer term study indicates that changing to a low fat diet (in the absence of weight loss) was associated initially with increased tiredness, even though after several weeks mood was actually improved overall (Lloyd and Rogers, unpublished observations). This is perhaps consistent with the adaptation of gastrointestinal and other physiological responses to food known to occur as a result of medium to longer term changes in diet composition (e.g. French *et al.*, 1995).

The appearance of certain mood and performance effects within half an hour of eating suggests that preabsorptive and/or early postabsorptive mechanisms are implicated. At present, however, the nature of these mechanisms is unclear, although one study has demonstrated substantial, dose-related effects of intravenously administered cholecystokinin on alertness and performance (Stacher *et al.*, 1979), with subjects feeling more relaxed, drowsy, sluggish and inert with increasing doses of the hormone. This is of particular interest, because fat and protein are more potent releasers of cholecystokinin than carbohydrate (Liddle *et al.*, 1983). Therefore cholecystokinin may provide part of a mechanism mediating effects of meal composition.

9.2.6 Chocolate

Speculation as to what makes chocolate special abounds in the popular media. Some of this speculation originates from the scientific literature. However, relatively little empirical research has actually investigated the various claims of chocolate 'addiction' and psychoactive effects of chocolate and chocolate constituents.

Of all foods, chocolate is certainly the most frequent focus of food cravings, especially among women (Weingarten and Elston, 1991; Rodin *et al.*, 1991). It is also clear that eating chocolate can have significant influences on mood, generally leading to an increase in pleasant feelings and a reduction in tension, although increased 'guilt' may be a penalty for some individuals (e.g. Hill *et al.*, 1991; Hill and Heaton-Brown, 1995; Macdiarmid and Hetherington, 1995). Furthermore, craving for various foods, including chocolate, is associated with negative moods (e.g. boredom, tension, anger, depression and tiredness) (Schlundt *et al.*, 1993; Rogers *et al.*, 1994). However, this and other circumstantial evidence, for example the existence of groups such as Chocoholics Anonymous, is not sufficient to demonstrate addiction to chocolate (Rogers, 1994). Indeed, it is obvious that very many people eat chocolate regularly without becoming addicted.

In fact, serious reviews have generally found little or no support for the suggestion that the liking for chocolate is related to the presence of psychoactive constituents (Tarka, 1982; Max, 1989; Rozin *et al.*, 1991). For instance, although chocolate can contain relatively high concentrations of theobromine, this is a relatively weak central nervous system stimulant and does not have strong subjective effects (Mumford *et al.*, 1994). The related methylxanthine, caffeine, is also found in chocolate but in much lower concentrations. Compared with coffee and tea, chocolate is an insignificant source of dietary caffeine (Gilbert, 1984). There also appears to be a lack of any generally significant relationship between reported chocolate craving and the liking and consumption of other xanthine-containing substances (Rozin *et al.*, 1991). Other substances present in chocolate which have been discussed as potentially pharmacologically significant, include histamine, serotonin, tryptophan, phenylethylamine, tyramine, salsolinol and magnesium (Cockcroft, 1993; Michener and Rozin, 1994), though many of these exist in higher concentrations in other foods with less appeal than chocolate (Robinson and Ferguson, 1992). The same is true for certain bioactive peptides, such as casomorphins which can act as opioid agonists and occur in a variety of food sources, for example milk and gluten (Morley *et al.*, 1983; Gardner, 1984).

Very few studies have attempted to directly investigate the psychoactive effects of these various substances administered orally, either alone or in combination, in amounts relevant to dietary intakes of chocolate. A notable exception is a recent report by Michener and Rozin (1994), who provided chocolate cravers with sealed boxes containing either a milk chocolate bar, a bar of white chocolate, capsules containing cocoa (and therefore many of the presumed psychoactive ingredients of chocolate), placebo capsules, white chocolate plus cocoa capsules, or nothing. The subjects consumed, in random order, the contents of one of these boxes when they experienced a craving for chocolate, and they rated the intensity of their chocolate craving just before, just after and 90 minutes after this. The results showed that only consumption of chocolate itself, either white or brown, substantially reduced the craving, suggesting that there is 'no role for pharmacological effects in the satisfaction of chocolate craving' (p. 419). As the authors themselves point out, this cannot be a definite conclusion because the subjects may have had different expectations for the effects of the substances they were asked to consume, nor does it exclude the possibility that pharmacological reinforcement is a significant factor influencing the acquisition of liking for chocolate (Michener and Rozin, 1994). Nevertheless, it would appear that current evidence points to the more mundane conclusion that liking for chocolate and its effects on mood are due primarily to its principal constituents sugar and fat, and their related sensory and physiological effects (Rogers, 1994).

9.2.7 Dieting and dietary restriction

One of the findings of the Minnesota study of semi-starvation (Keys *et al.*, 1950) was that long-term food restriction resulted in lethargy, tiredness, depression and greatly decreased feelings of 'energy'. The subjects also complained of a number of changes in their intellectual functioning, including an inability to concentrate, impaired judgement and poorer memory, although this was not confirmed by the results of a battery of 'objective' tests of performance. Dieting to lose weight, although usually involving less severe degrees of food restriction, can be also be regarded as a form of semi-starvation (anorexia nervosa is of course an extreme example of dieting, but is relatively uncommon). The effects of dieting on cognitive performance have, however, only recently begun to be investigated.

In an initial study (Rogers and Green, 1992) women undergraduates were assessed for their current dieting behaviour, their concerns about eating, body weight and body shape, and were also tested on a demanding, rapid information processing task. In this procedure a continuous stream of single digits was presented on a computer VDU at the rate of 100 digits per minute. The subjects were required to press a response button whenever they detected a sequence of 3 odd or 3 even digits. A total of 40 such targets were presented over 5 minutes, each separated by a minimum of 5 and a maximum of 30 digits. Thirteen of the 55 volunteers reported that they were currently dieting to lose weight. These subjects were compared with the remaining subjects divided into low-to-medium restraint and high restraint groups on the basis of their scores on the dietary restraint factor of the English version of the Dutch Eating Behaviour Questionnaire (van Strien *et al.*, 1986). The dietary restraint scale from this questionnaire is a measure of concerns about eating in relation to body weight. The currently dieting subjects did not differ from the high restraint subjects on dietary restraint or body mass index, although other results indicated greater dissatisfaction with body shape and weight among the dieters. The dieters also tended to perform less well than the non-dieters on the cognitive task. This last result was confirmed and extended in a second study (Green *et al.*, 1994) in which current dieters were found again to be relatively impaired on the rapid information processing task, and also impaired on simple reaction time and immediate memory tasks (Table 9.2). Furthermore, performance was poorest in subjects who had been on the current diet the longest and had lost the most weight.

The above results indicate a deterioration of cognitive performance associated with dieting. This could be due to a variety of factors, including the effects of food restriction and temperamental differences between dieters and non-dieters. Dietary records revealed a substantially lower energy intake in the dieters, but whether this was due to actual differences in intake or to underreporting by the dieters was unclear (Green *et al.*, 1994). The dieters though did not appear to be less motivated to perform the tasks, because if anything they were better than the non-dieters at a tapping task (for which subjects were required simply to tap the '1' and '2' keys on a computer keyboard as fast as possible) (Table 9.1). This also shows that their fine motor responses were not impaired. However, in order to investigate whether the performance deficits in spontaneous dieters are related to dieting itself, rather than to pre-existing differences between dieters and non-dieters, we carried out a further study in which the same subjects were tested twice—once when they were dieting and once when they were not dieting (Green and Rogers, 1995). Again dieting was found to be associated with impaired performance, giving results very similar to those shown in Table 9.2. Nevertheless, weight loss during dieting in these subjects was found to be negligible. This suggests that for example the effort of attempting to restrict eating or stresses related to the perceived need to diet might be more important than the actual level of food restriction in determining the effects of dieting on cognitive performance.

Table 9.2 Cognitive performance and motor speed in dieters and non-dieters[a]

	Dieters (n = 18)	Non-dieters (n = 45)	p (2-tail)[b]
Rapid information processing: % correct hits (in first 3 min)	37 ± 5	48 ± 3	0.041
number of false alarms	5.1 ± 1.5	5.2 ± 0.7	>0.5
Immediate memory (words recalled[c])	8.6 ± 0.9	10.5 ± 0.5	0.050
Simple reaction time (msec)	382 ± 18	338 ± 10	0.026
Tapping rate (taps/sec)	8.8 ± 1.0	6.9 ± 0.4	0.050

[a] Figures are means ± SEM. [b] t-test. [c] Maximum is 20. Adapted from Green *et al.* (1994).

9.2.8 Are there net benefits from caffeine use for mood and cognitive performance?

Caffeine is probably the most widely 'used' psychoactive substance in the world. This is based on an estimated global consumption of 120 000 tonnes of caffeine per year (Gilbert, 1984), the main sources of this caffeine being coffee and tea. In part, the consumption of caffeine-containing drinks is influenced by the recognition of their potential psychoactive effects. For instance, the coffee drinker may choose to consume strong coffee at breakfast for its expected alerting effects, but avoid coffee late in the evening because of the expectation that consumption of caffeine will lead to difficulty in getting to sleep.

A psycho-stimulant action is also supported by extensive research on the effects of caffeine. For example, caffeine has been reported to quicken reaction time, improve vigilance and concentration, and increase feelings of alertness and energy (reviewed by James, 1991). However, other behavioural effects of caffeine are clearly less desirable, including increased anxiety at higher doses or in certain vulnerable individuals, and decreased hand steadiness (James, 1991). Furthermore, with sustained exposure, tolerance develops to at least some of the effects of caffeine, and in many regular users cessation of caffeine consumption is followed temporarily by adverse changes such as increased incidence of headache, drowsiness and fatigue (e.g. Griffiths and Woodson, 1988; van Dusseldorp and Katan, 1990; Höfer and Bättig, 1994).

This is very clearly demonstrated in a study in which we compared morning mood in caffeine users and nonusers (Richardson *et al.*, 1995). The group of users was divided into 3 matched sub-groups who avoided all significant sources of caffeine for either 1.5 hours, 13 hours, or at least 7 days before testing. Two patterns were apparent in the results. First, the overnight (13 hours) deprived group showed markedly increased levels of tiredness and drowsiness, and were more angry and dejected compared with all of the other groups, who did not differ significantly on these moods. The second pattern was less definite, but tended to be characterised by poorer mood (e.g. decreased clearheadedness and cheerfulness) and increased headache in both the 13-hour and 7-day groups.

These results show that overnight caffeine deprivation is sufficient to induce significant negative effects, including tiredness, headache and depressed mood. Although it is possible that pre-existing differences in personality or other factors unaffected by caffeine use could explain differences in mood between users and nonusers (Smith *et al.*, 1991; Rogers *et al.*, 1995a), this study also found that increased tiredness, drowsiness, anger and dejection were present in caffeine users after overnight caffeine deprivation but not after prolonged deprivation. In other words, the only factor which can reasonably account for the presence or absence of these particular symptoms is the subjects' recent history of caffeine consumption. Similar findings from a smaller study were reported by Bruce *et al.* (1991), who found increased tiredness in 24-hour compared with 7-day caffeine-deprived subjects. Although nonusers were not tested, a further result was that high doses (250 and 500 mg) of caffeine reduced tiredness, and also headache, only in the 24-hour group.

Such findings illustrate the difficulty of determining the net effects of caffeine use. In a typical experiment most if not all of the subjects have a history of regular caffeine consumption, and they are tested on caffeine and a placebo after a period of caffeine deprivation (usually no longer than 24 hours). The problem with relying solely on this approach is that it leaves open the question as to whether the results obtained are due to beneficial effects of caffeine or to deleterious effects of caffeine deprivation (or a combination of both of these) (James, 1991; Rogers *et al.*, 1995a).

To add further to this complexity it appears that cognitive performance is less affected by caffeine deprivation than is mood. On reviewing the literature we could find no unequivocal examples of impaired cognitive performance associated with caffeine deprivation (Rogers *et al.*, 1995a), and this included our own recent study which was designed specifically to test for the effects of overnight caffeine deprivation on performance. The subjects in this experiment were divided into lower ($n = 37$) and higher ($n = 36$) caffeine user groups (mean ± s.e.m. caffeine intakes: 232 ± 17 and 522 ± 32 mg/d, respectively), equated for age and gender. They were tested mid-morning on a 25–30 minute long, simple reaction time task. Caffeine (1 mg/kg body weight) and placebo capsules were administered on separate occasions in a counterbalanced order with at least one week between test sessions. Testing began one hour after the capsule was consumed. The results, summarised in Figure 9.1, showed that this cup-of-coffee equivalent dose of caffeine significantly improved performance only in the higher consumers. Superificially, this is consistent with an action of caffeine confined to the reversal of performance degraded due to caffeine deprivation. That is, higher consumers experience greater deprivation effects and consequently benefit more from caffeine administration. However, other aspects of the results suggest a different interpretation. First, compared with the lower caffeine users, the higher users receiving placebo actually displayed significantly faster reaction times during the first block of trials. Furthermore, at no point during the test was their performance clearly worse than that of the lower users. Second, the higher users given caffeine displayed the fastest overall reaction times. Therefore, if anything, these data indicate a net beneficial effect of caffeine on cognitive performance in higher users.

The findings of some other recent studies also support this conclusion. Jarvis (1994), for example, recently reported results showing a strong positive dose-response relationship between habitual caffeine intake and cognitive performance. This relationship remained even 'after controlling

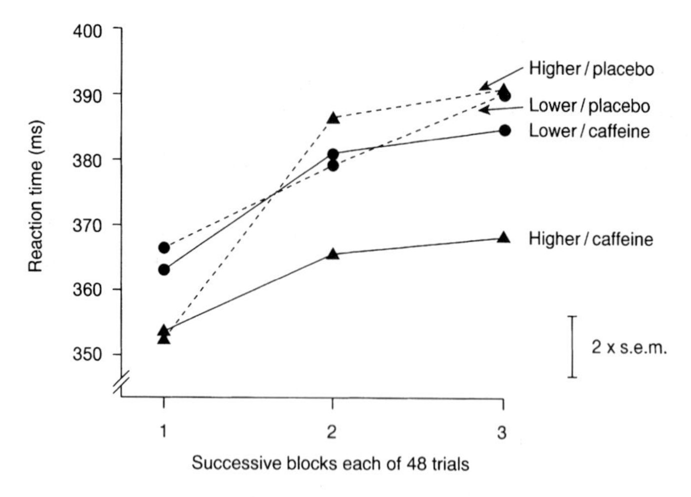

Figure 9.1 Performance of lower and higher caffeine users (see text for details) on a long-duration simple reaction time task following overnight caffeine deprivation. Testing began mid-morning one hour after administration of caffeine (1 mg/kg) or placebo. Each trial was preceded by a 'get ready' signal and, to prevent anticipatory responding, a fixed preparatory interval of 1, 3, 7 or 16 seconds. Analysis of variance revealed a significant 3-way interaction (Use x Treatment x Block), $F(2,142) = 3.84$, $p = 0.024$. The vertical bar is the least significant difference at the 5% significance level. (Rogers *et al.*, 1995a; reproduced with permission of the publisher, Karger: Basel.)

extensively for potential confounding variables' (p. 45). In addition cognitive performance-enhancing effects of caffeine have been found in subjects deprived of caffeine for only 3 to 4 hours (e.g. Smith *et al.*, 1994), where deprivation effects are likely to be minimal (Rogers *et al.*, 1995a).

Although there is a very large literature on the psychoactive effects of caffeine, this does not provide precise evidence as to the benefit of caffeine use in everyday life. Therefore, a priority for future research is to evaluate the psychoactive effects of caffeine in relation to the nature of the task and extraneous factors affecting performance, and using doses and a pattern of administration of caffeine which more closely model the spontaneous behaviour of typical caffeine users. In particular, we have suggested that there is a case for investigating the cumulative effects of caffeine administered on repeated occasions throughout the day, and relating these effects to biological variables indicative of the systemic activity of caffeine (Rogers *et al.*, 1995a).

9.3 Relationship between the mood and performance effects of foods and food choice

The discussion above has reviewed some of the evidence demonstrating the existence of psychoactive effects of common dietary constituents. It has also

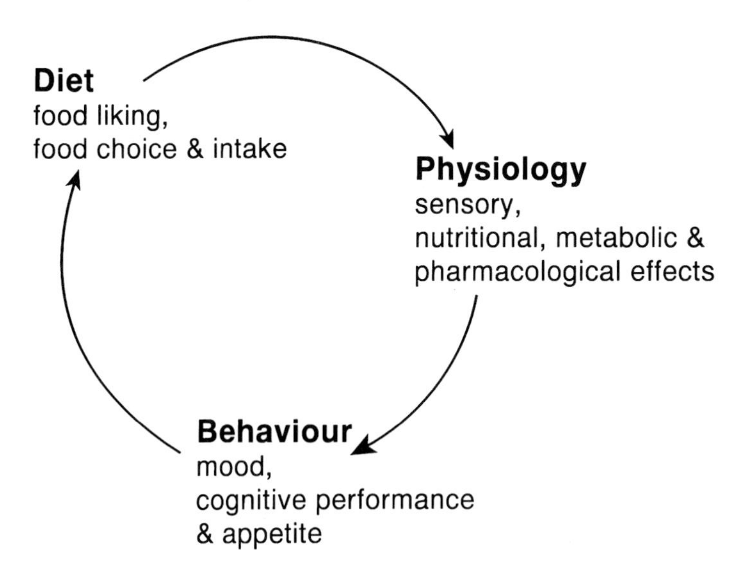

Figure 9.2 Interrelationships between diet and behaviour: dietary influences on mood, cognitive performance and appetite are mediated by physiological mechanisms and can feed back to influence food preference, food choice and food intake. (Adapted from Rogers *et al.*, 1994.)

pointed to certain mechanisms which have been suggested to underlie these effects. A further possibility is that the mood and certain other effects of food can in turn influence food choice. These interrelationships are illustrated in Figure 9.2, and the purpose of the final sections of this chapter is to examine examples of the possible bases for the connection going from 'behaviour' to 'diet'.

9.3.1 Learned preferences reinforced by the psychoactive effects of food and drinks

Beliefs about the supposed psychoactive effects of a food or drink will sometimes influence a person's decision to consume that food or drink. As suggested above (§9.2.8), coffee and tea drinking may be to a significant extent motivated by anticipation of benefiting from the stimulant properties of caffeine (Rogers and Richardson, 1993). However, if people are asked why they drink coffee and tea, they are more likely to say that this is because they like the 'taste' of the drink. Typically, it is not the case that people consume coffee as if it were a medicine, being prepared to tolerate its taste in the expectation of a benefit. On the other hand, it is fairly certain that human beings are not born with a liking for the taste and flavour of either coffee or tea, at least partly, because these drinks contain bitter constituents (including caffeine) and bitterness is innately aversive (Cines and Rozin, 1982). This then raises the question of how people come to acquire a liking for the sensory qualities of these drinks.

One way in which preferences are modified is through the association of the orosensory and postingestive effects of eating and drinking. The most dramatic example of this is the strong and specific aversions which can develop when consumption of a food is followed by gastrointestinal illness (Garcia *et al.*, 1974). Similarly, there is now good evidence that association of a taste or flavour paired with positive postingestive consequences can result in increased preference for that specific taste or flavour (e.g. Booth, 1978; Sclafani, 1990). For example, a small number of studies on human subjects have demonstrated conditioned increases in preference in children and adults for flavours associated with high carbohydrate content (Booth *et al.*, 1982; Birch *et al.*, 1990) and high fat content (Kern *et al.*, 1993). These changes appear to persist over time, and also tend to be state-dependent, such that preference is more readily modified if initial exposure occurs during a state of hunger, and is subsequently expressed more strongly when the individual is hungry compared with when he or she has recently eaten. Another feature of conditioned preferences and aversions is that they appear to be characterised by a change actual liking for the food or drink. For instance, after aversive conditioning the food tastes unpleasant, it is not simply that it is avoided because it is expected to cause harm (Booth, 1978; Rogers, 1990).

Studies on conditioned preferences have demonstrated the existence of a capacity to adapt preference for a food according to the benefit or otherwise derived from consuming that food. Not surprisingly, the main focus of this research has been on the 'nutritional' effects (e.g. high versus low energy density) of eating and drinking. However, as shown by the previous sections of this review, certain constituents of foods and drinks can also have pharmacological activity and significant effects on mood. Caffeine and alcohol are the most obvious examples. Perhaps, therefore, preferences for coffee, tea, beer, wine etc. are reinforced by the psychoactive effects of caffeine and alcohol (see also Cines and Rozin, 1982; Zellner, 1991; Rogers and Richardson, 1993).

Recently, we attempted to test this suggestion directly in studies in which caffeine ingestion was paired with the consumption of novel-flavoured fruit juices. Caffeine was given either in the drink or in a capsule swallowed with the drink. A drink of a different flavour was given without caffeine or with a placebo capsule containing a non-pharmacologically active substance such as cornflour. The design of these studies is, therefore, similar in principle to methods used in the work on flavour preferences conditioned by nutrient manipulations, and the straightforward prediction is that if caffeine has beneficial effects on mood, then pairing the drink with the consumption of caffeine should promote increased preference for that drink. Initial results showed that under these circumstances caffeine was not a strong positive reinforcer, although it did have aversive effects (Rogers *et al.*, 1992) and, more importantly, it appeared to act as a negative reinforcer by removing or alleviating the negative effects of overnight caffeine withdrawal (Rogers *et al.*, 1995b). In this latter study caffeine was found to have no significant effects on drink preference or mood in subjects with habitually low intakes of caffeine, whereas moderate users of caffeine developed a relative dislike for the drink lacking caffeine (Table 9.3). The moderate users, moreover, showed lowered mood following overnight caffeine abstinence (e.g. less lively, clearheaded and cheerful), and this was significantly improved by caffeine.

Results of an earlier study indicated similar effects in rats (Vitiello and Woods, 1977). Rats which had been previously given caffeine for 12 consecutive days (by injection) developed a relative aversion for a novel taste (saccharin) paired with the absence of caffeine. In addition, aversion for saccharin was also seen when this taste was paired with the injection of caffeine in caffeine-naive rats. These contrasting effects occurred at doses of caffeine which, at least based on a simple body-weight calculation, can be equated to the intakes of caffeine in humans drinking, for example, no more than 6 cups of instant coffee per day. The rats, however, received the caffeine in a single daily injection.

An implication of these findings is that caffeine will have a positive influence on the consumption of caffeine-containing drinks only after a pattern of fairly frequent, perhaps daily, intake of these drinks has already been established. Initially, therefore, other factors must operate to promote the habit (Zellner, 1991). Thirst may motivate consumption on some occasions, but in general most people consume tea and coffee, together with other beverages, well in excess of what is needed to maintain adequate fluid balance. Alternatively, preferences for coffee and tea could be acquired through association with the nutritional benefit derived from added milk,

Table 9.3 Development of preferences in low and moderate caffeine users for non-caffeinated and caffeinated fruit-juice drinks consumed at breakfast

	Subjects' usual level of caffeine intake	
Caffeine content of fruit juice	Low (<120 mg/day) $n = 25$	Moderate (≥120 mg/day) $n = 24$
Zero	1.43 ± 0.53 (a)	−1.08 ± 0.38 (a)
70 mg	0.40 ± 0.60	0.92 ± 0.47 (b)

The data are mean ± SEM changes in ranked preference measured before and after 10 conditioning trials. The maximum change in preference is 3 or −3, representing respectively an increase or decrease in preference. Means denoted by different letters are significantly different ($p < 0.05$, least significant difference test). For each subject a conditioning trial consisted of the consumption of 180 ml of the fruit-juice immediately after eating their normal breakfast. From the evening before this until mid-morning they were not allowed to consume any other drinks except water. Adapted from Rogers *et al.* (1995b).

cream and/or sugar, or through a flavour–flavour conditioning process (Baeyens *et al.*, 1990). If nothing else, the use of sweeteners, milk and cream may provide an immediate way of improving the sensory appeal of the beverage for the novice coffee or tea drinker.

Yet a further possibility is that situational influences on mood play a role in reinforcing preferences for certain foods and beverages. Coffee, tea and indeed many drinks are typically consumed in social contexts and during, for example, breaks from work or other activities. Pairing the positive shifts in mood occurring in such situations with consumption of a drink could result in a conditioned increase in preference for that drink. This process is sometimes referred to as evaluative conditioning (Zellner, 1991).

The reinforcing effects of caffeine, alcohol and a variety of other psychoactive substances used by human beings have been studied extensively (e.g. Stolerman, 1992; Giffiths and Mumford, 1995; O'Brien *et al.*, 1995). However, these effects have not generally been considered in relation to their impact on orosensory preference (liking), and at present rather little is known about how preferences for coffee, beer and cigarettes, for example, are learned. Further study of the role of conditioning in the acquisition of dietary habits is therefore clearly warranted. This can be expected to confirm a significant influence of the mood effects of dietary constituents on preferences for particular foods and drinks.

9.3.2 Carbohydrate craving and the self-medication of depression

Another hypothesis which explicitly links the effects of food on mood to food choice is Wurtman and Wurtman's (1989) proposal concerning the relationship between dysphoric mood and carbohydrate craving. This is based on the possibility that increased carbohydrate intake might augment brain serotonergic activity (see §9.2.2) and in turn ameliorate depression. Some support for this comes from results showing improvements in mood following the consumption of high-carbohydrate meals (see §9.2.3). In addition, several studies indicate a specific increase in carbohydrate consumption associated with depression in CCO, SAD and PMS (e.g. Fernstrom *et al.*, 1987; Rosenthal *et al.*, 1987; Krauchi and Wirz-Justice, 1988; Wurtman *et al.*, 1989).

Serotonergic-mediation of these effects has also been tested. For example, Wurtman *et al.* (1987) identified a group of obese carbohydrate cravers and a group of obese noncarbohydrate cravers and compared their responses to the drug *d*-fenfluramine which increases serotonergic neurotransmission. Carbohydrate craving was defined on the basis of the frequent consumption of snacks high in carbohydrate and fat, but low in protein (average of 7 such snacks/day). The smaller number of noncarbohydrate cravers had a similar frequency of snacking, divided about equally between high protein and high carbohydrate snacks. Compared with placebo, treatment with *d*-fenfluramine had a more immediate and larger effect on the frequency of snacking and snack intake in the carbohydrate cravers, although there was also a significant reduction in snacking in the noncarbohydrate cravers during the third and final month on *d*-fenfluramine. These results were interpreted as being consistent with the proposal that *d*-fenfluramine decreases hunger for carbohydrates in the carbohydrate craver by mimicking the effects of carbohydrate consumption on serotonergic functioning. However, although *d*-fenfluramine also reduced meal time energy intake in the carbohydrate cravers (but not in noncarbohydrate cravers), this was due to a similar reduction in intake of all macronutrients. In addition, *d*-fenfluramine did not appear to be superior to placebo in its effects on mood. On the other hand, *d*-fenfluramine has been found to be effective in the treatment of depression in SAD and PMS (O'Rourke *et al.*, 1989; Brzezinski *et al.*, 1990), and in the latter of these studies the improvement in mood was accompanied by a suppression of premenstrual increases in carbohydrate and fat intake.

An aspect of the mood, carbohydrate craving and serotonin hypothesis which has not been widely discussed is the mechanism by which depressed mood is supposed to give rise to carbohydrate craving. One suggestion is that the depressed individual recognises the beneficial effects of carbohydrates and accordingly deliberately increases carbohydrate intake in order to improve his or her mood (Wurtman, 1988). Wurtman and Wurtman (1989) have also proposed that the increased carbohydrate consumption may occur because the normal feedback regulation of carbohydrate and protein intake is disrupted in CCO, SAD and PMS. Normally, it is argued, consumption of carbohydrate would be expected to increase brain serotonin activity and inhibit further carbohydrate intake. In depression, when brain serotonin levels are low this response is reduced and the desire to eat carbohydrates persists. Other variants of these ideas have also been suggested (e.g. Møller, 1992), but none of them appears to have been tested directly.

A further possibility is that appetite for carbohydrates is amplified because of an increased liking for high-carbohydrate foods reinforced by their effects on mood (Rogers *et al.*, 1992; and §9.3.1). Strong conditioned preferences are unlikely to develop, however, if there is any substantial delay between the ingestion of carbohydrate and improved mood. A similar suggestion was made to account for the observation that the carbohydrate to fat ratio normally chosen by subjects at either breakfast or lunch tended to correspond with the macronutrient composition favouring optimal post-meal alertness (Lloyd and Rogers, 1994; §9.2.5). These effects occurred relatively soon after eating.

One important criticism of Wurtman and Wurtman's (1989) carbohydrate-craving hypothesis has been the use and definition of the term craving. Although there may be an association between lowered mood and increased carbohydrate intake or increased preference for high carbohydrate foods, it is far from clear that these changes in eating behaviour can be characterised as arising from cravings for carbohydrates. Craving suggests a particular intensity as well as specificity of appetite, and while carbohydrate craving is obviously an appealing term, its existence as a distinct form of appetite remains unproven. The most significant difficulty for the hypothesis, however, is the evidence suggesting that the actual effect of carbohydrate on brain serotonin synthesis is at best very small and therefore probably functionally insignificant (see §9.2.4).

9.3.3 *Sugar, fat and endogenous opioids*

The notion of carbohydrate craving has also been challenged on the grounds that the preferences appear to be for foods high in both fat and (sweet) carbohydrate (Drewnowski *et al.*, 1992b). This, moreover, has been linked with activity of the endogenous opioid system rather than serotonergic mediation (Drewnowski, 1992). Recent evidence shows that opioid peptides play a significant role in the regulation of food and fluid intake, and specifically in the mediation of hedonic responses to orosensory stimuli (reviewed by Kirkham and Cooper, 1991; Rogers, 1993). Other findings indicate that opioids are released in response to the ingestion of palatable foods (Fullerton *et al.*, 1985; Blass, 1991).

These observations have led, in turn, to proposals linking effects of eating on endogenous opioids with food craving and effects on mood. For example, Drewnowski (1992) has argued that overeating and reported cravings for foods high in sugar and fat may share a common mechanism with opiate drug addiction. Earlier, Blass (1987) suggested that consumption of sweet and other palatable foods may be motivated in part by their capacity to relieve stress, this effect being mediated by the release of endogenous opioids in response primarily to sweet taste. These ideas are supported by results showing that the opioid antagonist drug naloxone is highly effective in reducing food intake in binge eaters, due largely to an effect on sweet, high-fat foods such as chocolate and biscuits (Drewnowski *et al.*, 1992a). It is also claimed that opioid antagonists suppress stress-induced overeating (Fullerton *et al.*, 1985).

9.3.4 *Possible motivational effects of learned associations between mood and eating*

A different analysis of food craving is suggested by results obtained by Weingarten (1983) demonstrating the influence of learned associations on basic motivational processes. In these studies rats were fed liquid meals on a preprogrammed schedule giving a total daily intake of 70% of their *ad libitum* intake. During this training phase each meal was signalled with a tone and light (the conditioned stimuli, CS+) presented for 4 minutes before and 30 seconds after the food was made available. In the test phase the CS+ was again presented, but the rats were no longer food restricted since a bottle of the liquid food was available *ad libitum*. The rats, nonetheless, responded 'robustly and rapidly' to the CS+ by taking a substantial meal, and this behaviour was maintained throughout many subsequent days of testing. This shows that external stimuli previously associated with food consumption can reliably motivate eating in the absence of immediate nutritional 'need'.

These observations would appear to be very important for understanding the control of food intake and food choice, and other appetitive activities including drug use and abuse (*cf.* Stolerman, 1992). The basic findings have been replicated in a study on preschool children (Birch *et al.*, 1989), and further studies on rats have investigated suggestions for physiological mechanisms mediating external stimulus control of eating (Weingarten, 1984; Weingarten and Martin, 1989). It also appears that a relatively lengthy presentation of the conditioned stimuli is necessary to enhance eating (Weingarten, 1985). However, many questions about this phenomenon remain unanswered.

Of particular interest here is the possibility that presentation of a CS+ which has become associated with consumption of a food may elicit a desire to eat or 'craving' for that specific food or reinforcer (Weingarten, 1985). If this is the case, then external stimulus control of eating, which Weingarten refers to as conditioned meal initiation, might more accurately be said to be an example of a conditioned or learned specific appetite. An extension of this idea is that specific appetites might also become conditioned to salient internal stimuli accompanying, for example, particular emotional states. In fact, more probably this would involve associations formed between eating and a configuration of both internal and external stimuli evoking the emotional response (*cf.* Robbins and Fray, 1980; Booth, 1994). Therefore, as well as providing a mechanism linking mood and food craving, this could help explain effects of mood on eating at individual level, since these relationships would be shaped according a person's own unique learning history.

9.4 Conclusions

The examples of research findings reviewed in this chapter show that eating and drinking can have substantial effects on mood and cognitive performance. These effects appear to be mediated by sensory and predigestive as well as postabsorptive influences of the substances consumed. Some dietary constituents, such as caffeine and alcohol, have relatively direct actions on the central nervous system, although of course ultimately all changes in mood and performance will be encoded by changes in brain activity. In turn, the aftereffects of ingestion, including changes in mood, can be expected to influence food choice on future occasions. This may involve a modification of liking, that is, an alteration in the hedonic response to the taste, flavour, texture, etc. of the food or drink, through the association of the orosensory and postingestive effects of eating and drinking. Formation of associations between mood and eating may also underlie some instances of food 'craving'. However, in contrast to possible dietary effects on, for example, brain serotonin and endogenous opioids, these mechanisms have received very little attention. This, together with the potential for understanding the relationships between food choice, mood and appetite at an individual level, suggests that the investigation of the role of basic conditioning processes in the control of normal and excessive appetites should be a priority for future research.

Acknowledgement

The preparation of this review was supported by the Biotechnology and Biological Sciences Research Council.

References

Baeyens, F., Eelen, P., Van den Bergy, O. and Crombez, G. (1990) Flavour–flavour and colour–flavour conditioning in humans. *Learning and Motivation* **21**, 434–455.

Benton, D. and Sargent, J. (1992) Breakfast, blood glucose and memory. *Biological and Psychology* **33**, 207–210.

Birch, L.L., McPhee, L., Sullivan, S. and Johnson, S. (1989) Conditioned meal initiation in young children. *Appetite* **13**, 105–113.

Birch, L.L., McPhee, L., Steinberg, L. and Sullivan, S. (1990) Conditioned flavour preferences in young children. *Physiology and Behavior* **47**, 501–505.

Blass, E.M. (1987) Opioids, sweets and a mechanism for positive affect: Broad motivational implications. In *Sweetness*, pp. 115–126 [J. Dobbing, ed.]. Berlin, Springer-Verlag.

Blass, E.M. (1991) Suckling: Opioid and nonopioid processes in mother-infant bonding. In *Chemical Senses*: Volume 4, *Appetite and Nutrition*, pp. 283–302 (M.I. Friedman, M.G. Tordoff, M.R. Kare, eds). New York: Marcel Dekker.

Booth, D.A. (1978) Acquired behaviour controlling energy intake and output. *Psychiatric Clinics of North America* **1**, 545–579.

Booth, D.A. (1994) *Psychology of Nutrition*. London: Taylor and Francis.

Booth, D.A., Mather, P. and Fuller, J. (1982) Starch content of ordinary foods associatively conditions human appetite and satiation, indexed by intake and eating pleasantness of starch-paired flavours. *Appetite* **3**, 163–184.

Brody, S. and Wolitzky, D.L. (1983) Lack of mood changes following sucrose loading. *Psychosomatics* **24**, 155–162.

Bruce, M., Scott, N., Shine, P. and Lader, M. (1991) Caffeine withdrawal: a contrast of withdrawal symptoms in normal subjects who have abstained from caffeine for 24 hours and for 7 days. *Journal of Psychopharmacology* **5**, 129–134.

Brzezinski, A.A., Wurtman, J.J., Wurtman, R.J., Gleason, R., Greenfield, J. and Nader, T. (1990) d-Fenfluramine suppresses the increased calorie and carbohydrate intakes and improves the mood of women with premenstrual depression. *Obstetrics and Gynecology* **76**, 296–301.

Christensen, L. and Redig, C. (1993) Effect of meal composition on mood. *Behavioral Neuroscience* **107**, 346–353.
Cines, B.M. and Rozin, R. (1982) Some aspects of liking for hot coffee and coffee flavour. *Appetite* **3**, 23–34.
Cockcroft, V. (1993) Chocolate on the brain. *The Biochemist* Apr/May, 14–16.
Cowen, P.J., Anderson, I.M. and Fairburn, C.G. (1992) Neurochemical effects of dieting: Relevance to changes in eating and affective disorders. In *The Biology of Feast and Famine: Relevance to Eating Disorders*, pp. 269–284. (G.H. Anderson and S.H. Kennedy, eds). San Diego: Academic Press.
Craig, A., Baer, K. and Diekmann, A. (1981) The effects of lunch on sensory–perceptual functioning in man. *International Archives of Occupational and Environmental Health* **49**, 105–114.
de Castro, J.M. (1987) Macronutrient relationships with meal patterns and mood in the spontaneous feeding behaviour of humans. *Physiology and Behavior* **39**, 561–569.
Deijen, J.M., Heemstra, M.L. and Orlebeke, J.F. (1989) dietary effects on mood and performance. *Journal of Psychiatric Research* **23**, 275–283.
Department of Health (1992) *Health of the Nation*. London: HMSO.
Dickie, N.H. and Bender, A.E. (1982) Breakfast and performance in schoolchildren. *British Journal of Nutrition* **48**, 483–496.
Drewnowski, A. (1992) Food preferences and the opioid peptide system. *Trends in Food Science and Technology* **3**, 97–99.
Drewnowski, A., Krahn, D.D., Demitrack, M.A., Nairn, K. and Gosnell, B.A. (1992a) Taste responses and preferences for sweet high-fat foods: evidence for opioid involvement. *Physiology and Behavior* **51**, 371–379.
Drewnowski, A., Kurth, C., Holden-Wiltse, J. and Saari, J. (1992b) Food preferences in human obesity: carbohydrate versus fats. *Appetite* **18**, 207–221.
Editor. (1957) Physiologic results of breakfast habits. *Nutrition Reviews* **15**, 196–198.
Fernstrom, J.D. (1994) Dietary amino acids and brain function. *Journal of the American Dietetic Association* **94**, 71–77.
Fernstrom, M.H., Krowinsk, R.L. and Kupfer, D.J. (1987) Appetite and food preference in depression: effects of imipramine treatment. *Biological Psychiatry* **22**, 529–539.
Finnigan, F. and Hammersley, R. (1992) The effects of alcohol on performance. In *Handbook of Human Performance, Volume 2, Health and Performance*, pp. 73–126 (A.P. Smith and D.M. Jones, eds). London: Academic Press.
French, S.J., Murray, B., Rumsey, R.D.E., Fadzlin, R. and Read, N.W. (1995) Adapation to a high-fat diets: effect on eating behaviour and plasma cholecystokinin. *British Journal of Nutrition* **73**, 179–189.
Fullerton, D.T., Getto, C.J., Swift, W.J. and Carlson, I.H. (1985) Sugar, opioids and binge eating. *Brain Research Bulletin* **14**, 673–680.
Garcia, J., Hankins, W.G. and Rusiniak, K.W. (1974) Behavioural regulation of the milieu interne in man and rat. *Science* **185**, 824–831.
Gardner, M.L.G. (1984) Intestinal assimilation of intact peptides and proteins from the diet– a neglected field? *Biological Review* **59**, 289–331.
Gilbert, R.M. (1984) Caffeine consumption. In *The Methylxanthine Beverages and Foods: Chemistry, Consumption and Health*, pp. 185–214 (G.A. Spiller, ed.). New York: Alan R. Liss.
Goodwin, G.M., Cowne, P.J., Fairburn, C.G., Parry-Billings, M., Calder, P.C. and Newsholme, E.A. (1990) Plasma concentrations of tryptophan and dieting. *British Medical Journal* **300**, 1499–1500.
Goodwin, G.M., Fairburn, C.G. and Cowne, P.J. (1987) Dieting changes serotonergic function in women, not men: implications for the aetiology of anorexia nervosa? *Psychological Medicine* **17**, 839–842.
Green, M.W. and Rogers, P.J. (1995) Impaired cognitive functioning in dieters during dieting. *Psychological Medicine*, **25**, 1003–1010.
Green, M.W., Rogers, P.J., Elliman, N.A. and Gatenby, S.J. (1994) Impairment of cognitive performance associated with dieting and high levels of dietary restraint. *Physiology and Behavior* **55**, 447–452.
Green, M.W., Elliman, N.A. and Rogers, P.J. (1995) Lack of effect of short-term fasting on cognitive function. *Journal of Psychiatric Research*, **29**, 245–253.

Griffiths, R.R. and Woodson, P.P. (1988) Caffeine physical dependence: a review of human and laboratory animal studies. *Psychopharmacology* **94**, 437–451.
Griffiths, R.R. and Mumford, G.K. (1995) Caffeine—A drug of abuse? In *Psychopharmacology: The Fourth Generation of Progress*, pp. 1699–1713 (F.E. Bloom and D.J. Kupfer, eds). New York: Raven Press.
Hassmén, P., Blomstrand, E., Ekblom, B. and Newsholme, E.A. (1994) Branched-chain amino acid supplementation during 30-km competitive run: Mood and cognitive performance. *Nutrition* **10**, 405–410.
Hill A.J. and Heaton-Brown, L. (1995) The experience of food craving: A prospective study in healthy women. *Journal of Psychosomatic Research*, **38**, 801–814.
Hill, A.J., Weaver, C.F.L. and Blundell, J.E. (1991) Food craving, dietary restraint and mood. *Appetite* **17**, 187–197.
Höfer, I. and Bättig, K. (1994) Cardiovascular, behavioural, and subjective effects of caffeine under field conditions. *Pharmacology, Biochemistry and Behavior* **48**, 899–908.
Hrboticky, K., Leiter, L.A. and Anderson, G.H. (1985) Effects of L-tryptophan on short term food intake in lean men. *Nutrition Research* **5**, 595–607.
James, J.E. (1991) *Caffeine and Health*. Academic Press, London.
Jarvis, M. (1993) Does caffeine intake enhance absolute levels of cognitive performance? *Psychopharmacology* **110**, 45–52.
Kanarek, R.B. and Marks-Kaufman, R. (1991) *Nutrition and Behavior: New Perspectives*. New York: Van Nostrand Reinhold.
Kaplan, J.R., Shively, C.A., Fontenot, M.B., Morgan, T.M., Howell, S.M., Manuck, S.B., Muldoon, M.F. and Mann, J.J. (1994) Demonstration of an association among dietary cholesterol, central serotonergic activity, and social behaviour in monkeys. *Psychosomatic Medicine* **56**, 479–484.
Kern, D.L., McPhee, L., Fisher, J., Johnson, S. and Birch, L.L. (1993) The postingestive consequences of fat condition preferences for flavours associated with high dietary fat. *Physiology and Behavior* **54**, 71–76.
Keys, A., Brozek, J., Henschel, A., Mickelsen, O. and Taylor, H.F. (1950) *The Biology of Human Starvation*, Vols 1 and 2. Minneapolis: University of Minnesota Press.
Kirkham, T.C. and Cooper, S.J. (1991) Opioid peptides in relation to the treatment of obesity and bulimia. In *Peptides: A Target for New Drug Development*, pp. 28–44 (S.R. Bloom and G. Burnstock, eds). London: IBC.
Krauchi, K. and Wirz-Justice, A. (1988) The four seasons: food intake frequency in seasonal affective disorder in the course of a year. *Psychiatric Research* **25**, 232–338.
Leathwood, P.D. (1987) Tryptophan availability and serotonin synthesis. *Proceedings of the Nutrition Society* **46**, 143–156.
Liddle, R.A., Goldfine, I.D. and Williams, J.A. (1983) Bioassy of circulating CCK in rat and human plasma. *Gastroenterology* **84**, 1231–1236.
Lieberman, H.R., Corkin, S., Spring, B.J., Growden, J.H. and Wurtman, R.J. (1983) Mood, performance, and pain sensitivity: Changes induced by food constituents. *Journal of Psychiatric Research* **17**, 135–145.
Lieberman, H., Wurtman, J. and Chew, B. (1986a) Changes in mood after carbohydrate consumption among obese individuals. *American Journal of Clinical Nutrition* **45**, 772–778.
Lieberman, H.R., Caballero, B. and Finer, N. (1986b) The composition of lunch determines plasma tryptophan rations in humans. *Journal of Neural Transmission* **65**, 211–217.
Lloyd, H.M. and Rogers, P.J. (1994) Acute effects of breakfasts of differing fat and carbohydrate content on morning mood and cognitive performance. *Proceedings of the Nutrition Society* **53**, 239A.
Lloyd, H.M., Green, M.W. and Rogers, P.J. (1994) Mood and cognitive performance effects of isocaloric lunches differing in fat and carbohydrate content. *Physiology and Behavior* **56**, 51–57.
Macdiarmid, J.I. and Hetherington, M.M. (1995) Mood modulation by food: An exploration of affect and cravings in 'chocolate addicts'. *British Journal of Psychology* **34**, 129–138.
Maes, M. and Meltzer, H.Y. (1995) The serotonin hypothesis of major depression. In *Psychopharmacology: The Fourth Generation of Progress*, pp. 933–944 (F.E. Bloom and D.J. Kupfer, eds). New York: Raven Press.
Max, B. (1989) This and that: chocolate addiction, the dual pharmacogenetics of

asparagus eaters and the arithmetic of freedom. *Trends in Pharmacological Science* **10**, 390–393.
Michener, W. and Rozin, P. (1994) Pharmacological versus sensory factors in the satiation of chocolate craving. *Physiology and Behavior* **56**, 419–422.
Møller, S.E. (1992) Serotonin, carbohydrates, and atypical depression. *Pharmacology and Toxicology* **71** (Supplement 1), 61–71.
Morley, J.E., Levine, A.S., Yamada, T., Gebhard, R.L., Prigge, W.F., Shafer, R.B., Goetz, F.C. and Silvis, S.E. (1983) Effect of exorphins on gastrointestinal function, hormonal release and appetite. *Gastroenterology* **84**, 1517–1523.
Muldoon, M.F., Manuck, S.B. and Matthews, K.A. (1991) Mortality experience in cholesterol reduction trials. *New England Journal of Medicine* **324**, 922–923.
Mumford, G.K., Evans, S.M., Kaminski, B.J., Preston, K.L., Sannerud, C.A., Silverman, K. and Griffiths, R.R. (1994) Dicriminative stimulus and subjective effects of theobromine and caffeine in humans. *Psychopharmacology* **115**, 1–8.
Newsholme, E.A., Blomstrand, E. and Ekblom, B. (1992) Physical and mental fatigue: Metabolic mechanisms and importance of plasma amino acids. *British Medical Bulletin* **48**, 477–495.
O'Brien, C.P., Eckardt, M.J. and Linnoila, V.M. (1995) Pharmacotherapy of alcoholism. In *Psychopharmacology: The Fourth Generation of Progress*, pp. 1745–1755 (F.E. Bloom and D.J. Kupfer, eds). New York: Raven Press.
O'Rouke, D., Wurtman, J.J., Wurtman, R.J., Chebli, R. and Gleason, R. (1989) Treatment of seasonal depression with d-fenfluramine. *Journal of Clinical Psychiatry* **50**, 343–347.
Pijl, H., Koppeschaar, H.P.F., Cohen, A.F., Iestra, J.A., Schoemaker, H.C., Frölich, M., Onkenhout, W. and Meinders, A.E. (1993) Evidence for brain serotonin-mediated control of carbohydrate consumption in normal weight and obese humans. *International Journal of Obesity* **17**, 513–520.
Pivonka, E.E.A. and Grunewald, K.K. (1990) Aspartame- or sugar-sweetened beverages: Effects on mood in young women. *Journal of the American Dietetic Association* **90**, 250–254.
Pollitt, E. (1987) Effects of iron deficiency on mental development: Methodological considerations and substantive findings. In *Nutritional Anthropology*, pp. 225–254 [F.E. Johnson, ed.]. New York: Alan R. Liss.
Pollitt, E., Leibel, R.L. and Greenfield, D. (1981) Brief fasting, stress, and cognition in children. *American Journal of Clinical Nutrition* **34**, 1526–1533.
Pollitt, E., Lewis, N.L., Garza, C. and Shulman, R.J. (1983) Fasting and cognitive function. *Journal of Psychiatric Research* **17**, 169–174.
Richardson, N.J., Rogers, P.J., Elliman, N.A. and O'Dell, R.J. (1995) Mood and performance effects of caffeine in relation to acute and chronic caffeine deprivation. *Pharmacology, Biochemistry and Behavior*, **52**, 313–320.
Robbins, T.W. and Fray, P.J (1980) Stress-induced eating: fact, fiction or misunderstanding? *Appetite* **1**, 103–133.
Robinson, J. and Ferguson, A. (1992) Food sensitivity and the nervous system: hyperactivity, addiction and criminal behaviour. *Nutrition Research Reviews* **5**, 203–223.
Rodin, J., Mancuso, J., Granger, J. and Nelbach, E. (1991) Food cravings in relation to body mass index, restraint and estradiol levels: a repeated measures study in healthy women. *Appetite* **17**, 177–185.
Rogers, P.J. (1990) Why a palatability construct is needed. *Appetite* **14**, 167–170.
Rogers, R.J. (1993) The experimental investigation of human eating behaviour. In *Human Psychopharmacology: Measures and Methods*, Volume 4, pp. 123–142 (I. Hindmarch and P.D. Stonier, eds). Chichester: Wiley.
Rogers, P.J. (1994) Mechanisms of moreishness and food craving. In *Pleasure: the Politics and the Reality*, pp. 38–49 (D.M. Warburton, ed.). Chichester: Wiley.
Rogers, P.J. and Green, M.W. (1993) Dieting, dietary restraint and cognitive performance. *British Journal of Clinical Psychology* **32**, 113–116.
Rogers, P.J. and Jas, P. (1994) Menstrual cycle effects on mood, eating and food choice. *Appetite* **23**, 289.
Rogers, P.J. and Lloyd, H.M. (1994) Nutrition and mental performance. *Proceedings of the Nutrition Society* **53**, 443–456.

Rogers, P.J. and Richardson, N.J. (1993) Why do we like drinks that contain caffeine? *Trends in Food Science and Technology* **4**, 108–111.

Rogers, P.J., Edwards, S., Green, M.W. and Jas, P. (1992) Nutritional influences on mood and cognitive performance: The menstrual cycle, caffeine and dieting. *Proceedings of the Nutrition Society* **51**, 343–351.

Rogers, P.J., Anderson, A.O., Finch, G.M., Jas, P. and Gatenby, S.J (1994a) Relationships between food craving and anticipatory salivation, eating patterns, mood and body weight in women and men. *Appetite* **23**, 319.

Rogers, P.J., Green, M.W. and Edwards, S. (1994b) Nutritional influences on mood and cognitive performance: Their measurement and relevance to food acceptance. In *Measurement of Food Preferences*, pp. 227–252 (H.J.H. MacFie and D.M.H. Thomson, eds). Glasgow: Blacke A&P.

Rogers, P.J., Richardson, N.J. and Dernoncourt, C. (1995a) Caffeine use: Is there a net benefit for mood and psychomotor performance? *Neuropsychobiology*, **31**, 195–199.

Rogers, P.J., Richardson, N.J. and Elliman, N.A. (1995b) Overnight caffeine abstinence and negative reinforcement of preference for caffeine-containing drinks. *Psychopharmacology*, **120**, 457–462.

Rosenthal, N.E., Genhart, M.J., Jacobson, F.M., Skwerer, R.G. and Weht, T.A. (1987) Disturbances of appetite and weight regulation in seasonal affective disorder. *Annals of the New York Academy of Sciences* **499**, 216–223.

Rosenthal, N.E., Genhart, M.J., Caballero, B., Jacobsen, F.M., Skwerer, R.G., Coursey, R.D., Rogers, S. and Spring, B.J. (1989) Psychobiological effects of carbohydrate- and protein-rich meals in patients with seasonal affective disorder and normal controls. *Biological Psychiatry* **25**, 1029–1040.

Schlundt, D.G., Virts, K.L., Sbrocco, T., Pope-Cordle, J. and Hill, J.O. (1993) A sequential behavioural analysis of craving sweets in obese women. *Addictive Behaviors* **18**, 67–80.

Schweiger, U., Laessle, R., Kittle, S., Dickhaut, B., Schweiger, M. and Pirke, K.M. (1986) Macronutrient intake, plasma large neutral amino acids and mood during weight reducing diets. *Journal of Neural Transmission* **67**, 77–86.

Sclafani, A. (1990) Nutritionally based learned flavour preferences in rats. In *Taste, Feeding and Experience*, pp. 139–156 (E.D. Capaldi and T.L. Powley, eds). Washington, DC: American Psychological Association.

Smart, J.L. (1993) 'Malnutrition, learning and behaviour': 25 years on from the MIT symposium. *Proceedings of the Nutrition Society* **52**, 189–199.

Smith, A.P. and Kendrick, A.M. (1992) Meals and performance. In *Handbook of Human Performance, Volume 2, Health and Performance*, pp. 2–23 (A.P. Smith and D.M. Jones, eds). London: Academic Press.

Smith, A.P. and Miles, C. (1986) Acute effects of meals, noise and nightwork. *British Journal of Psychology* **77**, 377–387.

Smith, A.P., Leekam, S., Ralph, A. and McNeill, G. (1988) The influence of meal composition on post-lunch performance efficiency and mood. *Appetite* **10**, 195–203.

Smith, A.P., Rusted, J.M., Savory, M., Eaton-Williams, P. and Hall, S.R. (1991) The effects of caffeine, impulsivity and time of day on performance, mood and cardiovascular function. *Journal of Psychopharmacology* **5**, 120–128.

Smith, A., Maben, A., and Brockman, P. (1994) Effects of evening meals and caffeine on cognitive performance, mood and cardiovascular functioning. *Appetite* **22**, 57–65.

Spring, B. (1986) Effects of foods and nutrients on the behaviour of normal individuals. In *Nutrition and the Brain*, Volume 7, pp. 1–47 (R.J. Wurtman and J.J. Wurtman, eds). New York: Raven Press.

Spring, B., Chiodo J. and Bowen, D.J. (1987) Carbohydrates, tryptophan, and behavior: A methodological Review. *Psychological Bulletin* **102**, 234–256.

Stacher, G., Bauer, H. and Steinringer, H. (1979) Cholecystokinin decreases appetite and activation evoked by stimuli arising from the preparation of a meal in man. *Physiology and Behavior* **23**, 325–331.

Stolerman, I. (1992) Drugs of abuse: behavioural principles, methods and terms. *Trends in Pharmacological Sciences* **13**, 170–176.

Tarka, S.M. (1982) The toxicology of cocoa and methylxanthines: a review of the literature. *CRC Critical Reviews of Toxicology* **9**, 275–312.

Teff, K.L., Young, S.N. and Blundell, J.E. (1989a) The effect of protein or carbohydrate breakfasts on subsequent plasma amino acid levels, satiety and nutrient selection in normal males. *Pharmaclogy, Biochemistry and Behavior* **34**, 829–837.

Teff, K.L., Young, S.M., Marchand, L. and Botez, M.I. (1989b) Acute effect of protein or carbohydrate breakfasts on human cerebrospinal fluid monoamine precursor and metabolite levels. *Journal of Neurochemistry* **52**, 235–241.

van Dusseldorp, M. and Katan, M.B. (1990) Headache caused by caffeine withdrawal among moderate coffee drinkers switched from ordinary to decaffeinated coffee: a 12-week double-blind trial. *British Medical Journal* **300**, 1558–1559.

van Strein, T., Frijters, J.E.R., Bergers, G.P.A. and Defares, P.B. (1986) The Dutch Eating Behaviour Questionnaire (DEBQ) for assessment of restrained, emotional, and external eating behaviour. *International Journal of Eating Disorders* **5**, 295–315.

Vitiello, M.V. and Woods, S.C. (1977) Evidence for withdrawal from caffeine in rats. *Pharmacology Biochemistry and Behavior* **6**, 553–555.

Weingarten, H.P. (1983) Conditioned cues elicit eating in sated rats: A role for learning in meal initiation. *Science* **220**, 431–433.

Weingarten, H.P. (1984) Meal initiation controlled by learned cues: Effects of peripheral cholinergic blockade and cholecystokinin. *Physiology and Behavior* **32**, 403–408.

Weingarten, H.P. (1985) Stimulus control of eating: implications for a two-factor theory of hunger. *Appetite* **6**, 387–401.

Weingarten, H.P. and Elston D. (1991) Food cravings in a college population. *Appetite* **17**. 167–175.

Weingarten, H.P. and Martin, G.M. (1989) Mechanisms of conditioned meal initiation. *Physiology and Behavior* **45**, 735–740.

Wurtman, J.J. (1988) Carbohydrate craving, mood changes, and obesity. *Journal of Clinical Psychiatry* **49**, (Supplement), 37–39.

Wurtman R.J. and Wurtman J.J. (1989) Carbohydrates and depression. *Scientific American* **260**, 50–57.

Wurtman, R.J., Hefti, F. and Melamed, E. (1981) Precursor control of neurotransmitter synthesis. *Pharmacological Reviews* **32**, 315–335.

Wurtman, J., Wurtman, R., Reynolds. S., Tsay, R., and Chew, B. (1987) Fenfluramine suppresses snack intake among carbohydrate cravers but not among noncarbohydrate cravers. *International Journal of Eating Disorders* **6**, 687–699.

Wurtman, J.J., Brzezinski, A., Wurtman, R.J., and Laferrere, B. (1989) Effect of nutrient intake on premenstrual depression. *American Journal of Obstetrics and Gynecology* **161**, 1228–1234.

Young, S.N. (1991) Some effects of dietary components (amino acids carbohydrate, folic acid) on brain serotonin synthesis, mood and behaviour. *Canadian Journal of Physiology and Pharmacology* **69**, 893–903.

Zellner, D.A. (1991) How foods get to be liked. In *The Hedonics of Taste*, pp. 199–217 (R.C. Bolles, ed.). Hillside New Jersey: Erlbaum.

10 Attitudes and beliefs in food habits

RICHARD SHEPHERD and MONIQUE M. RAATS

10.1 Models of food choice

The choice of foods by free-living individuals is an area of concern for many people involved in the production and distribution of foods, but also for those concerned with nutrition and health education. Despite a great deal of knowledge gained on the impacts of diet on health and on specific diseases (Committee on Medical Aspects of Food Policy, 1994), relatively little is known about how and why people choose the foods that make up their diets or about how to influence their choices in an effective way. Given recommendations, for example, to reduce fat in the diet or increase the consumption of fruit and vegetables, it is then necessary to understand what determines people's choices of foods and what obstacles there might be to such changes. Although official recommendations have been in place in the UK since the report by COMA in 1984 (Committee on Medical Aspects of Food Policy, 1984) for a reduction in the energy in the diet derived from fat there has been relatively little change (Committee on Medical Aspects of Food Policy, 1994).

Food choice, like any complex human behaviour, will be influenced by many interrelating factors. A number of models seeking to delineate the effects of likely influences have been put forward in the literature (e.g. Khan, 1981; Pilgrim, 1957; Randall and Sanjur, 1981; Shepherd, 1985; reviewed by Shepherd, 1989). However, few of these models present any indication of likely mechanisms of action of the multitude of factors identified, nor do they quantify the relative importance of, or interactions between, factors. Likewise they do not allow any quantitative tests which are predictive of food choice. To date many such models are really only catalogues of the likely influences although as such they are useful in pointing to the variables to consider in studies in this area.

An example of such a model is shown in Figure 10.1. The factors influencing food choice are categorised as those related to the food, to the person making the choice and to the external economic and social environment within which the choice is made. Some of the chemical and physical properties of the food will be perceived by the person in terms of sensory attributes, e.g. flavour, texture or appearance. Simply perceiving these sensory attributes in a particular food does not necessarily mean that a person will or will not choose to consume that food. Rather it is the person's liking

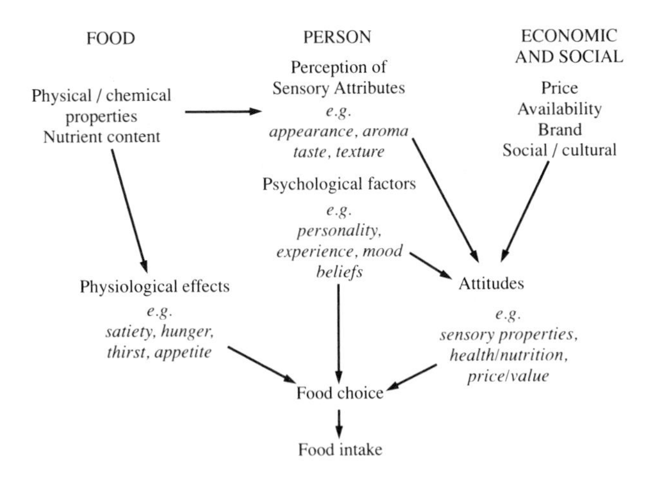

Figure 10.1 Some factors affecting food choice and intake (from Shepherd, 1985).

for that attribute in that particular food which will be the determining factor. Other chemical components in the foods, such as the amount of protein or carbohydrate, will have effects upon the person, e.g. reducing hunger, and the learning of the association between the sensory attributes of a food and its post-ingestional consequences appears to be a major mechanism by which preferences develop. Psychological differences between people, such as personality, may also influence food choice (Shepherd and Farleigh, 1986). There are also many factors in the context within which the choice is made that are likely to be very important. These include marketing and economic variables as well as social, cultural, religious or demographic factors (Murcott, 1989; Shepherd, 1989).

10.2 Theory of planned behaviour

Many of the influences on food choice are likely to be mediated by the beliefs and attitudes held by an individual. Beliefs about the nutritional quality and health effects of a food may be more important than the actual nutritional quality and health consequences in determining a person's choice. Likewise various marketing, economic, social, cultural, religious or demographic factors may act through the attitudes and beliefs held by the person. Thus the study of the relationship between choice and the beliefs and attitudes held by a person offers one possible route towards a better understanding of the influence of different factors on food choice.

The idea behind measuring attitudes is that they are thought to be causally related to behaviour. This is true both in the common use of the term attitude and in the research literature in social psychology (Eagly and

Chaiken, 1993), but the empirical evidence for this link has not always been clear. In the nutrition literature, for example, many studies have attempted to measure the degree of association between attitudes and consumption of foods. Axelson, *et al.* (1985) performed a meta-analysis of such studies and found evidence for small (although statistically significant) correlations between attitudes and behaviour ($r = 0.18$); there was an even lower association between nutrition knowledge and behaviour ($r = 0.10$). Thus a superficial survey of this area might lead to the conclusion that attitudes are not related to behaviour to an important degree. The same type of finding in social psychology led to a crisis in attitude research in the late 1960s (Wicker, 1969) which resulted in the generation of a number of structured attitude models (e.g. Ajzen and Fishbein, 1980). One example of such a structured attitude model is the theory of reasoned action (TRA) (Ajzen and Fishbein, 1980) and its extension in the form of the theory of planned behaviour (TPB) (Ajzen, 1988). These models have been widely applied in the area of social psychology and more recently have been applied to food choice issues.

The TRA seeks to explain rational behaviour which is under the control of the individual. The TPB, however, in addition seeks to be applicable to non-volitional behaviours, goals and outcomes which are not entirely under the control of the person. With volitional behaviours it is argued that intention to perform a behaviour is the best predictor of behaviour. Intention, in turn, is predicted by two components: the person's own attitude (e.g. whether the person sees the behaviour as good, beneficial, pleasant, etc.) and perceived social pressure to behave in this way (termed the subjective norm). These relationships are shown schematically in Figure 10.2. The TPB includes a component of perceived control, in addition to attitude and subjective norm in the prediction of behavioural intentions, and as a possible influence on the intention–behaviour link (see Figure 10.2).

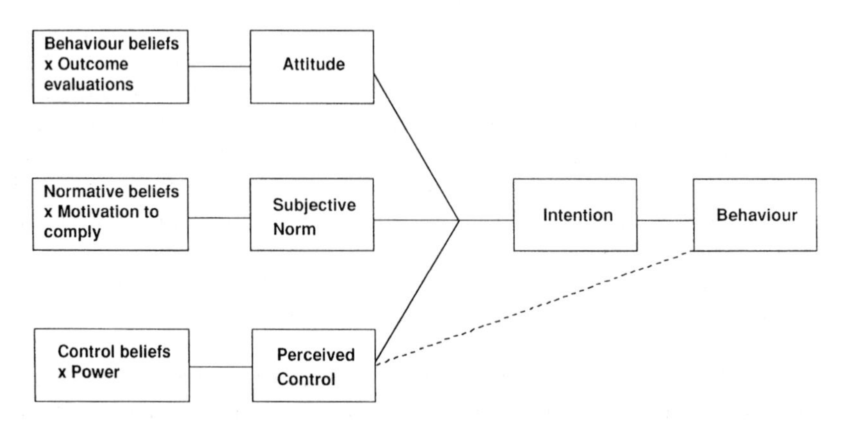

Figure 10.2 Schematic representation of the theory of planned behaviour.

In turn, attitude is predicted by the sum of products of beliefs about outcomes of the behaviour and the person's evaluations of these outcomes as good or bad. The subjective norm is predicted by the sum of products of normative beliefs, which are perceived pressure from specific people or groups (e.g. doctors, family) and the person's motivation to comply with the wishes of these people or groups. In a similar fashion perceived control is determined by specific control beliefs.

An important point in the original conception of the TRA is that influences other than beliefs, attitudes, social pressure and intention should act through these variables (Ajzen and Fishbein, 1980); this would also apply to the TPB, although here perceived control would form part of the model. Thus demographic variables, such as age or social class, should influence behaviour only through the model variables and not act as independent influences on behaviour.

The TRA has been widely applied to many issues in social psychology (Ajzen and Fishbein, 1980; Tesser and Shaffer, 1990), and more recently also successfully applied to a range of food choice issues (Axelson *et al.*, 1983; Shepherd, 1988, 1989; Shepherd and Stockley, 1985, 1987; Shepherd and Farleigh, 1986; Tuorila, 1987; Tuorila and Pangborn, 1988). Sheppard *et al.* (1988) carried out a meta-analysis of 87 studies using this model in the area of general consumer choice (not specifically related to foods), which met certain criteria for inclusion. They found an estimated correlation of 0.53 between intention and behaviour and a multiple correlation of 0.66 between attitude plus subjective norm against intention (Sheppard *et al.*, 1988). Thus this model has validity both in the study of general consumer choice and the study specifically of food choice.

The incorporation of perceived control has received some support in social psychological applications (e.g. Beale and Manstead, 1991), and in studies of weight loss (Schifter and Ajzen, 1985) and dietary health behaviours (Ajzen and Timko, 1986), although not in all applications (Fishbein and Stasson, 1990). In a study of biscuit and bread consumption (Sparks *et al.*, 1992), intentions to consume wholemeal bread were not influenced by perceived control but intentions to consume biscuits were. Thus, inclusion of a measure of perceived control may be important in predicting choices of some, although not all, foods. It is important to remember that a failure to find a significant effect for perceived control does not invalidate the TPB because it would be argued that in those cases the behaviour is volitional.

10.3 Extensions of the theory of planned behaviour

Although the TRA and TPB have proved successful in many applications in the food choice area there are a number shortcomings in their conceptualisation and implementation. This has led to a number of suggested modifications and extensions. Two such extensions will be described here applied to issues of food choice.

10.4 Self-identity and organic food consumption

One recent suggested modification to the TPB is that a person's self-identity may influence behaviour independently of his or her attitudes (Biddle *et al.*, 1987; Charng *et al.*, 1988). An example of evidence for such a modification is in the field of blood donation, where a person's identification of him or herself as a blood donor is a predictor of intention to donate blood over and above the effects of the person's attitude towards blood donation (Charng *et al.*, 1988). These findings are surprising since if a person identifies him or herself as a blood donor this would be expected to be reflected in favourable attitudes towards blood donation, rather than giving an effect on intention independent of attitude. The inclusion of self-identity in the TPB was investigated in a study of the consumption of organic vegetables (Sparks and Shepherd, 1992). Two hundred and sixty-one subjects completed questionnaires which included measures of the components of the TPB, in addition to measures of self-identity.

Attitude and subjective norm were measured as specified in Ajzen and Fishbein (1980). All responses were on 9-category scales. There were five attitude items: 'eating organic vegetables is . . .' with response scales of 'extremely foolish—extremely wise', 'extremely bad—extremely good', 'extremely harmful—extremely beneficial', 'extremely unenjoyable—extremely enjoyable', 'extremely unpleasant—extremely pleasant'. The responses to these items were summed to give an overall measure of attitude. Cronbach's alpha, a measure of how well the individual items relate to such a total, was high at 0.90. There was one subjective norm item in the form:

> 'Most people who are important to me think I should . . . I should not eat organic vegetables'.

Belief items were generated from a short pre-test questionnaire given to 75 people not involved in the main study. Seven salient beliefs were derived from this pre-test and were included in the main questionnaire:

> 'eating organic vegetables entails eating vegetables (i) that have a good flavour, (ii) that are healthy, (iii) that are expensive to purchase, (iv) that are easy to obtain in the shops, (v) that are associated with an "alternative" lifestyle image, (vi) that have an attractive appearance, (vii) that are "environmentally friendly".'

The response scales were marked 'extremely unlikely' to 'extremely likely'. Outcome evaluations were assessed by asking subjects to rate each of the seven characteristics of vegetables from 'extremely bad' to 'extremely good'.

Perceived control was measured with three items: (i) 'How much control do you have over whether you do or do not eat organic vegetables?', with the response scale end-points marked 'very little control' and 'complete control'; (ii) 'For me to eat organic vegetables is . . .', with response scale endpoints marked 'extremely difficult' and 'extremely easy'; (iii) 'If I wanted to, I could easily eat organic vegetables whenever I eat vegetables', with response scale endpoints marked 'extremely unlikely' and 'extremely likely'. Responses were summed to give an overall measure of 'perceived control'. Two measures of identification with green consumerism were constructed: 'I think of myself as a green consumer' and 'I think of myself as someone who is very concerned with green issues' (with responses from 'disagree very strongly' to 'agree very strongly'). These two measures were summed to create the self-identity measure of 'green identity'.

Correlations between the components of the TPB were reasonably high, confirming the basic applicability of the model in this case: the summed products of beliefs and evaluations correlated significantly with attitudes ($r = 0.44$, $p < 0.001$), attitudes correlated significantly with intentions to eat organic vegetables the following week ($r = 0.38$, $p < 0.001$), as did subjective norm ($r = 0.30$, $p < 0.001$) and perceived control ($r = 0.27$, $p < 0.001$). Self-identity was also significantly correlated with intentions ($r = 0.37$, $p < 0.001$).

A multiple regression tested the hypothesis that 'green identity' would not add to the prediction of intentions over and above the contributions of the other components. As expected, attitudes, subjective norm and perceived control all revealed independent effects but there was also a highly significant effect for 'green identity' (see Figure 10.3).

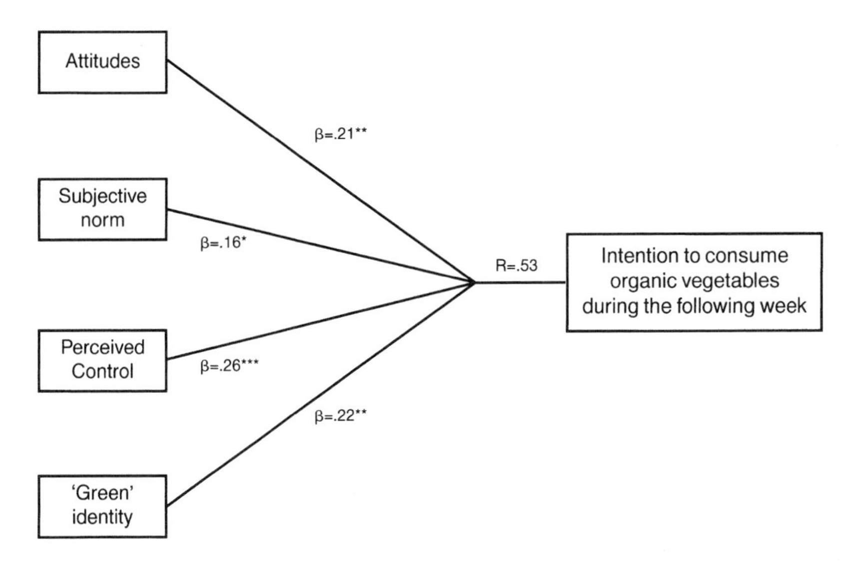

Figure 10.3 Standardised (beta) regression coefficients from a multiple regression of intentions to buy organic vegetables on to attitude, subjective norm, perceived control and self-identity. *$p < 0.05$; **$p < 0.01$; ***$p < 0.001$.

One possible reason for 'green identity' having such an independent effect is that the people were using this as a proxy for past behaviour, i.e. inferring their self-identity response from an examination of past behaviour, and past behaviour has been found to be a powerful predictor of both intention and future behaviour (Bentler and Speckart, 1979). However, adding responses on past behaviour into the above regression showed that the measure of self-identity was still an independent predictor of intention thus arguing against this interpretation of the effect.

The reason why self-identity should have such an independent effect is not entirely clear (Sparks and Shepherd, 1992), although it may reflect the inability of current measures of attitude adequately to assess various symbolic and emotional factors which might nonetheless influence intention and behaviour. Extensions of this kind to the basic model offer a potential means towards a better understanding of the different factors determining food choice.

10.5 Moral obligation

One aspect of behaviour not well covered by existing components of the TPB is that of moral or ethical concerns. In its basic form the model is purely utility or instrumentally based with behaviour leading to outcomes which are seen as beneficial or not. However, there are some behaviours where moral considerations might also have a significant impact, irrespective of beliefs about outcomes. Moral obligation has been investigated in studies outside the food area, primarily on behaviours such as stealing or cheating which have a very obvious moral component (e.g. Beck and Ajzen, 1991). In such cases it has been found to add significantly to the predictive power of the basic TPB. Although food choice is a less obvious domain within which such factors might operate there are nonetheless particular instances where moral or ethical issues might be of some importance. These include those instances where decisions are made on behalf of others, and the application of new techniques to food production (e.g. genetic engineering) which might be considered to have a significant moral dimension. Applications in these fields will be described in the following sections.

10.5.1 Choice of milks

Not all decisions on foods affect only ourselves. In many cases foods are chosen for other people, in particular within families. In such cases there is a responsibility which rests with the person making the choice, for example, for the health and well-being of other family members. This would therefore appear to be an example where moral obligation might play a significant role in the choice of foods. Moral obligation for the health of the family has been examined in a study of the consumption of different types of milks (Raats *et al.*, 1995).

Two hundred and fifty-seven people completed a questionnaire on the consumption of whole, semi-skimmed and skimmed milk. In addition to questions assessing the components of the TPB there was one question for each type of milk: 'I feel obliged to use (skimmed) milk for my family's health' which was taken as a measure of moral obligation.

For each type of milk the basic model was shown to give a reasonable degree of prediction of intention. Perceived control added significantly to the prediction of skimmed milk ($p < 0.001$) consumption but was only marginally significant for the consumption of semi-skimmed milk ($p = 0.06$) and not significant for whole milk. This confirms the applicability of the component of perceived control over and above the impact of attitude and subjective norm for some choices.

When the responses on obligation for family's health were added into the regression predicting intention, this led to small but significant increases in prediction for whole and semi-skimmed milk ($p < 0.05$) but not for skimmed milk. If, however, we consider the obligation component as feeding into attitude as an addition to beliefs and evaluations then we get the results shown in Table 10.1. As expected, belief-evaluations are important predictors of attitude in each case but the measure of obligation for family's health adds an independent effect to this and is highly significant for each type of milk.

It would therefore appear that moral obligation is important but that its position within the model does not lie at the level of being an extra predictor of intention but rather is a stage further back in feeding into attitude in

Table 10.1 Results from multiple regressions predicting attitude to consuming milk of different fat levels

Milk	R	Regression coefficient (beta)
Whole milk ($n = 224$)		
Belief evaluations	0.83	0.71***
Obligation for family's health	0.85	0.24***
Semi-skimmed milk ($n = 229$)		
Belief evaluations	0.78	0.68***
Obligation for family's health	0.82	0.25***
Skimmed milk ($n = 225$)		
Belief evaluations	0.84	0.74***
Obligation for family's health	0.85	0.19***

The multiple correlation coefficients (R) are shown for entering first the belief evaluations and then the measures of moral obligation; the standardised regression coefficients (beta) are from the final regression equation. ***$p < 0.001$.

additon to belief-evaluations about the outcomes of the behaviour. Such questions on moral obligation appear to tap something which is not measured using the questions on beliefs about instrumental outcomes of actions. While this has previously been shown to be true for behaviours with an obvious moral component such as stealing or cheating, this is a demonstration of the importance of this type of component in the choice of foods where issues of moral obligation are less clear. Like perceived control or self-identity it is unlikely to be important for the choice of all types of foods in all contexts but it may be particularly important where food choices are made on behalf of others. This issue is explored further in the following section.

10.5.2 Attitudes towards additives in foods

The amounts of additives found in foods are of concern to the public. Certain additives have been publicised in a negative light, some particularly in relation to children. Mothers have some degree of influence over the foods their children eat through purchase and preparation but also they may feel that this carries with it an element of responsibility for the diets of their children. For this reason, a study was conducted examining the issue of mothers' attitudes towards giving their children foods containing certain types of chemicals. Again the TPB was used as the basis for designing the questions to be asked and again measures of moral obligation were included in the model.

An additional measure was also included which will be referred to as negative affect; this is really a measure of the anticipated worry, concern or regret that the mother might feel about having performed the behaviour. There are a number of reasons for including such a measure within the TPB. Affect and cognition have long been considered to be distinct components of attitude (e.g. McGuire, 1985; Rosenberg and Hovland, 1960); the cognitive component of attitude incorporates attributes or beliefs about the attitude object (e.g. Fishbein and Ajzen, 1975), while feelings created by the attitude object are referred to as the affective component (e.g. Breckler and Wiggens, 1989). The affective component has been relatively neglected in recent research although affective factors are thought to be highly influential in people's behavioural choices (Wilson *et al.*, 1989), including those related to health behaviours (Ajzen and Timko, 1986). Specific assessment of affective factors in food choice is noticeable by its absence. Given the type of behaviour considered here it was felt that negative affect might be potentially a very important predictor of behaviour and hence such measures were also included in the study.

Before the main study, a preliminary study was carried out in order to discover the type of issues which concerned mothers about foods for their children. Fifteen mothers were interviewed in this preliminary study. Each one was presented with a range of products (fruit squashes, flavoured yoghurts, table jellies, sweets, breakfast cereals, fishfingers, crisps, breads, jams) with varying levels of additives to aid in the elicitation of the beliefs.

A questionnaire was developed on the basis of the responses from the preliminary study and this was completed by 172 mothers of children aged 5–11. The questionnaire covered perceptions of four groups of foods, two concerning additives (food with artificial sweeteners and food with synthetic colourings) and two concerning nutritional issues (high-fat food and high-sugar food). The questionnaire was designed according to the TPB but also included questions on perceived obligation for family's health (3 items) and negative affect (3 items, worry, regret and concern).

The following examples show the wording for the items on foods high in sugar but there were equivalently worded items for foods high in fat, foods with artificial sweeteners and foods with synthetic colourings. For each type of food there were eight belief items which were generated from the interviews held before the study. Each one was worded: 'Giving my children food that *is high in sugar* means giving them food that': (i) 'is inexpensive', (ii) 'they find tasty', (iii) 'is natural', (iv) 'damages their teeth', (v) 'is healthy', (vi) 'causes behaviour problems', (vii) 'helps control their weight', (viii) 'is processed'. The responses were made on 7-category rating scales labelled at the extremes with 'extremely unlikely' and 'extremely likely'. There were corresponding outcome evaluation questions for each of the beliefs. Attitudes were assessed using four items: 'Giving my children food that *is high in sugar* is', with responses on the following scales: 'extremely bad/extremely good', 'extremely unpleasant/extremely pleasant', 'extremely harmful/extremely beneficial' and 'extremely foolish/extremely wise'. There were two subjective norm items. The first was 'Most people who are important to me think I should give my children food that *is high in sugar*' with response scales from 'extremely unlikely' to 'extremely likely' and the second was 'Most people who are important to me approve/disapprove of my giving my children food that *is high in sugar*' with the response scales labelled 'approve very strongly' to 'disapprove very strongly'. A third normative question was added specifically addressing the mother's perceptions of what her child would want. This was worded 'My children think I should give them food that *is high in sugar*', with responses from 'extremely unlikely' to 'extremely likely'.

Perceived control was assessed with three items. The first two were: 'How easy is it for you to give your children food that *is high in sugar*' and 'How easy is it for you to avoid giving your children food that *is high in sugar*', both with responses on scales from 'extremely easy' to 'extremely difficult'. The third perceived control item was 'How much control do you feel you have over whether or not you give your children food that *is high in sugar*', with responses from 'no control at all' to 'complete control'. The responses on these three items were summed to give a measure of perceived control.

Moral obligation was assessed with the following items: (i) 'It would be wrong for me to give my children food that *is high in sugar*' (disagree very strongly/agree very strongly); (ii) 'I would feel guilty if I gave my children food that *is high in sugar*' (extremely true/extremely false); (iii) 'It goes against my principles to give my children food that *is high in sugar*' (disagree very strongly/agree very strongly). Again the responses on these three items were summed to give an overall measure of moral obligation.

Negative affect was the sum of responses on the following items: (i) 'How worried would you feel after giving your children food that *is high in sugar*' (not at all worried/extremely worried); (ii) 'How concerned would you feel after giving your children food that *is high in sugar*' (not at all concerned/extremely concerned); (iii) 'How much regret would you feel after giving your children food that *is high in sugar*' (no regret at all/extreme regret).

The mean ratings on these components are shown in Table 10.2 and it can be seen that the groups of foods studied were perceived differently by the mothers. Attitudes and behavioural scores were more negative for the nutritional issues than for the additives. Thus the chemical additives studied here are of less importance to the mothers than are the nutritional issues. Levels of negative affect (worry and concern) were significantly higher for the nutritional issues.

As in the previous studies, regression analyses were used to test whether perceived control added significantly to the prediction of intention. Perceived control was significant in the regression equations for three of the four groups of foods (food with artificial sweetener, food with artificial colouring and high-sugar food) but not for high-fat food. Thus this component of the TPB was again shown to be of importance.

Hierarchical regressions of attitudes on to the sum of belief-evaluations

Table 10.2 Mean scores and standard errors for intention (to present children with) (range 1 to 7), attitude (range −12 to +12), subjective norm (range 2 to 14), perceived control (range 2 to 14), moral obligation (range 3 to 21) and negative affect (range 3 to 21)

	Food with artificial sweeteners		Food with synthetic colourings		High-fat food		High-sugar food	
	Mean	SE	Mean	SE	Mean	SE	Mean	SE
Intention	3.5[a]	0.1	3.2[b]	0.1	2.9[c]	0.1	2.8[c]	0.1
Attitude	−2.6[a]	0.3	−4.0[b]	0.3	−6.1[c]	0.3	−6.3[c]	0.2
Subjective norm	6.8[a]	0.2	6.3[b]	0.1	5.8[c]	0.2	5.4[d]	0.2
Perceived control	8.7[a]	0.2	8.1[b]	0.2	9.6[c]	0.2	9.4[c]	0.2
Moral obligation	13.5[a]	0.3	14.2[b]	0.3	15.7[c]	0.3	16.3[d]	0.2
Negative affect	10.0[a]	0.4	11.0[b]	0.3	13.2[c]	0.3	14.2[d]	0.3

[abcd] Values with same superscript in same row do not differ from each other ($p < 0.05$; paired t-test).

Table 10.3 Results from multiple regressions predicting attitude towards eating various types of foods ($n = 172$)

	R	Regression coefficient (beta)
Food with artificial sweeteners		
Belief evaluations	0.57	0.19***
Moral obligation	0.85	–0.74***
Food with synthetic colourings		
Belief evaluations	0.50	0.16**
Moral obligation	0.81	–0.73***
High-fat food		
Belief evaluations	0.37	0.15**
Moral obligation	0.81	–0.75***
High-sugar food		
Belief evaluations	0.36	0.16**
Moral obligation	0.73	–0.67***

The multiple correlation coefficients (R) are shown for entering first the belief evaluations and then the measures of moral obligation; the standardised regression coefficients (beta) are from the final regression equation. *$p < 0.05$; **$p < 0.01$; ***$p < 0.001$.

and perceived obligation for family's health showed that the addition of perceived obligation for family's health significantly improved the prediction of attitude for all four behaviours (see Table 10.3). Thus moral obligation has again been shown to be important in determining attitudes in addition to the more cognitively based beliefs about outcomes of the behaviour. The evidence in Table 10.3 is even more clear than in the previous study (Table 10.2) with the beta coefficients for moral obligation being far greater than those for belief-evaluations. A further addition of negative affect again increased the degree of prediction (Table 10.4) but negative affect did not simply replace moral obligation in the regression equation, since moral obligation was still highly statistically significant.

As in the previous study, the inclusion of measures of moral obligation significantly increased the predictive power of the model, again underlining the importance of this component when a person is involved in buying and preparing foods for other people, and perhaps particularly for children.

10.5.3 Genetic engineering and ethical concerns

While the inclusion of additives in foods represents a technology which is now relatively familiar to people there are other technologies which are currently not widely applied to food production and manufacturing but are likely to be applied very widely in the future. One such technology is genetic engineering. This involves a set of techniques which allow scientists:

Table 10.4 Results from multiple regressions predicting attitude towards eating various types of foods ($n = 172$)

	R	Regression coefficient (beta)
Food with artificial sweeteners		
Belief evaluations	0.57	0.16***
Moral obligation	0.85	–0.58***
Negative affect	0.86	–0.23***
Food with synthetic colourings		
Belief evaluations	0.50	0.13**
Moral obligation	0.81	–0.43***
Negative affect	0.86	–0.41***
High-fat food		
Belief evaluations	0.37	0.10*
Moral obligation	0.81	–0.57***
Negative affect	0.83	–0.28***
High-sugar food		
Belief evaluations	0.36	0.12*
Moral obligation	0.73	–0.51***
Negative affect	0.76	–0.26***

The multiple correlation coefficients (R) are shown for entering first the belief evaluations, then the measures of moral obligation and finally measures of negative affect; the standardised regression coefficients (beta) are from the final regression equation. *$p < 0.05$; **$p < 0.01$; ***$p < 0.001$.

'. . . to cut and splice DNA, to move genes from one organism to another and to get them to work there' (Advisory Council on Science and Technology, 1990, p. 1).

It probably represents the most contentious example of a range of biotechnological processes which have diverse potential applications in medicine, agriculture and food production.

There are already a number of applications in the pharmaceuticals area but there are currently relatively few commercial examples in food. One example is the use in cheese making of genetically engineered chymosin replacing rennet extracted from the stomachs of calves. The first genetically engineered complete food to go on sale is Calgene's Flavr-Savr tomato which was marketed in the USA in 1994. This has been genetically modified in order to extend shelf life and, it is argued, improve the flavour through not having to pick and transport the fruit under-ripe. There are, however, many more possible applications where crops are being genetically engineered often to provide some form of agricultural advantage such as pest resistance or herbicide resistance. The intention is that many of these crops will be used in food production in the future. Genetic engineering does not have the same limitations as conventional cross breeding, which has been used for centuries to change the genetic makeup of animals and plants. Whereas cross breeding relies on transferring genetic characteristics within a species, genetic engineering can be used to transfer genetic material across species boundaries and between very different types of organisms (e.g. from microorganisms to animals, or from plants to animals).

Although some evidence suggests that the general public knows or understands little about the technology (Hamstra, 1991; Sparks *et al.*, 1994), the direction of public opinion is thought likely to be a major factor in influencing its future role in society (Advisory Council on Science and Technology, 1990). The application of this type of technology raises a number of issues of potential concern to the public, including issues of safety and what benefits are likely to derive from the application of the technology but it also raises moral and ethical concerns in the minds of many people. The Advisory Council on Science and Technology (1990) report emphasises this in its recommendation that 'government departments need to be sufficiently alert to the ethical issues that are likely to cause concern in the future' (p. 28).

As part of an extensive series of studies examining public attitudes towards the application of genetic engineering to food production (Frewer *et al.*, 1994; Sparks *et al.*, 1994), two studies were conducted applying an extended form of the TPB to this issue.

In the first study (Sparks *et al.*, 1995), 334 members of the public completed a questionnaire designed following the basic structure of the TPB. Since public awareness and knowledge of the technology had previously been shown to be very low (Sparks *et al.*, 1994), the questionnaire began with a simple short description of the technology. Instead of intention, two expectation questions were included, one related to eating food produced using genetic engineering and the other to supporting the use of the technology. In addition to the TPB questions, two questions measuring perceived ethical obligation were included: 'I feel that I have an ethical obligation to avoid eating food produced by gene technology' and 'I feel that I have an ethical obligation to support the use of gene technology in food production' (both with responses from 'disagree very strongly' to 'agree very strongly'). One was framed positively and the other negatively so as not to suggest that ethical and moral issues can only be negative and also to determine whether ethical issues have the same impact when presented positively.

Ten possible outcomes of the application of genetic engineering to food production were used to assess beliefs and outcome evaluations. These were selected from a cross-section of the literature on genetic engineering (e.g. Pimentel *et al.*, 1989; Straughan, 1991). These are not necessarily the salient beliefs of a random sample of the population but because people generally profess little knowledge about the technology, it was not considered feasible in this first study to assess salient beliefs in the standard way. The selected outcomes were:

(i) increased food production,
(ii) ecological damage,
(iii) cheaper food,
(iv) animal and/or plant extinction,
(v) reduced use of pesticides,
(vi) detrimental effects on animal welfare,
(vii) improvements in food quality,
(viii) unpredictable animal and/or plant mutations,
(ix) economic growth, and
(x) increased use of weedkillers.

Multiple regression analysis showed that people's ($n = 334$) expectations of (i) supporting the technology and (ii) consuming food produced by the technology could be predicted from their attitudes (respectively: beta = 0.77, $p < 0.001$ and beta = 0.23, $p < 0.05$). These attitudes, in turn, could be predicted, not only from a consideration of the perceived tangible advantages and disadvantages of the technology but also from a consideration of people's levels of ethical concern (Table 10.5). The positive beta coefficient for ethical obligation for supporting the technique and negative coefficient for eating the foods is because of the opposite wording of the question in the two cases (see above).

A second study (Sparks and Shephered, in preparation) essentially replicated these findings but looked at specific applications of the technology. These were pork produced using genetically engineered growth hormone and genetically engineered tomatoes.

In the previous study, the beliefs were generated from the literature rather than from a representative sample of the population and hence the

Table 10.5 Standardised regression coefficients (beta) from multiple regressions predicting attitudes

	Number of subjects	Belief evaluations	Ethical obligation
Study 1			
Eating genetically engineered foods	284	0.46***	–0.41***
Supporting genetic engineering	284	0.33***	0.60***
Study 2			
Pork produced with genetically engineered growth hormone	60	0.30***	–0.51***
Genetically engineered tomatoes	61	0.56***	–0.37***

In Study 1 these are attitudes towards eating foods produced using genetic engineering and supporting the use of the technology and in Study 2 the purchase of two foods produced using genetic engineering. ***$p < 0.001$.

beliefs included may not have been truly salient for the people taking part. This might then have underestimated the contribution from the belief component in the prediction of attitude, and hence possibly overestimated the contribution of ethical concerns. In this second study, a method was used to generate the beliefs directly from the subjects themselves (Towriss, 1984). That is, each person generated his or her own set of beliefs and it was this set of beliefs that the person then received in the questionnaire rather than an overall set being derived from pre-interviews and then everyone receiving this same set of beliefs. The method used here would then be expected to give the highest estimate of the impact of belief-evaluation on attitude.

Following a similar pattern to that found in the previous study, it was apparent that respondents' perceived ethical obligation had a significant independent predictive effect on both their attitudes towards purchase of the pork and their attitudes towards purchasing the tomatoes (Table 10.5). Respondents' intentions to purchase these products were very highly correlated with their attitudes: $r = 0.80$ ($p < 0.001$) in the case of pork, and $r = 0.70$ ($p < 0.001$) in the case of the tomatoes.

Thus ethical and moral concerns seem to play a major role in determining attitudes, and hence intention, when considering technologies such as genetic engineering. This extends the findings discussed previously, where moral obligation was important in those instances where people had responsibility for others, to demonstrate its potential importance in the acceptance of novel technologies. It would thus appear that moral and ethical concerns may play an important role across a whole range of food decisions but the range of its applicability remains to be determined.

10.6 Conclusions

Food choice is influenced by a large range of potential factors. Many models put forward in this area involve merely listing the likely influences rather than offering a framework for empirical research and practical application. Although there is general agreement on the types of influences likely to be important, the integration of these factors into a coherent and quantitative model of food choice remains an area in need of development.

The attempt to model food choice via an understanding of people's beliefs and attitudes requires a structured framework within which to measure and relate the variables of interest. One model from social psychology for achieving this is the TPB. This model generally reveals good prediction of behaviour and can be used to determine the relative importance of different factors in influencing food choice.

Various extensions of this model, including self-identity, moral obligation and negative affect, offer a means for developing a clearer understanding of the factors influencing the choice of particular types of foods in particular contexts. In particular they pave the way for exploring the more emotional and feeling elements potentially important in food choice rather than simply addressing the rational cognitive issues prevalent in the literature.

Acknowledgements

The work reported here was funded by the Biotechnology and Biological Sciences Research Council and the UK Ministry of Agriculture, Fisheries and Food.

References

Advisory Council on Science and Technology (1990) *Developments in Biotechnology*, HMSO, London.

Ajzen, I. (1988) *Attitudes, Personality, and Behavior*, Open University Press, Milton Keynes.

Ajzen, I. and Fishbein, M. (1980) *Understanding Attitudes and Predicting Social Behavior*, Prentice-Hall, Englewood Cliffs, New Jersey.

Ajzen, I. and Timko, C. (1986) Correspondence between health attitudes and behavior. *Basic and Applied Social Psychology*, **7**, 259–76.

Axelson, M.L., Brinberg, D. and Durand, J.H. (1983) Eating at a fast-food restaurant—a social-psychological analysis. *Journal of Nutrition Education*, **15**, 94–8.

Axelson, M.L., Federline, T.L. and Brinberg, D. (1985) A meta-analysis of food and nutrition-related research. *Journal of Nutrition Education*, **17**, 51–4.

Beale, D.A. and Manstead, A.S.R. (1991) Predicting mothers' intentions to limit frequency of infants' sugar intake: testing the theory of planned behavior. *Journal Applied Social Psychology*, **21**, 409–31.

Beck, L. and Ajzen, I. (1991) Predicting dishonest actions using the theory of planned behavior. *Journal of Research in Personality*, **25**, 285–301.

Bentler, P.M. and Speckart, G. (1979) Models of attitude-behavior relations. *Psychological Review*, **86**, 452–64.

Biddle, B.J., Bank, B.J. and Slavinge, R.L. (1987) Norms, preferences, identities and retention decisions. *Social Psychology Quarterly*, **50**, 322–37.

Breckler, S.J. and Wiggens, E.C. (1989) Affect versus evaluation in the structure of attitudes. *Journal of Experimental Social Psychology*, **25**, 253–71.

Charng, H.-W., Piliavin, J.A. and Callero, P.L. (1988) Role identity and reasoned action in the prediction of repeated behavior. *Social Psychology Quarterly*, **51**, 303–17.

Committee on Medical Aspects of Food Policy (1984) *Diet and Cardiovascular Disease*, HMSO, London.

Committee on Medical Aspects of Food Policy (1994) *Nutritional Aspects of Cardiovascular Disease*, HMSO, London.

Eagly, A.H. and Chaiken, S. (1993) *The Psychology of Attitudes*. Harcourt, Brace and Jovanovich, San Diego.

Fishbein, M. and Ajzen, I. (1975) *Belief, Attitude, Intention and Behavior. An Introduction to Theory and Research*, Addison-Wesley, Reading, MA.

Fishbein, M. and Stasson, M. (1990) The role of desires, self-predictions, and perceived control in the prediction of training session attendance. *Journal of Applied Social Psychology*, **20**, 173–98.

Frewer, L.J., Shepherd, R. and Sparks, P. (1994) The interrelationship between perceived knowledge, control and risk associated with a range of food related hazards targeted at the self, other people and society. *Journal of Food Safety*, **14**, 19–40.

Hamstra, A. (1991) *Biotechnology in Foodstuffs: Towards a Model of Consumer Acceptance*, Institute for Consumer Research, The Hague.

Khan, M.A. (1981) Evaluation of food selection patterns and preferences. *CRC Critical Reviews in Food Science and Nutrition*, **15**, 129–53.
McGuire, W. (1985) Attitudes and attitude change, in *The Handbook of Social Psychology. Vol. 2. 3rd edn* (eds G. Lindzy and E. Aronson), Random House, New York, pp. 233–346.
Murcott, A. (1989) Sociological and social anthropological approaches to food and eating, in *World Review of Nutrition and Dietetics, 55*, (ed. G.H. Bourne), Karger, Basel, pp. 1–40.
Pilgrim, F.J. (1957) The components of food acceptance and their measurement. *American Journal of Clinical Nutrition*, **5**, 171–5.
Pimentel, D., Hunter, M.S., LaGro, J.A., Efroymson, R.A., Landers, J.C., Mervis, F.T., McCarthy, C.A. and Boyd, A.E. (1989) Benefits and risks of genetic engineering in agriculture. *Bioscience*, **39**, 606–14.
Raats, M.M., Shepherd, R. and Sparks, P. (1995) Including moral dimensions of choice within the structure of the Theory of Planned Behavior. *Journal of Applied Social Psychology*, **25**, 484–94.
Randall, E. and Sanjur, D. (1981) Food preferences—their conceptualization and relationship to consumption. *Ecology of Food and Nutrition*, **11**, 151–61.
Rosenberg, M.J. and Hovland, C.I. (1960) Cognitive, affective, and behavioral components of attitudes, in *Attitude Organization and Change: An Analysis of Consistency among Attitude Components* (eds C.I. Hovland and M.J. Rosenberg), Yale University Press, New Haven, pp. 1–14.
Schifter, D.E. and Ajzen, I. (1985) Intention, perceived control, and weight loss: an application of the theory of planned behavior. *Journal of Personality and Social Psychology*, **49**, 843–51.
Shepherd, R. (1985) Dietary salt intake. *Nutrition and Food Science*, **96**, 10–11.
Shepherd, R. (1988) Belief structure in relation to low-fat milk consumption. *Journal of Human Nutrition and Dietetics*, **1**, 421–8.
Shepherd, R (1989) Factors influencing food preferences and choice, in *Handbook of the Psychophysiology of Human Eating* (ed. R. Shepherd), Wiley, Chichester, pp. 3–24.
Shepherd, R, and Farleigh, C.A. (1986) Preferences, attitudes and personality as determinants of salt intake. *Human Nutrition: Applied Nutrition*, **40A**, 195–208.
Shepherd, R. and Stockley, L. (1985) Fat consumption and attitudes towards food with a high fat content. *Human Nutrition: Applied Nutrition*, **39A**, 431–42.
Shepherd, R. and Stockley, L. (1987) Nutrition knowledge, attitudes, and fat consumption. *Journal of the American Dietetic Association*, **87**, 615–19.
Sheppard, B.H., Hartwick, J. and Warshaw, P.R. (1988) The theory of reasoned action: a meta-analysis of past research with recommendations for modifications and future research. *Journal of Consumer Research*, **15**, 325–43.
Sparks, P. and Shepherd, R. (1992) Self-identity and the theory of planned behavior: assessing the role of identification with 'green consumerism'. *Social Psychology Quarterly*, **55**, 388–99.
Sparks, P. and Shepherd, R. (in preparation) The moral dimension of attitudes towards genetic engineering in food production.
Sparks, P., Hedderley, D. and Shepherd, R. (1992) An investigation into the relationship between perceived control, attitude variability and the consumption of two common foods. *European Journal of Social Psychology*, **23**, 55–71.
Sparks, P., Shepherd, R. and Frewer, L.J. (1994) Gene technology, food production, and public opinion: a UK study. *Agriculture and Human Values*, **11**, 19–28.
Sparks, P., Shepherd, R. and Frewer, L.J. (1995) Assessing and structuring attitudes toward the use of gene technology in food production: the role of perceived ethical obligation. *Basic and Applied Social Psychology*, **16**, 267–85.
Straughan, R. (1991) Genetic manipulation for food production: social and ethical issues for consumers. *British Food Journal*, **92** (7), 13–26.
Tesser, A. and Shaffer, D.R. (1990) Attitudes and attitude change. *Annual Review of Psychology*, **41**, 479–523.
Towriss, J.G. (1984) A new approach to the use of expectancy value models. *Journal of the Market Research Society*, **26**, 63–75.
Tuorila, H. (1987) Selection of milks with varying fat contents and related overall liking, attitudes, norms and intentions. *Appetite*, **8**, 1–14.
Tuorila, H. and Pangborn, R.M. (1988) Behavioural models in the prediction of consumption of selected sweet, salty and fatty foods, in *Food Acceptability* (ed. D.M.H. Thomson), Elsevier Applied Science, London, pp. 267–279.

Wicker, A.W. (1969) Attitude versus actions: the relationship of verbal and overt behavioral responses to attitude objects. *Journal of Social Issues*, **25**, 41–78.

Wilson, T.D., Dunn, D.S., Kraft, D. and Lisle, D.J. (1989) Introspection, attitude change and attitude-behavior consistency: the disruptive effects of explaining why we feel the way we do, in *Advances in Social Psychology*, Vol. 22 (ed. L. Berkowitz), Academic Press, New York, pp. 287–343.

11 Dietary change: changing patterns of eating

DAVID A.T. SOUTHGATE

11.1 Introduction

One characteristic of the human diet is the variety of foods eaten and the dietary patterns that can be seen. The capacity of the human species to accept and flourish on the foods that were available for consumption is possibly one reason why the human species has been able to succeed in such a wide range of environments (Andrews and Martin, 1991). While it is customary to assume that dietary changes are taking place at an increasing rate at the present time under the twin influences of developing agronomic practices and food technology, it is possibly misleading to think of periods when the human diet was stable, in fact one suspects that dietary change, whether due to human curiosity or imposed by the natural environment, has been a dominant factor in human evolution (Southgate, 1991a; Ulijaszek, 1991).

The evidence for changes in eating patterns comes from measurements of food consumption and these provide a variety of types of information especially in the level of detail of the types of foods consumed and the temporal patterns of their consumption. It is first necessary to outline the types of method which are in use (Cameron and van Staveren, 1988) and then useful to consider how one should characterise eating patterns, before moving on to discuss the range of factors which may influence dietary change.

In recent years much prominence has been given to the relationships between diet and the incidence of disease and as a consequence dietary changes have been advocated as health promotion measures (US Department of Health and Human Services, 1988; National Research Council, 1989; Department of Health, 1994). It is important to consider these suggested changes against the dietary changes that are taking place for other reasons.

The final section of this chapter considers some future directions that research might take to lead to a better understanding of the factors that influence dietary change, concluding that this will require the integration of all the factors involved and will almost certainly require the conceptual development of the ways in which we measure and evaluate dietary patterns. This will be particularly important if we wish to have a predictive understanding of the factors that influence dietary change.

11.2 Sources of evidence for dietary change

In analysing dietary changes it is necessary first to have serial measurements in time of food consumption coupled with a range of other types of information concerning the populations where dietary changes are being analysed. Thus one needs evidence on the socio-economic characteristics of the population studied and the changes that have occurred over the time periods studied together with information about cultural changes that have occurred, such as changes in immigration policies. Alongside this, information relating to the availability of foods at the times studied is required, including information on changes in food production at the primary, agricultural and horticultural levels and developments in food production technology.

In many cases where sound food consumption data are available, the subsidiary, but essential information which characterises the populations studied and the 'food environment' in all its facets is not available and the inferences drawn regarding the factors affecting dietary changes are in many ways intuitive. It is also important to recognise the limitations of the information on food consumption because the level at which the data were obtained limits the types of dietary change which can be observed and the conclusions regarding the factors involved.

11.2.1 The measurement of food consumption

Food consumption is measured in many different ways and for different purposes. Each of the methods used provides differing types of information particularly in the level of detail concerning the types, amounts and temporal patterns of consuming the foods (Southgate, 1991b). This means that the different approaches give different types of evidence relating to changes that are taking place in the diet and determine the conclusions which can be drawn regarding the rate and extent of dietary changes. These differences are therefore important when discussing dietary changes and in drawing conclusions from the different estimates of food consumption. Table 11.1 summarises the major types of measurements available for studying dietary changes. One of the most important differences in the methods in use is the population level at which they are applied, and the different methods will be described in relation to the level at which they are used.

11.2.1.1 Measurements at the country level

(a) Measurement of food production and consumption. These measurements are widely used for making international comparisons of food consumption in different countries and within countries over time. The Food Balance Sheets of the Food and Agricultural Organisation of the United

Table 11.1 Characteristics of the levels at which food consumption is measured

Level of measurement	Type of measurement	Measurements	Period of study	Primary purpose
National	Food balance sheet	Food production, food imports, food exports, non-food use, changes in food stocks	Typically annually	Economic Food planning Food relief programmes
Households	Household budget surveys	Food purchases by cost (all foods or specfic items)	Weekly, typically over 7 days, occasionally longer	Socio-economic
	House food consumption	Food purchases, foods coming into the house, amounts and costs, meals outside home	Typically over 7 days	Socio-economic Nutritional surveillance
Individual	Dietary recall	Foods as consumed in household portions	24 hours	Nutritional surveillance
	Weighed inventory	Foods as consumed weighed	7 or 3–4 days	
	Food frequency questionnaire	Foods as consumed by frequency of consumption	One month	Nutritional evaluation of dietary intake Nutritional epidemiology

Nations (FAO) are an example of this type of measurement (FAO, 1983; Kelly et al., 1991; Sizaret, 1992). Most national governments compile this type of data, in fact the FAO relies on the data submitted by national governments for the basic data used to compile the Food Balance Sheet Estimates.

In this method estimates are made of food production and imports, from which food exports, non-food use and changes in food stocks are deducted to give the food consumed (Kelly *et al.*, 1991). In the United Kingdom these estimates are called the Consumption Level Estimates and measure food moving into consumption. Another term used for this type of measurement is 'food disappearance'. The estimated amounts are divided by the population numbers to give estimates of per capita consumption.

The measurements of foods are made at the food supplies level and therefore are in terms of bulk quantities of the major foods. These quantities are converted into nutrients using a database that at present includes adjustments that allow for wastage as inedible material and which are weighted for the types of losses of foods during processing of, for example, cereals or dairy products which may differ significantly between countries. This type of measurement therefore gives a view of food consumption and dietary patterns that only permits analysis in terms of the major food commodities and on an annual basis.

The quality of the food estimates varies with the state of development, clearly where much food is produced and consumed locally the estimates tend to be less reliable than for countries with more centralised food markets where there are often fiscal influences which exert controls on the quality of the measurements.

(b) Population estimates by aggregation. These are national estimates of food consumption based on measurements made at a lower level, either at the household level or on individuals, where the population has been selected to be nationally representative. The findings may only be published in an aggregated form as the national values. The outlines of these methods are discussed below.

11.2.1.2 Measurements at the household level

(a) Household budget surveys (Cameron and van Staveren, 1988; van Staveren et al., *1991).* These surveys are usually undertaken to provide financial information and include measures of expenditure on food purchases at the household level. The populations studied may be nationally representative or they may be weighted to focus on specific socio-economic groups of the population. In some countries the range of foods for which expenditure is recorded is restricted whereas in other countries the expenditure on all foods is recorded. The number of food categories is usually limited but seasonal and socio-economic variables are often well documented. Conversion into amounts of food purchased requires information on food prices and further conversion into nutrients can be made using a conventional nutritional database that gives values for foods as purchased.

Although international comparisons with this type of survey are limited because of differences in the design of national studies, within countries these types of measurement provide information on dietary changes in the patterns of food purchases and the effects of seasonal and socio-economic variables on these patterns.

(b) Household food consumption. These studies, in which food consumption is measured directly in terms of both the amounts of foods purchased and expenditure on food, may be conducted for economic and nutritional surveillance purposes. The National Food Survey (NFS) carried

out by the Ministry of Agriculture, Fisheries and Food (MAFF) in the UK is an example of this type of study (MAFF, 1991, 1993, 1994; Nelson, 1991). The measurements are made at the household level and are usually restricted to foods consumed in the home. The UK NFS has recently been expanded to include meals eaten outside the home (MAFF, 1992). The studies are planned to give a nationally representative sample of the population by stratified sampling and cover virtually the entire year. In the UK NFS is carried out annually but many countries carry out similar types of study at less frequent intervals. Series of measurements are an essential source of information on dietary change and considerable use of the UK NFS data will be made in this chapter to illustrate factors affecting dietary patterns. The value of these studies is, however, constrained by the size of the populations studied which often means that the regional, social, or economic sub-groups are rather small in size. Despite this, the regular annual records based on essentially the same techniques provide an exceptionally valuable source for the documentation of dietary changes. Comparisons between countries can be difficult because the study designs may often differ because of differing national objectives and priorities. Such studies do not allow comment to be made on dietary patterns at the level of the individual.

The food consumption data are converted into nutrients using conventional nutritional databases using compositional data for foods as purchased; adjustments for food wastage are usually made in an arbitrary way (Southgate, 1991b). Some adjustments for losses of nutrients due to domestic cooking may also be incorporated in the nutritional assessments.

11.2.1.3 Measurements on individuals. These provide the most detailed type of evidence for the amounts of foods consumed and evidence on patterns of food intake, provided sufficient numbers of subjects are studied to provide representative information on the chosen population (Cameron and van Staveren, 1988). These studies can give very useful information on dietary changes although comparable serial studies are, of course, essential. A range of methods is used.

(a) Retrospective methods. In retrospective studies individuals recall, in a dietary diary or in a structured dietary recall interview, their previous dietary intake, usually over a 24 h period (Nelson, 1991). The food consumption may be estimated in household measures or the portion sizes may be allocated in the recall interview using food models or photographs of different standard portions (Nelson *et al.*, 1994). The accuracy of such measurements is limited by the ability of the subjects to remember all foods eaten and to estimate portion sizes.

Food frequency questionnaires (FFQs) have been developed to provide retrospective information over a longer time span that the traditional dietary recall (Nelson, 1991). The FFQs are usually designed to cover the consumption of specific foods or structured to focus attention on foods that are key sources of a specific nutrient or group of nutrients. Portion sizes are either estimated within the questionnaire or assigned from databases of portion sizes. FFQs can give evidence of dietary changes in individuals but each FFQ needs to be validated before use and when serial FFQ measurements are made the FFQs need to be strictly comparable (Nelson, 1991). Like all retrospective measures FFQs rely on the accuracy of the memory of the subjects and have a limited quantitative predictive accuracy because of the estimated portion sizes.

(b) Prospective Methods (Bingham, 1991). Prospective methods require the subjects to record food consumption over a period of time. The units of measurement may be household ones or the subjects may be asked to weigh every item of foods. Such studies are very demanding of the subjects and study resources because of the level of training required and because supervision of the subjects is usually necessary (Howat *et al.*, 1994). In principle these methods should give the most accurate results for food intake and permit detailed analyses of dietary patterns and changes in serial studies (Stockley, 1985; Booth *et al.*, 1986). In practice the period of study has to be restricted to a few days or a week and so only gives information over a short period of time. Furthermore there is considerable anecdotal evidence and some objective evidence to show that food consumption patterns may be modified by the measurement itself. Such studies can provide detailed evidence of dietary and meal patterns, and serial studies give evidence for dietary changes which can be analysed in terms of foods and nutrient intakes.

(c) Consumer surveys. A wide range of measurements of food consumption is also made for commercial marketing purposes. These are often designed to measure the consumption of specific categories of food or eating patterns, for example breakfast cereals or eating out. The panel of subjects used may be chosen to represent a balanced population (by age, sex, socio-economic status, education, etc.) or be targeted at a specific socio-economic or regional group of the population. Such studies may be semi-quantitative in a conventional scientific sense but can provide useful qualitative information on dietary changes (King, 1988). Their interpretation must, however, be based on a sound understanding of the study design and methods used, particularly in relation to the questions asked because these may be specifically chosen for the immediate purpose of the study. Many of these studies are not in the public domain, being conducted for clients within the food industry or government.

11.3 Dietary patterns

11.3.1 Defining patterns

The *Oxford English Dictionary* defines a pattern as either 'a repeated decorative design' or 'regular or logical form, order or arrangement of parts'. The latter definition is closer to what students of human food intake and dietary practices normally imply when the term dietary pattern is used. Clearly, in the formal sense, a post hoc judgement is being assigned to observations of eating behaviour in seeking to establish a regular or logical form to food consumption.

There are a number of ways in which dietary patterns can be described, for example, in a temporal sense as seasonal patterns of consumption over the year or a weekly or daily pattern of meals. The patterns can be described in terms of the foods themselves or the nutrients (or other constituents) within the foods.

Each of the descriptions of the patterns requires different types of information from the estimations of food intake described above and the capacity to describe the patterns is limited by the available information. Equally, each of the ways in which the patterns are measured and described provides a different type of information and it is important to be wary of limiting exploration of the ways in which diets are described.

11.3.2 Describing dietary patterns

(a) As nutrient intakes. This is possibly the more common method in which food intakes are converted to nutrient intakes and then analysed in terms of time and their changes in response to other parameters. The nutrient intake in these patterns may be given as absolute rates of intake (weight of nutrient per unit of time) or more commonly as nutrient density (weight of nutrient per 1000 kcal or MJ). The energy-yielding macronutrients may also be expressed in terms of a proportion or percentage of the energy intake.

Formally the composition of any diet in absolute rate terms may be represented by a position in a multi-dimensional space defined by the number of nutrients minus one (Southgate, 1988). Dietary change can in principle take place along any one of these dimensions although in reality, because most foods contain more than one nutrient, it takes place along several dimensions. Where the dietary composition is expressed as nutrient densities the changes in the intake of an energy yielding component will alter the values of all other nutrient dimensions.

Few formal *a priori* analyses of dietary patterns in this way have been attempted (Patterson *et al.*, 1994) but post hoc approaches to this way of analysing dietary patterns can be made by cluster analysis techniques or by principal component analysis to determine the major dimensions which define the dietary intake (Gregory *et al.*, 1990; Fidanza, 1991; Hulsof *et al.*, 1992; Hulsof, 1993).

(b) In terms of foods. This represents a more direct way of describing dietary patterns since the primary information which we have is in terms of foods consumed. It therefore has the advantage that the uncertainties in nutrient intake values inherent in the use of nutritional databases in calculating intakes are not introduced (West and van Staveren, 1991). Since the consumer chooses to eat foods rather than nutrients it is arguably more useful to analyse dietary changes initially in terms of foods and subsequently to interpret them in terms of changes in the dietary patterns of nutrient intakes.

Descriptions in terms of foods can be made as absolute rates of intake (weight of food per unit time), as a proportion of the total food consumption or in terms of expenditure on food. It is also possible to combine the food and nutrient modes by expressing the foods consumed as a proportion of total energy intake. Any diet described in terms of foods can analogously be defined as a point in multi-dimensional space defined by the food dimensions (the number of separate food items measured minus one). Clearly such an analysis presents even more difficulties in formal analysis than was the case for nutrients where the number of dimensions is finite.

In analysing dietary changes useful information can be obtained by aggregating foods into nutritionally coherent food groups (Kelly *et al.*, 1991). Any grouping of foods on a purely nutritional basis must be modulated by consideration of whether these food groupings are coherent in respect of consumer perceptions and in relation to the ways in which the foods are consumed within the food and meal cultures (Stockley *et al.*, 1985). This approach also enables one to consider food consumption records that have been aggregated into groups either in published records or at the point when dietary data was collected and also permits analysis at several different levels of aggregation.

11.3.3 Measurement level and interpretation of dietary change

The level at which food consumption is measured acts as a constraint on the types of dietary change which can be observed and the interpretation that can be placed on the reasons for such changes. Table 11.2 summarises the major constraints of the different types of measurement.

(a) Measurements at the population level provide a broad picture where one can only see major trends which may disguise dietary changes that are taking place. For example, if foods are measured at the wholesale commodity level, changes in the types of products being eaten are not evident nor

Table 11.2 Interpretation of food consumption data obtained at different levels of measurement

Level of measurement	Type of measurement	Limitations	Advantages	Dietary changes
National	Food disappearence	Foods measured at commodity level Estimation of losses during food chain Quality of data variable Assumptions necessary when converting data into nutrients Interpretation depends on good demographic data	Large population Data available for many countries Continuity	Gross changes in food supply Indications of adequacy of food available Dietary changes at low magnification little detail
Household	Budgetary surveys record food purchases only	Foods measured at retail level Conversion into amounts requires good price/ amount data Conversion into nutrients	Conducted in many industrialised countries Continuity	Changes in patterns of purchases Economic analyses of changes
	House food consumption	No indication of actual consumption within household Estimation of edible waste Estimation of nutrient losses in preparation Conversion into nutrients	Continuity Often linked with socio-economic data Reasonable estimates of nutritional adequacy of diet.	Changes in foods consumed in household Economic analyses Medium level of magnification

are changes occurring in specific sectors of the population which may have profound significance. Population measurements therefore tend to give a false impression of the constancy of food consumption and, because they relate to measurements of foods at a high level in the food chain they overestimate per capita food ingestion. They should, by analogy, be regarded as providing a 'low magnification' view of dietary change.

(b) Measurements at the household level provide more detail but again, primarily because of the limited sample sizes of specific groups of the population, may hide significant dietary changes. Changes in temporal patterns of food consumption, meal frequency and the size of meals consumed by individuals within the household groups studied are also not distinguished and these studies, particularly when aggregated to the national level, may also give a false impression of dietary constancy.

Many studies of this kind include an economic characterisation of the household so that the effects of socio-economic factors can be deduced. The National Food Survey conducted in the UK is a good example of a study that is documented in this way. The accumulated evidence from the annual reports of the National Food Survey and comparable studies in other countries provide the basis for much of the discussion of the factors producing dietary changes (MAFF, 1991).

(c) Studies of the foods eaten by individuals provide the 'high magnification' pictures of dietary change where changes in the selection of individual foods and the meal patterns can be distinguished. The major limitation of these measurements is often the small size of the population, the length of the study and the separation of the data from descriptions of the wider food environment (Gregory *et al.*, 1990).

11.4 Factors involved in dietary change

In discussing the range of factors involved in dietary change it is important to recognise the fact that in many cases the identification of the factors and the strength of their effects are usually retrospective assessments often with a strong intuitive influence. This 'post hoc, prompter hoc' analysis of dietary changes has severe limitations if there is a desire or perceived need to introduce dietary changes as, for example, desirable for public health measures for the reduction of the risk of chronic diseases. The formal analysis of the factors affecting food choice has not as yet produced a unifying hypothesis which has predictive properties although there is clear evidence that the attitudes and beliefs of the consumer have a powerful influence (Towler and Shepherd, 1992) just as there are strong physiologically driven aspects of appetite coupled with hedonic factors in the choice of some foods (Rogers and Blundell, 1990). The evidence for these factors has often been obtained under conditions where food availability and economic considerations were controlled and their influence relative to economic factors is unclear (Meiselman, 1992).

While there is value in identifying the individual factors which influence dietary change and establishing the strengths of the various factors, for example by using multiple regression (Gregory *et al.*, 1990), there is also merit in considering how the factors may be integrated (Shepherd, 1990). Thus, observed food choice as expressed in dietary behaviour represents an integration of the various factors which act as the vectors of dietary change (Figure 11.1). In categorising the factors it is important to emphasise that there are two limitations of this type of analysis; first, that the boundary

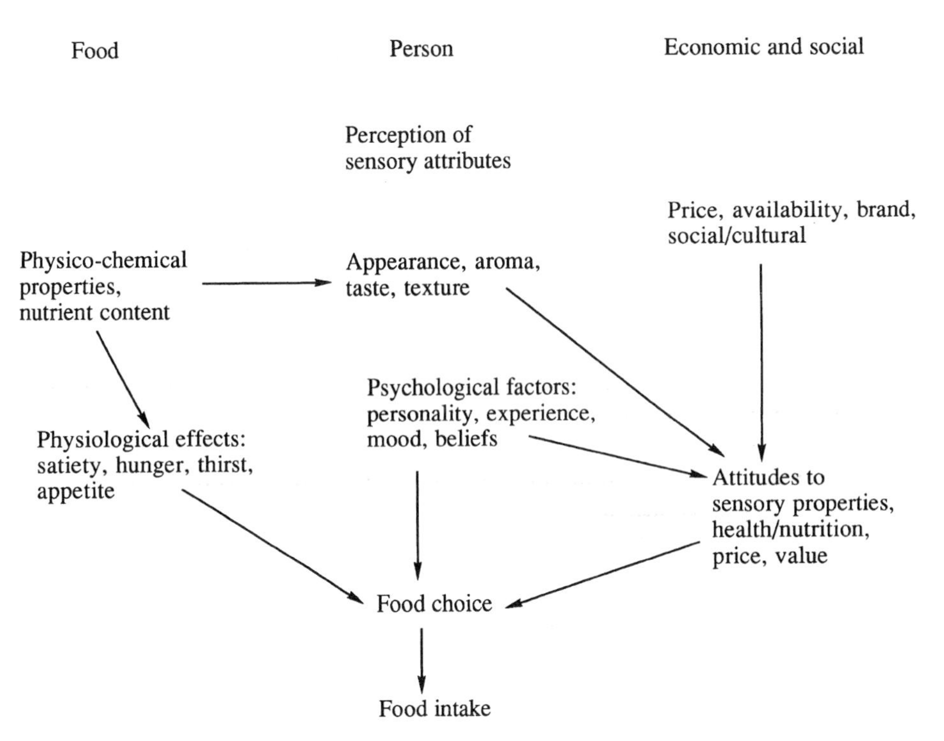

Figure 11.1 Interrelationships between the factors affecting food choice categorised into those relating to foods, the person and socio-economic factors (after Shepherd, 1990).

between the factors is not a clear one and that many of the factors operate at random and, second, that much of the evidence relating to the effects of these factors on dietary change is conjectural and not as yet tested in a formal sense. Moreover, such formal testing depends on chance observations of 'natural' experiments or the design and implementation of experimental studies with populations that few people would regard as ethical.

11.4.1 Cultural factors

The cultural factors involved in food consumption within a country or region are extremely important because they act as a constraint on dietary changes and form the background against which many attitudes and beliefs about foods arise (Fieldhouse, 1986; Wardle, 1977; Booth *et al.*, 1986; Terny, 1994). The overwhelming impression gained from measurements of food supplies and from Household Surveys, such as the UK NFS, is that food consumption patterns exhibit great conservatism. In primitive societies the food culture is passed down from mothers and the elders, and has a strong protective characteristic preventing the consumption of new and untested foods. There is some evidence that this transmission of food habits and attitudes continues in industrialised societies (Stafleu, 1994; Stafleu *et al.*, 1994). Many of the religious food laws also restrict the choice of foods to those known from experience to be safe and wholesome. Studies with immigrant communities provide evidence for the retention of known dietary choices possibly as a refuge against a foreign, and sometimes threatening, culture.

Cultural changes have a strong influence on food choices. Some of these are difficult to separate from socio-economic changes which in many cases are the driving force behind such cultural changes. Thus the changing role of women in most western societies, although stemming from the moral position of the justice of the emancipation of women, was, in many countries, the consequence of the economic need for increasing the proportion of women in employment. For example, in the UK there was a need in 1939–45 in munition factories to replace men involved in war. The increasing employment of women outside the home since the 1939–45 war has continued for personal economic reasons and is especially important in many European countries (King, 1988; Gofton, 1990). This has reduced the time available for the domestic preparation of food and therefore increased the demand for processed foods. For many of these foods the preparation has moved from the kitchen to the food factory (Den Hartog, 1992; Southgate and Johnson, 1993).

Changes in leisure activity have also resulted in an important cultural change; the development of the travel industry has permitted increasing numbers of people to experience other food cultures and the liking for different foods has resulted in increased demand for these foods both in the restaurants and as ingredients. In countries such as the UK, immigrants from the Asian sub-continent and the West Indies have also created a demand for their own foods and these have become available in the marketplace where they can be obtained by those sectors of any community which are active in seeking new experiences and enjoy trying new foods (Van Stryp, 1994).

11.4.2 *Socio-economic factors*

(a) *Economic status.* There is strong evidence from the Food Balance Sheet studies of the FAO of a relationship between patterns of food consumption at the national level and economic status. The classical study of Perisse and his colleagues (Perisse *et al.*, 1969) showed that the dietary pattern in terms of the sources of energy-yielding components was strongly related to gross national product and that diets could be categorised using a limited number of dimensions. Serial measurements confirm that improved economic status led to an increased proportion of energy from fats, a shift in the ratio between sugars and starches (FAO, 1980) and an increase in the proportion of protein from animal sources, although protein as a proportion of the energy supply is substantially constant.

In many cultures animal foods are associated with a higher prestige and price whereas starchy plant-based foods have an analogously lower status and price. Economic development in such communities is therefore accompanied by a shift in the patterns of food purchases towards the higher value, more prestigious foods. The series of studies carried out in the UK NFS provides clear evidence of such changes in the years following the abolition of food rationing in the 1950s (MAFF, 1991). This series of studies shows that with improved economic status the proportion of household budgets devoted to food declines. Although there was substantial inflation during the period 1950–1990, the prices of food showed a lower rate of inflation than the retail price index as a whole.

Studies of food purchases since the 1950s show a pronounced trend towards an increased proportion of fat in the diet and a reduction in the consumption of starchy foods such as bread and potatoes. This trend accompanied the improvement in economic status, thus reinforcing the position seen in the FAO analysis (Table 11.3). The changes took place at different rates in the different socio-economic groups used in the survey (Table 11.4). The analyses of the NFS in terms of foods show an increased consumption of processed foods which has been attributed to the influence of other social changes that were taking place over this period (MAFF, 1994).

Table 11.3 Changes in the consumption of the major food groups and energy intake in the United Kingdom between 1950 and 1990 (percentage change relative to 1950). Based on MAFF (1991).

Food group	1955	1960*	1970	1980	1990
Milk	–2.3	–1.4	–2.5	–12.1	–26.7
Cheese	+11.4	+19.7	+41.3	+53.1	+57.4
Meat	+14.4	+20.2	+32.1	+34.6	+14.2
Fish	–10.2	–11.5	–19.2	–27.5	–23.3
Eggs	+19.7	+32.6	+33.1	+5.4	–37.2
Fats and oils	+2.3	+3.1	+2.9	+3.5	–22.5
Sugars	–30.4	+27.6	+19.4	–19.5	–53.0
Vegetables	–1.6	–3.3	–6.7	–13.5	–19.3
Fruits	+21.1	+36.1	+41.1	+55.1	+74.2
Cereals	–30.1	–13.6	–22.6	–32.2	–36.6
Beverages	+30.1	+31.2	+32.7	+10.3	+9.2
Total energy	+6.8	+6.3	+3.5	–9.9	–24.3

*Between 1955 and 1960 all food rationing in the UK had been discontinued

Table 11.4 Comparison of the consumption of the major food groups by the highest (Group A) and lowest (Group D) income groups in the United Kingdom between 1955 and 1990 (1990 consumption as percentage of 1955 consumption). Based on MAFF (1991).

Income group	Group A	Group D	Group A adjusted for energy intake	Group D adjusted for energy intake
Food group				
Milk	61.6	82.3	86.9	116.0
Cheese	162.0	124.0	228.6	174.9
Meat	85.5	95.0	108.4	134.0
Fish	66.3	75.6	93.5	106.0
Eggs	39.7	97.4	55.9	137.0
Fats and oils	61.9	75.4	87.3	106.3
Sugars	24.8	42.5	35.0	60.0
Vegetables	79.0	89.0	111.5	125.6
Fruits	109.6	86.7	154.6	125.6
Cereals	62.3	65.0	87.8	91.0
Beverages	61.1	70.0	87.8	91.0
Total energy	+6.8	+6.3	+3.5	–9.9

(b) *Changes in working patterns.* The UK, like many developed societies, underwent substantial changes during the period 1950–1990. These, together with cultural changes, resulted in changes in working patterns, especially of women but also, to an increasing extent, for the whole community. The introduction of automation of production and the introduction of computers in many areas of work have resulted in increased unemployment and part-time working which have made the financial contribution of women to many households essential (Gofton, 1990).

In addition, in many communities commuting to work, often involving several hours in travelling, has produced changes in the structure of the working day which affect meal patterns and food choice. First, there has been an increase in eating out which has meant that the NFS is actually measuring a diminishing proportion of the total diet. Second, reductions in the physical effort required in many jobs and the increased use of the motor car for personal transport have led to reduced energy requirements and these are reflected in reduced food consumption, although obesity remains a problem. Third, the increased cost of labour and efficiencies of bulk marketing have led to changes in food retailing with the introduction of supermarkets and, latterly, hypermarkets. These also are seen as providing more freedom of consumer choice. While this development did not necessarily change food choice, it did alter food purchasing behaviour towards making bulk purchases once a week rather than the traditional daily shopping routine of the pre-war housewife.

(c) *Changes in meal patterns.* These cultural and socio-economic changes have resulted in changes in meal patterns (Wardle, 1977; King, 1988). The changes seem to be taking place at different rates in the different socio-economic groups and different geographical regions, and, as one might expect, show age-dependent effects.

First, there have been changes in the type of breakfast consumed, with the trend moving from the so-called traditional cooked breakfast to a meal based around breakfast cereals or, in many younger age groups, the omission of breakfast as a meal of any size. Second, the meals eaten at midday have become less likely to be provided in work place canteens, and the size of the meal and the type of foods consumed have changed to become less substantial. Third, a combination of the pressures imposed by working patterns on the housewife and her partner, coupled with leisure activities of all the members of the family, has resulted in a reduction in the importance of set family meals where the family eat together.

In the study of knowledge and beliefs about food conducted by the British Nutrition Foundation (King, 1988) there was strong evidence that the traditional UK meal pattern of three meals per day had largely disappeared amongst young adults and was being replaced by a disordered 'chaotic' style of taking meals. The authors were unable to say whether this represented a permanent change in the UK meal patterns or whether the traditional meal patterns would re-establish themselves as the subjects became older. A study of the food consumption of adults (Gregory *et al.*, 1990) suggests that the traditional 'three cooked meals a day' has virtually disappeared, probably for socio-economic and cultural reasons. In countries without this tradition of a warm breakfast the changes do not appear to be so significant (Ministerieren van Welzijn *et al.*, 1988, 1993).

11.4.3 Food availability

Food availability is a major factor influencing food choice since, *a priori*, food choice can only be made from the foods available for purchase, exchange or obtainable from the environment. In developing societies food availability and the related capacity to purchase the food are major determinants of food choice although these will always be modulated by cultural factors defining what are seen as proper or acceptable foods. In non-industrialised societies the availability of foods has a strong seasonal dependence and the patterns of food consumption therefore shows seasonal cycles. Seasonal influences on food availability, although they have been weakened in industrialised societies, are still present and can still be seen in the patterns of consumption of some foods where presumably some cultural determinants are operating. In the UK, for example, seasonal trends can still be seen for fresh vegetables, fresh fruit and to a certain extent fresh fish and shell fish. Eggs, which showed a pronounced seasonal pattern in 1950, have shown little seasonality since 1970 following the introduction of intensive egg production which increased their availability (MAFF, 1991).

Economic factors have a major influence an food availability because in many cases the market acts to control supply either directly, where there is management of food prices by the producers or governments through subsidies, or indirectly by the interplay of supply and demand.

In many developed countries populations are relatively stable and the reduced physiological need for the consumption of food energy has the effect of limiting the total volume of the food market. This places pressure on food producers, and especially food manufacturers, to focus attention on the development and marketing of value-added products and completely new products to capture an increased market share. Much of the innovation in the food industry stems from the developments in food technology. These have permitted the development of new products which meet an unfilled niche in consumer demand and the creation of demand for new products. All these activities expand the range of foods available and influence food choice (Den Hartog, 1992). The initial purchasers of these new foods are those consumers with adequate financial resources and who are seeking new experiences. These innovators establish these new products initially and create food fashion which then spreads to the more conservative segments of the population.

11.4.4 Changes in food production and food technology

It is possible to distinguish a number of developments which have had a profound influence on food patterns.

(a) *Food production.* The development of more efficient production methods, leading to increased availability at lower prices, can be seen as a major factor in determining dietary changes. In some cases this may result in profound changes in the foods consumed, for example the introduction of intensive methods for the production of poultry meat had a substantial effect on prices and resulted in a major change in the pattern of meat consumption in the UK over the period 1950–1990. The changes started in the late 1950s, leading to a doubling in the consumption of poultry. The intensive production of eggs, on the other hand, was not accompanied by an increase in consumption but the seasonal pattern of egg consumption was virtually abolished. Furthermore, the battery-produced eggs became the major proportion of total egg purchases over the course of the next five to six years.

This illustrates one of the important facets of changes in production methods, such as the introduction of a new process or variety of a food; if these changes offer substantial advantages to the producer they may became dominant very rapidly and produce radical changes in the types of food being eaten without appearing to produce dietary change. For example, the development of the Chorleywood Bread Process (Coppock *et al.*,

1958) had substantial benefits for the bread-making industry and permitted the use of an increased proportion of home-grown flour. Once introduced, the majority of white bread consumed was produced by this process over the course of about six years. Analogous effects flow from changes in the varieties of vegetables grown commercially which produce dietary changes in the types of foods eaten but no change in eating patterns.

(b) *Storage and preservation.* The development of refrigerated storage, initially at the industrial level and latterly in the widespread acquisition of domestic refrigerators and deep-freezes, has been an important factor in changing dietary patterns. At the industrial level the availability of crops and meats has been extended. The domestic deep freeze has altered purchasing patterns and increased the consumption of frozen food products. The consumption of conventionally processed canned products has not substantially declined, suggesting that frozen products have partially replaced the fresh produce or are effectively seen as new foods.

Controlled atmosphere storage has also extended the availability of fresh fruit and vegetables and is one of the factors contributing to the reduction of seasonal effects on the consumption of these foods. Seasonal effects are still evident but the amplitude of the changes in purchasing patterns has shown a progressive reduction for both fresh vegetables and fresh fruit (MAFF, 1991).

The development of refrigerated transport and the ability to maintain the quality of a chilled product have also served to increase the availability of products such as yoghurts.

(c) *Heat processing.* One major advance in food technology has been the application of extrusion cooking. This has extended the range of cereal foods available and has been particularly important in the development of snack foods. The changes in the meal patterns in many industrialised countries, leading to the reduction in the importance of formal, set meals and to grazing patterns of food consumption, have provided an excellent niche for snack foods.

The introduction of microwave cooking has increased opportunities for the development of convenience meals, often intended for a single consumer, and these have filled the niche produced by changing meal patterns and leisure activities, particularly in the younger age groups. Microwave cooking greatly increases the convenience of frozen foods because they allow safe defrosting and therefore the technology increases the attraction of these products.

Ultra-heat treatment technology has extended the shelf life of products such as milk and fruit drinks.

(d) *Packaging.* The advances in packaging technology have greatly improved sensory quality and extended availability. The development of aseptic packaging technology has been particularly valuable in expanding the shelf-life and quality of fruit juices. All these advances increase the range of consumer choice and thus provide the background against which other factors determining food choice operate (Swalwell, 1990).

(e) *Retailing and advertising.* The development of supermarkets, and more recently hypermarkets, has brought about major changes in food purchasing patterns in many industrialised societies. One can speculate that the presentation of a wide variety of foods directly to the consumer may have increased the variety of foods eaten. Purchasing patterns have undoubtedly been altered but it is not clear whether these have in themselves greatly influenced eating patterns directly.

Clearly, food advertising is another way in which the food industry seeks to modify food choice. It is arguable whether these affect eating patterns as such because the advertising is predominantly brand- or food group-based and may by analogy be compared with cigarette advertising which is said to be concerned with maintaining brand or food loyalty rather than encouraging consumption at the expense of other foods (Swalwell, 1990; Van Stryp, 1994). However, the observed dietary patterns, as remarked above, represent the integration of the different factors and at the present time quantitative models of that integration are not available.

11.4.5 Microbiological safety

While consumer concerns about the relationship between nutrition and health are relatively recent, concerns about the safety of food, especially microbiological safety, have been major influences on food purchasing. In most industrialised countries food legislation has for many years resulted in an essentially safe food supply and departures from this expected norm create considerable interest in the media. These incidents can have major effects on food purchasing patterns and produce dietary changes which may be transient (where the episode is soon no longer 'news') but in some cases can have persistent effects on dietary choices (Gofton, 1990). The identification of *Salmonella* in eggs in the UK in 1988 is one case where there was a substantial reduction in the consumption of eggs which has remained, even allowing for a general decline in egg consumption. The appearance of bovine spongiform encephalopathy in cattle has also contributed to the decline in carcass meat purchases which has been evident for many years, probably due to economic influences (MAFF, 1994).

11.5 Impact of recommendations on diet and health

In 1977 the report of the US Senate Select Committee for Health and Human Welfare (the McGovern report, US Senate Select Committee on

Nutrition and Human Needs, 1977) marked the beginning of a change in thinking about the relationships between diet and health. The report tried to identify the relationships between dietary intake and the incidence of many chronic diseases. It produced guidelines for changes in the US diet which it was believed would reduce the incidence of these diseases. The report initially produced estimates of the financial benefits which would flow from these changes apart from the benefits to health of the US population. Thus dietary changes were suggested as public health measures. The report called on consumers and both primary and secondary sectors of the food industry to contribute to the implementation of dietary changes. Although many previous publications had associated diet and disease and produced recommendations for a healthy diet this report, because of its provenance and approach, set in train a series of other reports in the USA (US Department of Health and Human Services, 1988; National Research Council, 1989) and from International Organisations (WHO, 1990). All these reports recognise the need, first, to guide the consumer in the choice of an appropriate healthy diet and, second, to encourage the food producers, retailers and catering industry to contribute to the process of bringing about dietary change.

11.5.1 Principal features of the recommendations

While there are many points of detail and emphasis in the different reports, the principles are to a considerable extent common ones:

1. It is desirable to eat a varied diet.
2. It is important to eat the amount required to maintain body weight within an ideal range (that is the body weight associated with lower mortality).
3. The proportion of total fat in the diet should be reduced to around 30–35% of total energy.
4. Of that fat no more than one third should be saturated fats.
5. The proportion of complex carbohydrates, starches and plant cell wall material (dietary fibre) should be increased to compensate for the reduction in fat energy and to replace some sugars. Some recommendations specifically call for a reduction in the proportion of energy from simple sugars especially those added to foods (Department of Health, 1989).
6. The consumption of vegetables and fruits should be increased to around five or more servings per day.

11.5.2 Implementation of dietary advice: effects on dietary patterns

The strategies adopted in various countries which are intended to produce these dietary changes have shown some differences in emphasis but all have recognised the need to first provide education to consumers and second to encourage the development by the food industry of modified or new products which would assist the consumer in the choice of an appropriate diet.

(a) *Nutritional labelling.* One element in the educational process has been the introduction of nutritional labelling of foods on either a voluntary or compulsory basis (MAFF, 1990; British Nutrition Foundation, 1990). The regulations differ between countries but most regulations insist on labelling when a nutritional claim is made for the food.

The effectiveness of nutritional labelling in educating consumers about foods has met with difficulties because of presentational issues and because the construction of diets on an energy basis requires rather sophisticated calculations and a certain level of numeracy and scientific knowledge. Furthermore, at the time of consuming food the labelling is not available and most individual consumers will not have read the label. The effectiveness of nutritional labelling is thus heavily dependent on the housewife or the 'gatekeeper' for the household food supply.

(b) *Educational programmes.* The problems experienced in using nutritional labels suggested that nutritional knowledge needed specific attention and studies by Fine and her colleagues (Fine *et al.*, 1994) have shown that by using a range of educational techniques it is possible to lay the foundations for dietary change. Few studies have shown that nutritional knowledge is necessarily correlated with actual changes in behaviour.

Food retailers and some sectors of the food industry have also produced educational material to guide consumers in their choice of foods. This has often been accompanied by the development of new or modified products (see section 11.5.3).

11.5.3 Development of modified food products

These have focused on a few major facets of food composition some of which appear to be having substantial effects on dietary patterns.

(a) *Reduced fat products.* The focus on reducing the consumption of fat, and especially saturated fats, has resulted in food producers developing lower-fat products. These products have in many cases replaced substantial proportions of the conventional product. Thus in the UK skimmed and semi-skimmed liquid milks have replaced a substantial part of whole milk purchases, now making up more than one third of liquid milk consumption (MAFF, 1991; 1994).

Yoghurt, which represented a rather exotic food until the 1970s, has shown a steady growth, although only a proportion of this is of the low fat varieties. Reduced fat cheeses have not been as successful.

The development of low fat spreads with comparable sensory properties has led to a substantial change and they now make up more than a fifth of fat purchases compared with a virtually zero level before 1980. Together with these low fat spreads the manufacturers have also developed margarines based on vegetable oils, and more recently based on monounsaturated oils, with high concentrations of polyunsaturated fats. The consumption of these margarines has increased since their introduction and they have replaced a substantial proportion of conventional margarines.

Since the late 1970s there has been a substantial decline in the consumption of butter and a rise in the consumption of vegetable oils (MAFF, 1991). The fall in butter consumption, however, started before nutritional advice began to be promoted and was probably of economic origin because of the relative price of butter compared with margarines. It is interesting to note that in the Netherlands, which was a high margarine consuming country, little change in butter consumption has occurred (de Graaf, personal communication). The increased consumption of vegetable oils also pre-dates nutritional advice and may be of cultural origin, representing the impact of foreign travel in the national cuisine. For both these foods it is probable that nutritional advice is serving to re-inforce changes that were taking place for other reasons.

(b) *Complex carbohydrates.* The hypothesis regarding the protective effects of the plant cell walls in foods (dietary fibre) was effectively promoted in the 1970s and resulted in an increased demand for high fibre foods. The food industry, especially the manufacturers of breakfast cereals, responded to this demand by expanding their range of products.

The consumption of wholemeal bread also showed a marked increase, although from a very low base. The rise did not continue beyond 1988 (MAFF, 1991) and there is a steady fall in the consumption of white bread so that total consumption of breads has fallen steadily in the UK since 1960.

Potato consumption has also fallen by a half over the period 1955–1990. There is little evidence that dietary advice in this area has had any influence on the dietary intake of the population as a whole.

(c) *Consumption of fruits and vegetables.* The consumption patterns of vegetables have shown only modest increases in the UK since 1960 although changes in the types of vegetables consumed have been evident with the traditonal vegetables being replaced by fresh green and salad vegetables, and exotic types. Fruit consumption has shown some increase although this is principally due to the rapid increase in the consumption of fruit juices. There is also evidence of an increased consumption of exotic fruits. In the Netherlands, where the consumption of both vegetables and fruit has been actively promoted, their consumption shows a slight decline believed to be due to a perceived lack of convenience in consuming these foods.

(d) *Consumption of salt and sugars.* The food industry has responded to these elements of the dietary advice by expanding the range of products so that alternative products without added sugar or salt have become available. Thus, unsweetened fruit juices, fruits canned in juice rather than sugar syrups and vegetables canned in lower salt concentrations have appeared. The salt content of bread has also been reduced.

Estimates of salt intake are extremely difficult to make from diet measurements of food purchases and, at the individual level, from food data alone because of the salt added to foods at the table or in cooking but the estimates since 1985 suggest a modest decline.

The consumption of sugars and preserves has declined in the population as a whole but the changes within the population have differed by economic group with the higher income groups showing a substantial reduction and pensioner groups being very much above average.

11.5.4 Overall impact of dietary advice on eating patterns

The studies in the UK show that, for the population as a whole, energy intakes are declining slowly and fat intakes are declining substantially in parallel with energy so that the proportion of energy from fat has remained virtually constant since the late 1970s. There has, however, been a significant change in the polyunsaturated/saturated fatty acid (P/S) ratio which has increased from 0.19 in 1975 to 0.43 in 1992. In the Netherlands analogous changes in the P/S ratio have occurred and there has been a significant decline in the proportion of energy from fat (Ministerieren van Welzijn *et al.*, 1993).

In the UK the changes within the income groups have been different. In the two higher income groups the percentage energy from fat has declined slowly since 1983 whereas the lower income groups show little change in the proportion of energy from fat although all show comparable changes in the P/S ratio (MAFF, 1991). There is some slight evidence for a modest increase in starch and a reduction in the consumption of added sugars.

It is probable that the view provided by household purchases is too low a magnification to identify changes being made at the individual level and the income group differences suggest that dietary advice may be influencing the eating patterns of individuals who are likely to have a higher educational level and are also in a financial position to experiment with the new products. One would also expect these income groups to include the innovative, experimentally inclined, individuals.

11.6 Understanding dietary change: future directions

It is clear from the previous discussion that dietary changes can be produced by a wide range of factors, many of which are interrelated, and the observed

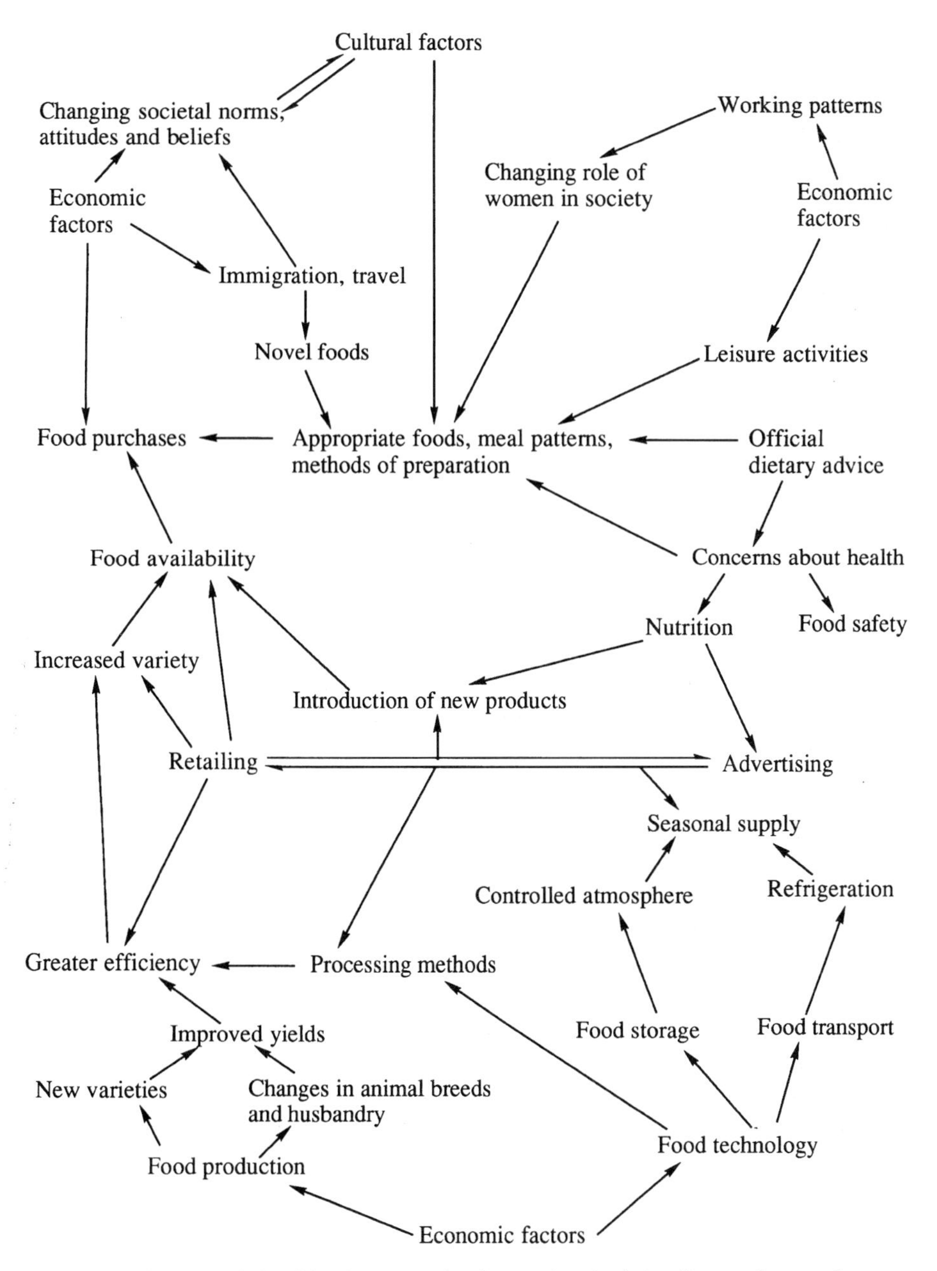

Figure 11.2 Interrelationships between the factors involved in dietary change. Some interactions have been omitted in the interests of clarity.

changes represent the integration of these many factors. Figure 11.2 illustrates how some of the many factors interact with one another. Understanding this integration is central to understanding dietary change and is especially important for public health nutritionists seeking to produce dietary changes. At the present time we are limited by a number of factors.

First, many measurements of food consumption are made without documentation of the other factors: cultural, socio-economic and those relating to the food supply of the population or groups of subjects being studied. Analogously, studies of sociological factors are often separated from the detailed nutritional measurements. Additionally, relatively few measurements of food consumption document the temporal patterns of food consumption. Understanding dietary change therefore depends, in the first instance, on the implementation of multidisciplinary studies which are designed to document all the factors known or believed to be involved in influencing food choice and dietary patterns (Terny, 1994).

Second, improving the measurements of food consumption is extremely important. The most widely available sets of data based on food supplies (food balance sheets) and from household budgetary surveys, while valuable, do not provide sufficent detail and there are serious doubts about their validity as measures of what people actually eat. Such doubts also extend to more detailed individual studies where under-reporting of a significant magnitude is common (Bingham, 1991). The effect of the measurements on eating behaviour is critical (Stockley, 1985).

Third, post hoc rationalisations of the effects of different factors on eating patterns are intrinsically unsatisfactory in that the development of predictive understanding is central to any public health policy for introducing dietary change and would be more satisfying intellectually. Designing experimental studies where all the factors are under control is, as mentioned earlier, difficult to accomplish in an ethical environment and therefore the one avenue that is open is the development of hypothetical, predictive models which can be tested. The features of such a model require considerable research but one could envisage that the biological determinants of appetite represent the core of the model upon which cultural factors operate as long-term effectors. Socio-economic and other factors related to the food supply would act on the cultural–biological axis. Such a model would need to be based on a multidimensional description of the diet. It would probably be wise initially to base this model on the consumption of foods rather than nutrients because such a basis would enable the model to be tested against a range of information from the different levels of measurement of food consumption. Without the development of such a model our understanding of dietary changes will remain incomplete and intuitive.

References

Andrews, P. and Martin, L. (1991) Hominoid dietary evolution. *Phil. Trans. Royal Society London B*, **334**, 199–209.

Bingham, S.A. (1991) Assessment of food consumption: Current, in *Design Concepts in Nutritional Epidemiology* (eds B.M. Margetts and M. Nelson), Oxford Medical Publications, Oxford, pp. 154–166.

Booth, D.A., Edema, J.M.P. and Stockley, L. (1986) Food habits, food preferences and their determinants: difficulties in social and behavioural measurements crucial to nutrition, in *Measurements and Determinants of Food Habits and Food Preferences* (eds J.M. Diehlr and C. Leitzman), Euro-Nut Report No 7., Wageningen, The Netherlands, pp. 45–61.

British Nutrition Foundation (1990) *Nutrition Labelling*, BNF Briefing Paper No. 21, BNF, London.

Cameron, M.E. and van Staveren, W.A. (eds) (1988) *Manual on Methodology for Food Consumption Studies*, Oxford University Press, Oxford.

Coppock, J.P.M., Knight, R.A. and Vaughan, M.C. (1958) The moisture content of white bread. *Nutrition Lond.*, **12**, 63.

Den Hartog, A.P. (1992) Dietary change and industrialization: the making of the modern Dutch diet. *Ecology of Food and Nutrition*, **27**, 307–318.

Department of Health (1989) *Dietary sugars and human disease*, Report on Health and Social Subjects No. 37, HMSO, London.

Department of Health (1994) *Nutritional aspects of cardiovascular disease*. Report On Health and Social Subjects No. 46, HMSO, London.

Gofton, L. (1990) *Food fears and time famines*. British Nutrition Foundation Bulletin 15 (Supplement), pp. 78–95.

Gregory, J., Foster, K., Tyler, H. and Wiseman, M. (1990) *The dietary and nutritional survey of British adults*, HMSO, London.

Fidanza, F. (1991) Food patterns and health problems. *Annals Nutrition Metabolism*, **35** (supplement), 78–80.

Fieldhouse, P. (1986) *Food and Nutrition: Custom and Culture*, Croom Helm, London.

Fine, G.A., Conning, D.M., Firmin, C., De Looy, A.E., Losowsky, M.S., Richards, I.D.G. and Webster, J. (1994) Nutrition education in young women. *British J. Nutrition*, **71**, 789–798.

Food and Agriculture Organization (1980) Carbohydrates in human nutrition. *A Joint FAO/ WHO report. Food and Nutrition Paper No. 15*, FAO, Rome.

Food and Agriculture Organization (1983) Comparison of food consumption data from food balance sheets and household surveys. *FAO Economic and Social Development Paper No. 34*, FAO, Rome.

Howat, P.M., Mohan, R., Champagne, C., Monlezun, C., Wozniak, P. and Bray, G.A. (1994) Validity and reliability of reported dietary intake data. *J. American Diet Assoc.*, **94**, 169–173.

Hulsof, K.F.A.M. (1993) *Assessment of variety, clustering and adequacy of eating patterns*, Dutch National Food Consumption Survey (thesis), Maastrict.

Hulsof, K.F.A.M., Wedel, M., Lowik, M.R.H., Kok, F.J., Kistemaker, C., Hermus, R.J.J., ten Hoor, F. and Ockhuizen, T. (1992) Clustering of dietary variables and other lifestyle factors. *J. Epidemiology and Community Health*, **46**, 417–424.

Kelly, A., Becker, W. and Helsing, E. (1991) Food balance sheets, in *Food and Health Data*. (eds W. Becker and E. Helsing.), WHO Regional Publications, European Series, No. 43, WHO, Copenhagen, pp. 39–48.

King, S. (1988) *Eating behaviour and attitudes to food, nutrition and health*. Report prepared for the British Nutrition Foundation, London.

Meiselman, H.L. (1992) Methodology and theory in human eating research. *Appetite*, **19**, 49–53.

Ministerieren van Welzijn, Cultuur, Landbouw en Visserij (1988) *Wat eet Nederland. Resultaten van de voedselconsumptiepeiling 1987–1988*, Ministerie van Cultuuur, Rijswik, pp. 119.

Ministerieren van Welzijn, Volkesgezondheid, Landbouw, Natuurbeheer, en Visserij (1993) *Zo eet Nederland 1992. Resultaten van de Voedselconsumptiepeiling 1992*, Voorlichtlingbureau voor de Voeding, Den Haag, pp. 198.

Ministry of Agriculture, Fisheries and Food (1990) *Food Labelling Survey England and Wales*, HMSO, London.

Ministry of Agriculture, Fisheries and Food (1991) *Household Food Consumption and Expenditure 1990*, HMSO, London.

Ministry of Agriculture, Fisheries and Food (1993) *National Foods Survey 1992*, HMSO, London.

Ministry of Agriculture, Fisheries and Food (1994) *National Food Survey 1993*, HMSO, London.

National Research Council (1989) *Diet and Health: implications for reducing chronic disease risk*, National Academy Press, Washington DC.

Nelson, M. (1991) Assessment of food consumption: past, in *Design Concepts in Nutritional Epidemiology* (eds B.M. Margetts and M. Nelson.), Oxford Medical Publications, Oxford, pp. 167–191.

Nelson, M., Atkinson, M. and Darbyshire, S. (1994) Food photography: perception of food portion size from photographs. *British. J. Nutrition.*, **72**, 649–663.

Patterson, R.E., Hains, P.S. and Popkin, B.M. (1994) Diet quality index: capturing a multidimensional behaviour. *J. American Dietetic Association*, **94**, 57–64.

Perisse, J., Sizaret, F. and Francois, P. (1969) The effect of income on the structure of the diet. *FAO Nutrition Newsletter 7*, FAO, Rome, pp. 1–17.

Rogers, P.J. and Blundell, J.E. (1990) Psychobiological basis of food choice. *British Nutrition Foundation Bulletin 15 (supplement 1)*, pp. 31–40.

Shepherd, R. (1990) Overview of factors influencing food choice. *British Nutrition Foundation Bulletin 15 (Supplement 1)*, pp. 12–30.

Sizaret, F. (1992) Food balance sheets and food consumption surveys, in *Flair Eurofoods-Enfant Second Annual Meeting Killiney Bay* (eds J. Castenmiller and C.E. West.), Wageningen, The Netherlands, pp. 62–71.

Southgate, D.A.T. (1988) Dietary fibre and the diseases of affluence, in *A Balanced Diet* (ed J. Dobbing), Springer-Verlag, London, pp. 117–141.

Southgate, D.A.T. (1991a) Nature and variability of human food consumption. *Phil. Trans. Royal Society London B*, **334**, 281–288.

Southgate, D.A.T. (1991b) Database requirements for calculations from food balance sheet data and household budget surveys, in *Food and Health Data* (eds W. Becker and E. Helsing), *WHO Regional Publications European Series*, **34**, 85–89.

Southgate, D.A.T. and Johnson, I. (1993) Food processing, in *Human Nutrition and Dietetics. 9th edn.* (eds J.S. Garrow and W.P.T. James), Churchill Livingstone, Edinburgh, pp. 335–348.

Stafleu, A. (1994) *Family resemblence in fat intake, nutritional attitudes and beliefs: a study among three generations of women*, thesis, Wageningen, pp. 189.

Stafleu, A., de Graaf, C., van Staveren, W. and de Jong, M. (1994) Attitudes towards high fat foods and their low fat alternatives: reliability and relationship with fat intake. *Appetitie*, **22**, 183–196.

Stockley, L. (1985) Changes in habitual food intake during weighed inventory surveys and duplicate diet collections: a short review. *Ecology of Food and Nutrition*, **17**, 263–269.

Stockley, L., Faulks, R.M., Broadhurst, A.J., Jones, F.A., Greatorex, E.A. and Nelson, M. (1985) An abbreviated food table using food groups for the calculation of energy, protein and fat intake. *Human Nutrition: Applied Nutrition*, **39A**, 339–348.

Swalwell, S. (1990) Packaging and advertising. *British Nutrition Foundation Bulletin 15 (Supplement)*, pp. 96–101.

Terny, R.D. (1994) Needed a new appreciation of culture and food behaviour. *J. Amer. Diet Assoc.*, **94**, 501–503.

Towler, G. and Shepherd, R. (1992) Application of Fishbein and Aijzen's expectancy-value model in understanding fat intakes. *Appetite*, **18**, 15–22.

Ulijaszek, S.J. (1991) Human dietary change. *Phil. Trans. Royal Society London B*, **334**, 217–279.

US Department of Health and Human Services (1988) *The Surgeon General's Report on Nutrition and Health*, US Government Printing Office, Washington DC.

US Senate Select Committee on Nutrition and Human Needs (1977) *Dietary Goals for the United States*, US Government Printing Office, Washington DC.

van Staveren, W.A., Van Beem, I. and Helsing, E. (1991) *WHO Regional Publications European Series No. 34.*, WHO, Copenhagen, pp. 49–61.

Van Stryp, H.C.M. (1994) Product related determinants of variety seeking behaviour for food. *Appetite*, **22**, 1–10.

Wardle, C. (1977) *Changing Food Habits in the UK*, Earth Resources Research Ltd, London.

West, C.E. and van Staveren, W.A. (1991) Food consumption, nutrient intakes and the use of food composition tables, in *Design Concepts in Nutritional Epidemiology* (eds B.M. Margetts and M. Nelson), Oxford Medical Publications, Oxford, pp. 97–100.

World Health Organization (1990) Diet, nutrition and the prevention of chronic diseases. *WHO Technical Report No. 797*, WHO, Geneva.

Index